AF772983

Kurt Hingkeldey

Englisch für Ingenieure

Kurt Hingkeldey

Englisch für Ingenieure

Mit 371 Bildern

Springer Fachmedien Wiesbaden GmbH

Verlagsredaktion: *Alfred Schubert*

ISBN 978-3-663-01999-2 ISBN 978-3-663-01998-5 (eBook)
DOI 10.1007/978-3-663-01998-5

1970

Alle Rechte vorbehalten

Copyright © 1970 Springer Fachmedien Wiesbaden

Ursprünglich erschienen bei Friedr. Vieweg + Sohn GmbH, Verlag, Braunschweig 1970.
Softcover reprint of the hardcover 1 st edition 1970

Bestell-Nr. 4056

Vorwort

Das Beherrschen der englischen Sprache ist auch für Ingenieure unerläßlich. So wendet sich dieses Buch sowohl an die Studenten der Ingenieurschulen, Akademien, Fachhochschulen und Universitäten als auch an die in der Industrie stehenden Ingenieure. Das Buch vermittelt Grundkenntnisse im technischen Englisch. Es soll sowohl als Vorbereitung für das Lesen englischer bzw. amerikanischer technischer Fachliteratur dienen, als auch eine Grundlage für das Führen von Gesprächen über technische Probleme in englischer Sprache geben. Es werden die Grundlagenfächer Mathematik, technische Mechanik, Festigkeitslehre, Hydraulik, technische Wärmelehre, Werkstoffkunde, technisches Zeichnen und Maschinenelemente in kurzer, leicht verständlicher Weise gebracht. Der Leser wird mit den wichtigsten Fachausdrücken und dem Formulieren einfacher technischer Probleme vertraut gemacht. Zugrunde gelegt ist dabei das metrische Maßsystem und die deutschen DIN Normen. Wörter, deren Sinn nicht klar aus dem Zusammenhang zu entnehmen sind, sind in einem beigefügten Wörterverzeichnis angeführt, wobei bei den englischen Wörtern nur das entsprechende deutsche Fachwort angegeben ist und die allgemeine Bedeutung des englischen Wortes weggelassen wurde.

Mr. *N. Hiller*, B. Sc. (eng.) möchte ich für die sorgfältige Korrektur des Manuskriptes danken, ebenso gilt mein Dank der Redaktion des Vieweg Verlages für die Bemühungen um die äußere Ausgestaltung des Buches.

München, im November 1969 *Kurt Hingkeldey*

Contents

Symbols | Formelzeichen

A	area	Fläche
A	mechanical equivalent of heat	mechanisches Wärmeäquivalent
a	acceleration	Beschleunigung
a_n	normal acceleration	Normalbeschleunigung
a_t	tangential acceleration	Tangentialbeschleunigung
B	angular momentum	Drall
b	width	Breite
c	coefficient of discharge	Ausflußfaktor
c	service factor	Betriebsfaktor
c	specific heat	spezifische Wärme
c	spring stiffness	Federkonstante
c_p	plate scale	Plattenkennziffer
c_p	specific heat at constant pressure	spezifische Wärme bei konstantem Druck
c_s	bolt scale	Schraubenkennziffer
c_v	specific heat at constant volume	spezifische Wärme bei konstantem Volumen
d	diameter	Durchmesser
E	energy	Energie
E	modulus of elasticity	Elastizitätsmodul
e	eccentricity	Exzentrizität
F	force	Kraft
F_a	thrust load	Axiallast
F_a	unbalanced force	Beschleunigungskraft
F_f	frictional force	Reibungskraft
F_r	radial load	Radialkraft
F_u	tangential force	Umfangskraft
f	deformation of a spring	Federweg
G	modulus of rigidity	Schubmodul
G	weight	Gewicht
g	acceleration due to gravity	Erdbeschleunigung
h	height	Höhe
h	lead	Steigung
h_w	loss head	Verlusthöhe
I	enthalpy	Wärmeinhalt (Enthalpie)
I_A	axial moment of inertia	axiales Trägheitsmoment
I_m	moment of inertia of mass	Massenträgheitsmoment
I_p	polar moment of inertia	polares Trägheitsmoment
i	enthalpy of unit mass	Wärmeinhalt (Enthalpie) der Masseneinheit
i	radius of gyration	Trägheitsradius
$\boldsymbol{i}$	unit tolerance	Toleranzeinheit
k_n	specific sliding resistance	spezifischer Gleitwiderstand
L	service life	Lebensdauer
L_{12}	work output (input) at expansion (compression)	absolute Gasarbeit

L_t	useful work	technische Arbeit
l	length	Länge
M	molecular mass	Molekularmasse
M	moment	Moment
M_b	bending moment	Biegemoment
M_t	twisting moment	Drehmoment
m	mass	Masse
m	module	Modul
m	number of shearing sections	Schnittigkeitszahl
m	Poisson ratio	Poisson'sche Konstante
m_t	mass per unit time	Durchflußmasse in der Zeiteinheit
N	normal force	Normalkraft
n	exponent for polytropic process	Exponent der Polytrope
n	speed (r.p.m.)	Drehzahl (1/min)
p	absolute pressure	absoluter Druck
p	pitch	Teilung
p	power	Leistung
p	pressure	Druck
Q	heat	Wärme
Q	shear force	Scherkraft
Q	volume of liquid per unit time	Durchflußmenge
Q_s	radiated energy	Strahlungsverlust
q	heat added (rejected) to unit mass	spezifische Wärme
R	gas constant	Gaskonstante
R	resultant force	resultierende Kraft
R	Reynolds number	Reynolds'sche Zahl
r	radius	Radius
r_h	hydraulics radius	hydraulischer Radius
S	cable tension	Seilkraft
S	entropy	Entropie
s	displacement	Weg
s	entropy of unit mass	Entropie der Masseneinheit
s	thickness	Dicke
s	width across flats	Schlüsselweite
T	absolute temperature	absolute Temperatur
t	immersed depth	Eintauchtiefe
t	temperature	Temperatur
t	time	Zeit
U	internal energy	innere Energie
U	oversize	Übermaß
u	internal energy of unit mass	innere Energie der Masseneinheit
V	volume	Volumen
v	velocity	Geschwindigkeit
v	volume of unit mass	spezifische Volumen
v_n	normal velocity	Normalgeschwindigkeit
v_0	initial velocity	Anfangsgeschwindigkeit
v_t	tangential velocity	Tangentialgeschwindigkeit

W	energy	Energie
W	section modulus	Widerstandsmoment
w	horizontal range	Wurfweite
x	dryness fraction of steam	Dampfgehalt
z	number	Anzahl
α	angle	Winkel
α	factor of stength for welds	Festigkeitsfaktor für Schweißnähte
α_t	thermal coefficient of linear expansion	linearer Wärmeausdehnungskoeffizient
β_k	fatigue stress concentration factor	Kerbwirkungszahl
γ	specific gravity	Wichte
γ	shearing strain	Verschiebungswinkel
δ	coefficient of cyclic variation	Ungleichförmigkeitsgrad
ε	angular acceleration	Winkelbeschleunigung
ε	strain	Dehnung
ε_l	axial strain	Längsdehnung
ε_q	lateral strain	Querdehnung
ζ	loss coefficient	Widerstandsziffer
η	coefficient of performance	Wirkungsgrad einer Kältemaschine
η	coefficient of viscosity	Zähigkeitszahl
η	efficiency	Wirkungsgrad
$\varkappa$	relation between specific heat at constant volume and at constant pressure	Verhältnis zwischen der spezifischen Wärme bei konstantem Volumen und bei konstantem Druck
λ	friction factor for pipes	Rohrreibungszahl
λ	slenderness ratio	Schlankheitsgrad
μ	coefficient of friction	Reibungskoeffizient
ν	kinematic viscosity	kinematische Zähigkeit
ϱ	angle of friction	Reibungswinkel
ϱ	density	Dichte
ϱ	radius of curvature	Krümmungsradius
σ	stress	Spannung
σ_{all}	allowable stress	zulässige Spannung
σ_b	bending stress	Biegespannung
σ_d	compressive stress	Druckspannung
σ_{rel}	related stress	Bezugsspannung
σ_{weld}	stress in welds	Spannung in Schweißnähten
σ_z	tensile stress	Zugspannung
τ_s	shearing stress	Schubspannung
τ_t	torsional shearing stress	Torsionsspannung
Φ	mole volume	Molvolumen
ψ	friction factor	Reibungszahl
ω	angular velocity	Winkelgeschwindigkeit

1. Mathematics

Mathematics is the basis of engineering. It may be subdivided into elementary mathematics and higher mathematics. Elementary mathematics comprises algebra, geometry and trigonometry. Higher mathematics includes the differential and integral calculus.

There are two main mathematical methods: the analytical (algebraical) method and the graphical method. Generally the first method is employed. The second method is only used, if the analytical method is too complicated. Before establishing a formula for any physical process the problem will be simplified by appropriate hypotheses. Later on, experiments will prove, whether the errors resulting from these simplifications are within reasonable limits. Many formulae are only applicable within a limited range which must be specified for such formulae. When using a formula, at the beginning of the mathematical operation it must be checked that the given problem is within the limits of applicability of the formula. Besides the formulae derived by mathematical procedures there are empirical formulae obtained by experiment. When using the latter kind of formula it must be checked that the problem is within the limits of applicability of the formula.

1, 2, 3 ... are whole numbers, 2, 4, 6 ... are even numbers, 1, 3, 5 ... are odd numbers. 1/2, 1/3, 1/4 ... (one half or one over two, one third or one over three, one quarter or one over four) are common fractions.

0.35, 0.45 ... (zero point three five, zero point four five) are decimal fractions.

5/8 is a proper fraction, 8/8 and 9/8 are improper fractions. If the value is equal to or greater than unity the fraction is called improper.

$5\frac{3}{8}$ and 6.3 are mixed numbers.

1.1. Algebra

1.1.1. The Fundamental Operations

Addition

$4 + 2 = 6$ four plus two is equal to six

Subtraction

$6 - 2 = 4$ six minus two is equal to four

Subtraction may be also considered as addition with negative numbers.
e. g.

$6 - 2 = 6 + (-2)$ in brackets minus two.

If there are several terms in a sum, each term is provided with a positive or a negative sign.

e. g.

$a + b + (-c) + (-d) = A$ small a, capital A

Multiplication

$4 \times 2 = 8$ four multiplied by two is equal to eight

Division

$8 \div 2 = 4$ eight divided by two is equal to four

Division may be also considered as multiplication with a fraction.

e. g.

$$8 \div 2 = 8 \times \frac{1}{2} = 8 \times 0.5$$

$\frac{a}{b}$ a is above the fraction line and b is below the fraction line. a is the numerator and b ist the denominator.

$\frac{a}{b} = \frac{e}{d}$ a is to b as e is to d

Additional expressions

$a < b$ a is less than b, $a > b$ a is greater than b
a' a dash, a'' a double dash, $0 \rightarrow \infty$ from zero to infinity
1 cm $\triangleq$ 100 m/s 1 cm corresponds to 100 metres per second
(. . .) bracket open, bracket closed, [. . .] in square brackets
$\{\ldots\}$ in braces, $\lim x \rightarrow 0$ limit x tends to zero, ΣF_i sum of F_i

1.1.2. Complex Numbers

Since the squares of positive and negative numbers are all positive, a negative number cannot have a real square root. For this reason square roots of negative numbers are imaginary. Every imaginary number can be expressed as a product of a real number and of the imaginary unit operator. The imaginary unit operator is $\sqrt{-1}$ and is denoted by i. For instance: $\sqrt{-3} = 3 \times i$. The quantity $4 + \sqrt{-3} = 4 + i\sqrt{3}$ is a complex number.

1.1.3. Powers, Roots and Logarithms

$3^2 = 9$	three squared is equal to nine
$3^3 = 27$	three cubed is equal to twenty seven
$3^4 = 81$	three to the power of four is equal to eighty one
$\sqrt{9} = 3$	square root of nine is equal to three
$\sqrt[3]{27} = 3$	cube root of twenty seven is equal to three
$\sqrt[4]{81} = 3$	fourth root of eighty one is equal to three

How much is or what is the value of $\sqrt[5]{125}$?

Logarithms

The logarithm of any number to a given base is the exponent to which the base must be raised, to produce that number.

Thus, if $b^n = a$, then b is the base and n the exponent or the power to which the base must be raised to produce the number a.

The logarithm of unity to any base (except zero) is zero.

$$\log_a 1 = 0 \quad \text{since} \quad a^0 = 1 .$$

The logarithm of any number (except zero) which is equal to its base, is unity.

$$\log_a a = 1 \quad \text{since} \quad a^1 = a .$$

The logarithm of a product is equal to the sum of the logarithms of its factors.

$$\log (a\, b) = \log a + \log b .$$

The logarithm of a fraction is equal to the difference of the logarithms of the numerator and the denominator.

$$\log \left(\frac{a}{b}\right) = \log a - \log b .$$

The logarithm of a number to a given power is equal to the product of the exponent and the logarithm of the base.

$$\log (a^b) = b \log a .$$

As a root may be written in the form of a power, the logarithm of a root of a number is found in the same way as the logarithm of a power.

$$\log \sqrt[b]{a} = \log a^{\frac{1}{b}} = \frac{1}{b} \log a .$$

There are two main systems of logarithms:

The Common or Briggs system and the Natural or Napierian system.

In the common system the base is 10 and in the natural system the base is

$$e = 2.718 .$$

The relation between the two system is:

$$\ln a = \ln 10 \times \log a . \tag{1}$$

The whole number of a logarithm is called its "characteristic".

The decimal number of a logarithm is called its "mantissa".

When a number is greater than 1, the logarithm of the number is positive. Then the characteristic of the logarithm is a number one less than the number of digits before the decimal point.

Thus the characteristic of log 475.6 is 2.

When the number is less than 1, the logarithm of the number is negative. Then the characteristic of the logarithm is a negative number one more than the number of zeros between the decimal point and the first significant figure.

Thus the characteristic of log 0.00085 is -4.

The mantissa may be found in a table of logarithms. Ordinarily tables give the logarithms to four or five decimal places.

1.1.4. Equations

Equations are the basis of the derivation of formulae. There are equations of different degrees.

$a x + b = 0$	is an equation of the first degree
$a x^2 + b x + c = 0$	is an equation of the second degree or a quadratic equation
$a x^3 + b x^2 + c x + d = 0$	is an equation of the third degree or a cubic equation.

All these equations are written in the normal form.

For each unknown quantity one equation is necessary. Hence to solve for six unknown quantities six equations are necessary.

From the two equations

$$a_1 x + b_1 y + c_1 = 0$$
$$a_2 x + b_2 y + c_2 = 0 ,$$

the values of x and y may be calculated in the following way:

The two equations are combined either by adding or subtracting or by computing one unknown quantity from one equation and substituting it in the other one or by computing the same unknown quantity from the two equations and equating these equations. From the combined equation one unknown quantity may be calculated. By substituting this quantity in one of the original equations the other unknown quantity may be found.

Quadratic equations

Every complete quadratic equation may be written in the form

$$x^2 + a x + b = 0 .$$

When the second term i. e. the term containing x is missing, the equation is called a pure quadratic equation

$$x^2 + b = 0 .$$

A quadratic equation has two roots:

$$x_1 = -\frac{a}{2} + \sqrt{\left(\frac{a}{2}\right)^2 - b}$$
$$x_2 = -\frac{a}{2} - \sqrt{\left(\frac{a}{2}\right)^2 - b} . \qquad (2)$$

Parametric equations

An equation may be written in the form:

$$x = f(p)$$
$$y = f(p)$$

where p is denoted as a parameter.

By computing the parameter from one equation and substituting it in the other equation we get the usual form of an equation:

$$y = f(x) .$$

1.2. Plane and Solid Geometry

Plane Geometry is a science that deals with plane figures. Solid Geometry is a science that deals with solid figures.

A main field of plane geometry is the computation of areas and circumferences of planes. A main field of solid geometry is the computation of volumes and surfaces of solids.

The best known plane figures are the triangle, the quadrilateral and the circle. The angles in a triangle may be right, acute or obtuse. The sum of the angles in any plane triangle is equal to 180°. The square on the hypotenuse is equal to the sum of the squares on the other two sides. A triangle which has no right angle and no two sides equal is called a "scalene triangle". Fig. 1 shows the forms of the different triangles.

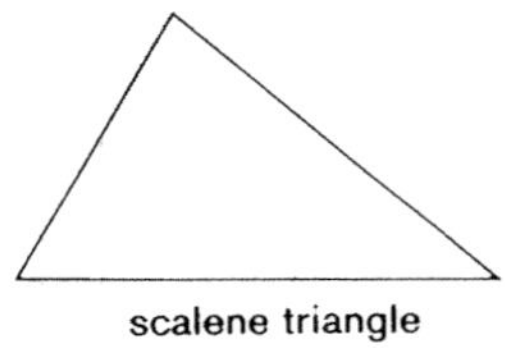

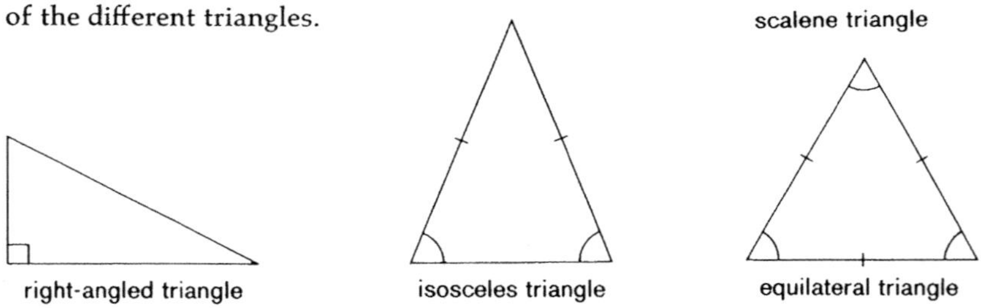

Fig. 1 Triangles

A quadrilateral, which has two sides parallel, is called a trapezium.
Fig. 2 shows the forms of the different quadrilaterals and a circle

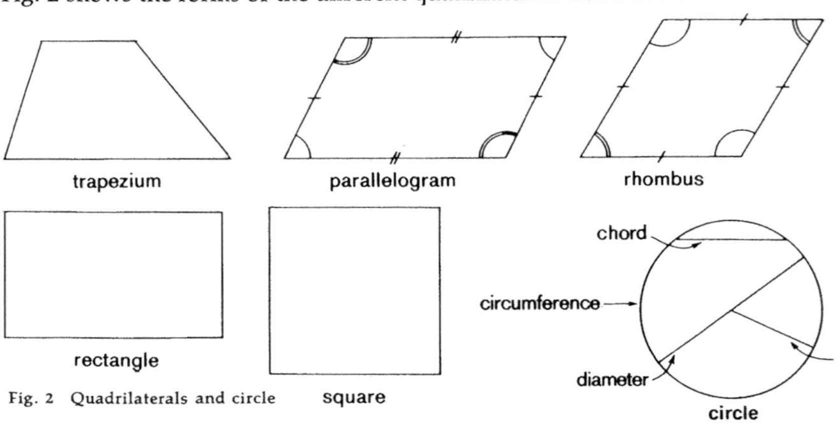

Fig. 2 Quadrilaterals and circle

In Fig. 3 some important solid figures are shown.

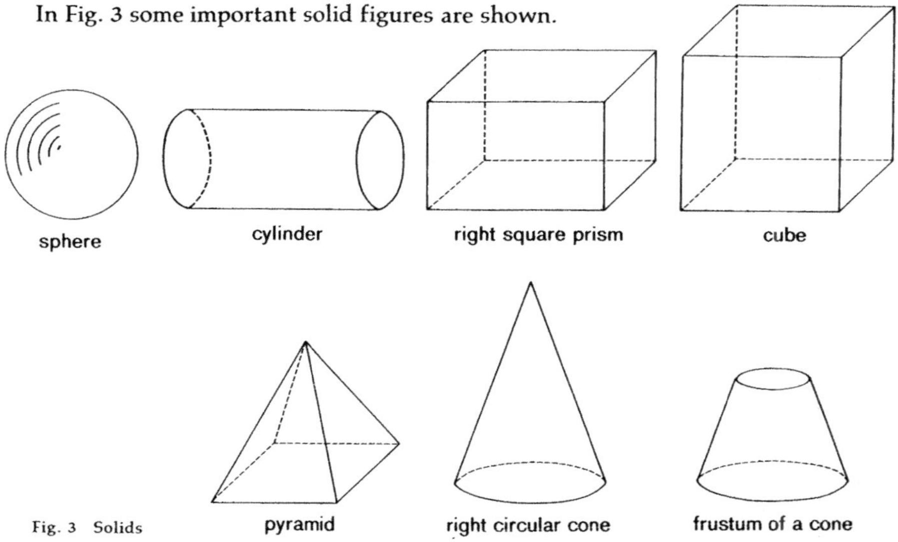

Fig. 3 Solids

1.3. Trigonometry

Trigonometry is that branch of mathematics which deals with measurements relating to angles.

There are two systems used in the trigonometric measurement of angles: the sexagesimal system and the circular system.

The unit of the sexagesimal system, 1 degree, is the angle at the centre of a circle subtended by an arc equal to $\frac{1}{360}$ of the circumference.

1 degree = 60 minutes and 1 minute = 60 seconds.

The unit in the circular system, 1 radian, is the angle at the centre of a circle subtended by an arc equal in length to the radius of the circle. It is equal to 57.3°.

The conversion of the sexagesimal to the circular system is given by the formula:

$$\varphi_{\text{radians}} = \varphi_{\text{degrees}} \times \frac{\pi}{180}. \tag{3}$$

The size of an angle may be determined from the ratio of two sides of a triangle. Generally four ratios are used.

These ratios or trigonometric functions are:

sine, cosine, tangent and cotangent.

They are defined by reference to a right-angled triangle:

$$\sin \alpha = \frac{\text{opposite side}}{\text{hypotenuse}} \qquad \cos \alpha = \frac{\text{adjacent side}}{\text{hypotenuse}}$$

$$\tan \alpha = \frac{\text{opposite side}}{\text{adjacent side}} \qquad \cot \alpha = \frac{\text{adjacent side}}{\text{opposite side}}$$

There are some important relationships between the trigonometric functions. For instance:

$$\cot \alpha = \frac{1}{\tan \alpha} \quad \text{or} \quad \frac{\sin \alpha}{\cos \alpha} = \tan \alpha \, .$$

Any plane triangle can be solved

1. When the three sides are given
2. When two sides and the included angle are given
3. When two sides and any other angle besides the included angle are given
4. When one side and any two angles are given.

1.4. Coordinate Geometry

There are two coordinate geometry systems:

The rectangular coordinate system and the polar coordinate system. In the first system a point is determined by the x and the y coordinates. In the latter system a point is determined by a radius and an angle.

Equations of straight lines

The equation of the x- axis is

$$y = 0 \, . \tag{4}$$

The equation of the y- axis is

$$x = 0 \, . \tag{5}$$

The equation of a straight line parallel to the x- axis is

$$y = b \, . \tag{6}$$

The equation of a straight line parallel to the y- axis is

$$x = a . \tag{7}$$

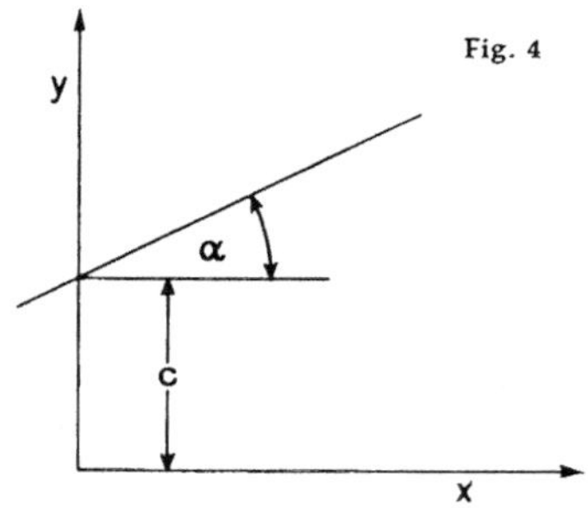

Fig. 4

The equation of a straight line which cuts off a given intercept (c) on the y- axis and includes a given angle with the x- axis, as shown in Fig. 4, is

$$y = m\,x + c$$

where $m = \tan\alpha$ (8)

Conic sections

A conic section is a curve traced out by a point in a plane so that the ratio of the distance from a fixed point to the perpendicular distance from a fixed straight line is constant (Fig. 5).

The fixed point is called the "focus" and the fixed straight line is called the "directrix". The constant ratio of the two distance is called the "eccentricity of the conic section" (e).

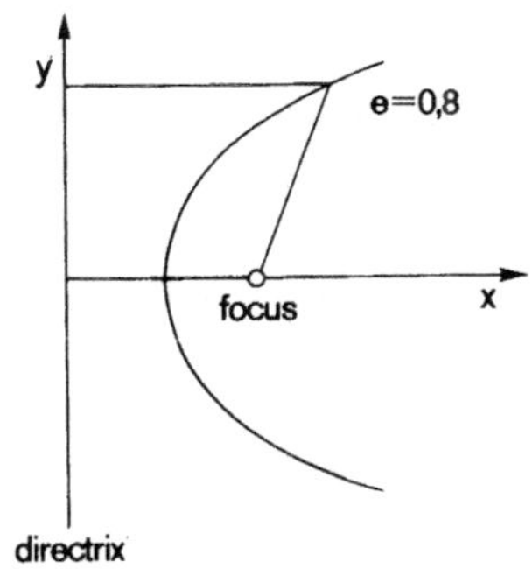

Fig. 5 Conic section (ellipse)

When $e = 1$ the conic section is a parabola
When $e < 1$ the conic section is an ellipse
When $e > 1$ the conic is a hyperbola
When $e = 0$ the conic section is a circle.

1.5. Differential and Integral Calculus

1.5.1. Differential Calculus

Every function $y = f(x)$ can be represented by a curve in a coordinate system.

The derivative of a function can be considered as the slope of the curve at any point (Fig. 6)

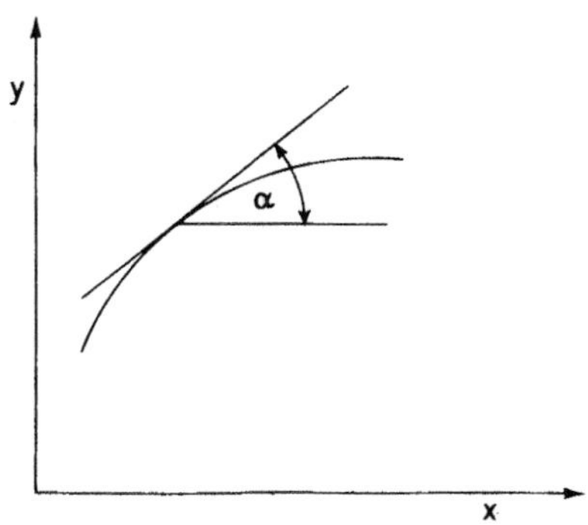

Fig. 6

$$\frac{dy}{dx} = \tan\alpha \qquad \text{first derivative of } y \text{ with respect to } x$$

The derivative of a function is a function of the same variable that occurs in the original function. The derivative of this second function is called the second derivative and the derivative of the second derivative is called the third derivative and so on.

A well known application of the differential calculus is the determination of the maxima and minima of a curve.

The condition for a maximum is: The first derivative is zero, the second derivative is negative.

The condition for a minimum is: The first derivative is zero, the second derivative is positive (Fig. 7).

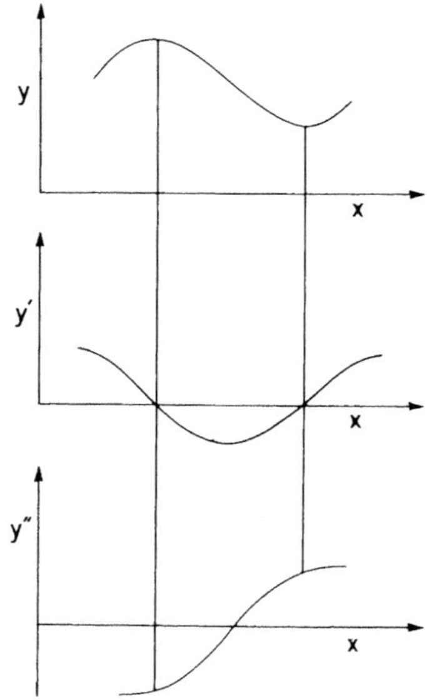

Fig. 7

1.5.2. Integral Calculus

The indefinite integral

The inverse process of differentiation, that is the process of finding the function, whose differential is given, is known as integration. Integration provides the means of solving problems in many technical fields.

If the derivative of a function $\frac{dy}{dx} = f(x)$ is given, then the original function is found from

$$y = \int f(x)\,dx = F(x) + c.$$

This integral is defined as an indefinite integral, because the constant term is not determined.

The definite integral

Integration is applicable to the determination of an area bounded by a curve, the abscissa and two ordinates y_1 and y_2 (Fig. 8)

$$F = \int_a^b y \, dx$$

integral $y \, dx$ between the limits $x = a$ and $x = b$.

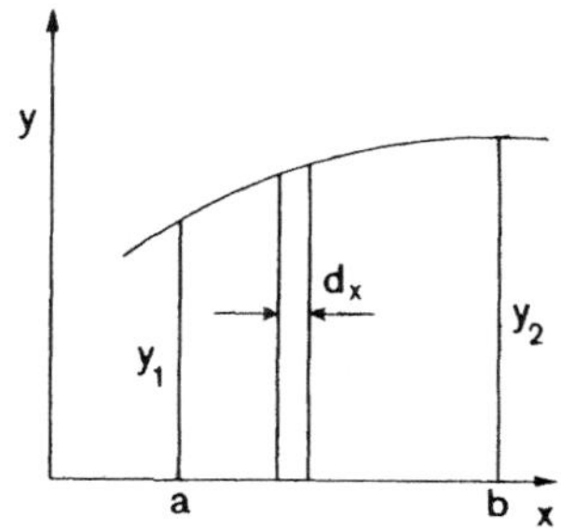

Fig. 8

This kind of integral is called a "definite integral".

The differential equation

An equation in which, with x and y, dx and dy also occurs is called a differential equation. The differential equations are described according to the order of the highest present derivative of the function. A differential equation in which the third derivative of the function occurs is a differential equation of the third order.

By integrating a differential equation an equation of the form

$y = f(x)$ is found.

One method of obtaining a technical formula is to set up a differential equation and then to integrate it. If the initial conditions are given the constants of integration may be determined and so a non-arbitrary formula obtained.

2. Mechanics

Mechanics is that branch of physics which deals with the action of forces on bodies and the resulting motions of bodies. Engineering mechanics or applied mechanics is the study of the principles of mechanics as applied to problems of engineering. Bodies may be at rest or in motion, therefore mechanics may be classified as:

Statics or the study of forces acting on bodies at rest and Dynamics or the study of the motions of bodies and of forces acting on moving bodies.

Mechanics also may be classified according to the state of aggregation of bodies as: Mechanics of solid bodies, Mechanics of liquids or Hydraulics and Mechanics of gases.

There are three fundamental laws which represent the basis of mechanics.

1. The law of the parallelogram of forces.

   A force may be resolved into two components and conversely two forces may be combined to a resultant force by means of a parallelogram of forces (Fig. 9).

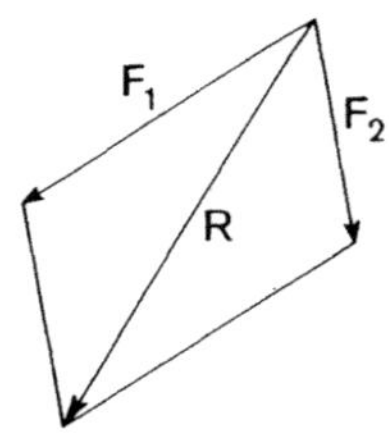

Fig. 9
Parallelogram of forces

2. The law of action and reaction.

   For every force there is an equal, opposite and collinear reaction force.

   (Newton's third law of motion)

3. The law of inertia.

   A body will remain at rest or continue to move in a straight line with uniform motion, unless an external force acts on it. (Newton's first law of motion).

2.1. Statics of Rigid Bodies

In the mechanics of solid bodies all bodies are considered as rigid, i. e. as bodies which do not endure distortions under load. Small distortions of bodies may be neglected. Plastic bodies, liquid and gases may not be considered as rigid bodies. The law of mechanics of these bodies are found both by calculus and by experiment. The results of experiments, represented by curves, are taken into consideration in the subsequent mathematical operations.

There are two kinds of physical quantities:

1. The vector quantity which possesses direction. Such quantities include: force, displacement, velocity, acceleration etc.
2. The scalar quantity which does not possess direction. Such quantities include: time, temperature, work etc.

A force has four characteristics:

1. point of application
2. magnitude
3. angle of direction (α)
4. angle of inclination (β) (Fig. 10)

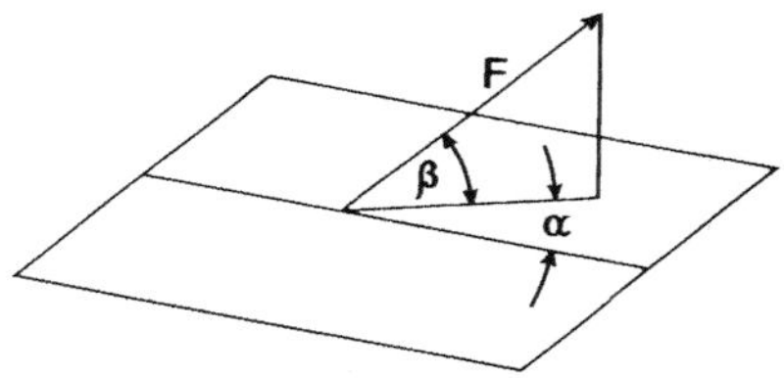

Fig. 10 Characteristics of a force

There are two basic systems of units:

1. The physical system which has mass as a fundamental quantity and
2. The technical system which has force as a fundamental quantity.

In this book the technical system will be used.

The unit of force in the technical system is 1 kilopond (1 kp). A multiple of this unit is the megapond (1 Mp = 1000 kp) and a sub-multiple is the pond (1 p = 1/1000 kp).

2.1.1. Composition, Resolution and Equilibrium of Coplanar, Concurrent Forces

A system of forces is coplanar, if the lines of action of all the forces lie in one plane. A system of forces is concurrent, if the lines of action of all the forces pass through a common point.

Composition of two forces

According to the parallelogram law, which may be applied to all vector quantities, two concurrent forces may be combined into a resultant force. A space diagram and a force diagram may be drawn.

The resultant of two forces may be found by drawing a triangle the two sides of which represent the given forces and the third and closing side represents the resultant force (Fig. 11).

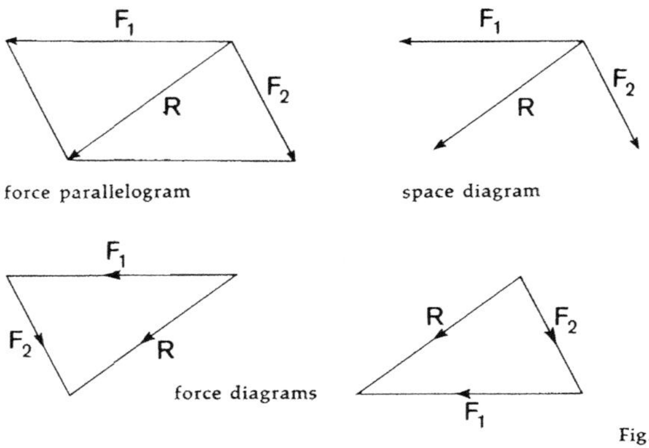

Fig. 11 Composition of two forces

Resolution of a force into two components

The components of a given force are also found by a triangle of vectors. Through the beginning and through the end of the vector of the given force the direction lines of the two components are laid off up to their point of intersection and thus the magnitude of the two components is obtained as Fig. 12 shows.

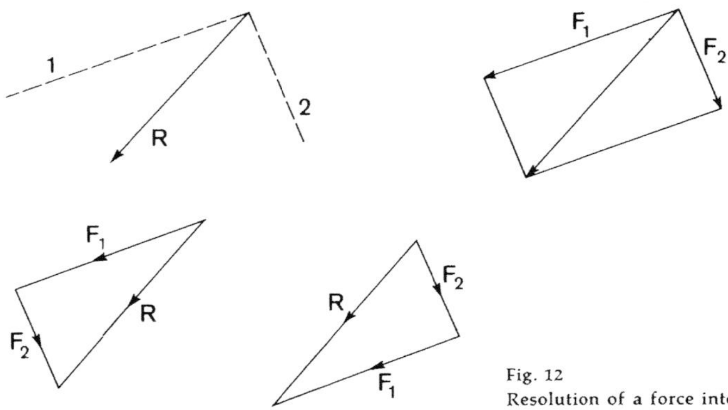

Fig. 12
Resolution of a force into two components

Composition of any number of coplanar, concurrent forces

Graphically the resultant of the given forces may be found in the following way. Any two of the forces are combined into their resultant, which is in turn combined with a third force to give a new resultant which is then combined with a fourth force and so on. The resultant extends from the beginning of the first force to the end of the last force. The order in which the forces are taken is arbitrary. The directional arrows follow round in sequence. The arrow of the resultant is in the opposite direction to the arrows of the forces (Fig. 13).

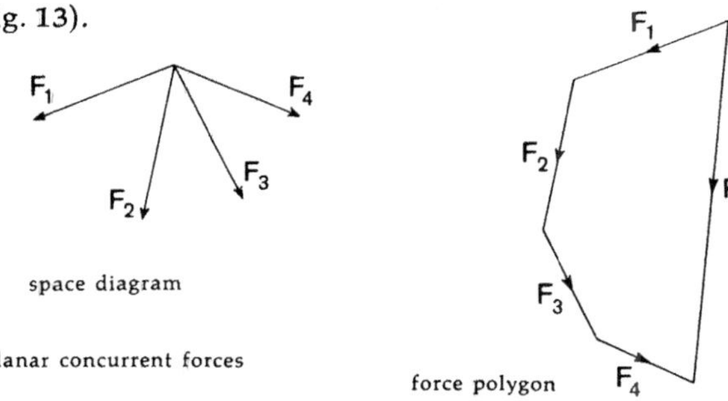

Fig. 13
Graphical composition of coplanar concurrent forces

When all the forces act in the same direction, their resultant is equal to the arithmetic sum of the forces and acts in the direction of the forces.

Analytically the resultant of a number of forces may be found thus: The vectors of all forces are resolved along two perpendicular directions as shown in Fig. 14.

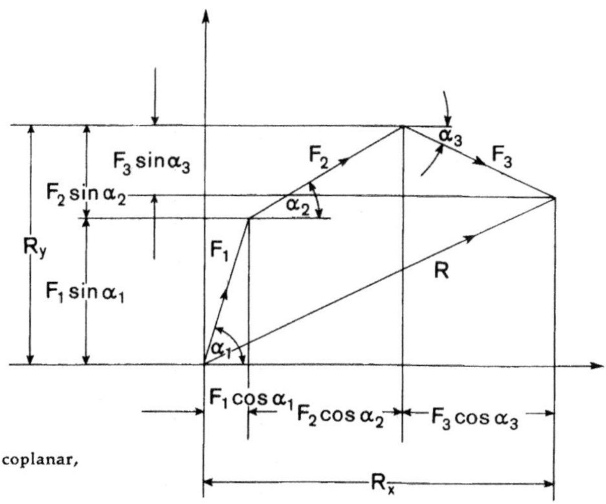

Fig. 14
Analytical composition of coplanar, concurrent forces

The sum of the resolved parts in the direction of x is:

$$R_x = F_1 \cos\alpha_1 + F_2 \cos\alpha_2 + \ldots$$

The sum of the resolved parts in the direction of y is:

$$R_y = F_1 \sin \alpha_1 + F_2 \sin \alpha_2 + \ldots$$

The magnitude of the resultant may be found from

$$R = \sqrt{R_x^2 + R_y^2}\,. \tag{9}$$

The inclination of the resultant is calculated from

$$\tan \varphi = \frac{R_y}{R_x}\,. \tag{10}$$

Equilibrium of coplanar, concurrent forces

Any number of forces are in equilibrium if their resultant is zero. In this case the forces form a closed polygon. All the arrows of the vectors follow round in sequence (Fig. 15).

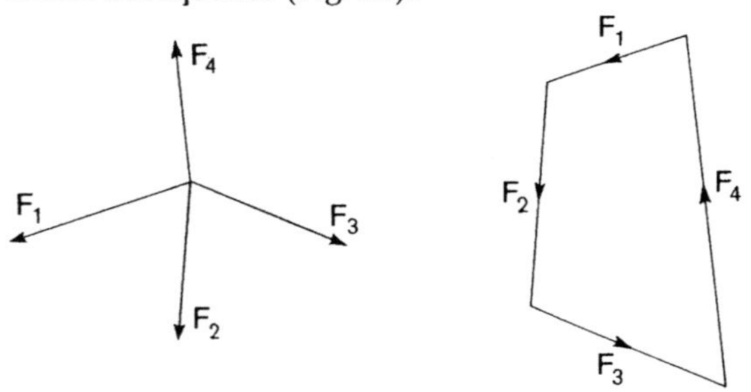

Fig. 15
Equilibrium of coplanar, concurrent forces

The x and y components of the resultant must be zero.

$$R_x = \Sigma F_x = 0 \qquad R_y = \Sigma F_y = 0\,. \tag{11}$$

Resolution of a force in several directions

If a force is resolved in more than two directions, any number of components is possible (Fig. 16).

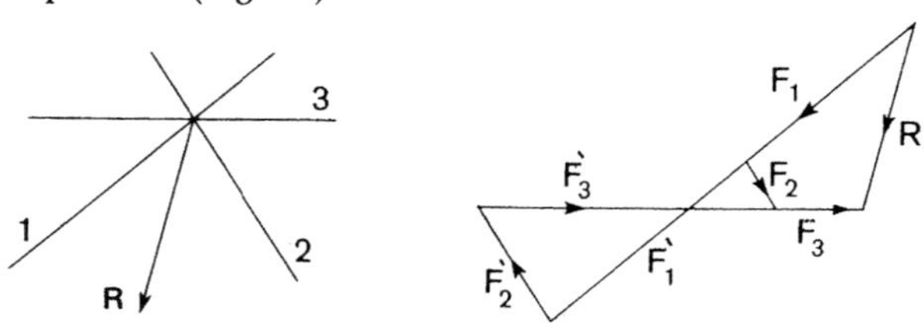

Fig. 16
Resolution of a force into three components

A force may be resolved into two components only, if one solution is desired.

2.1.2. Moments, Composition, Resolution and Equilibrium of Coplanar, Non-concurrent Forces

Equilibrium of forces

Two forces are in equilibrium, if they are equal, opposite in direction and are acting along the same line.

Three forces, not being parallel, are only in equilibrium, if they meet at a common point (Fig. 17).

If there are four forces in equilibrium, we can combine two forces to give one resultant and also the two other forces to give a second resultant. The two resultants must be equal, opposite in direction and lie along the same line.

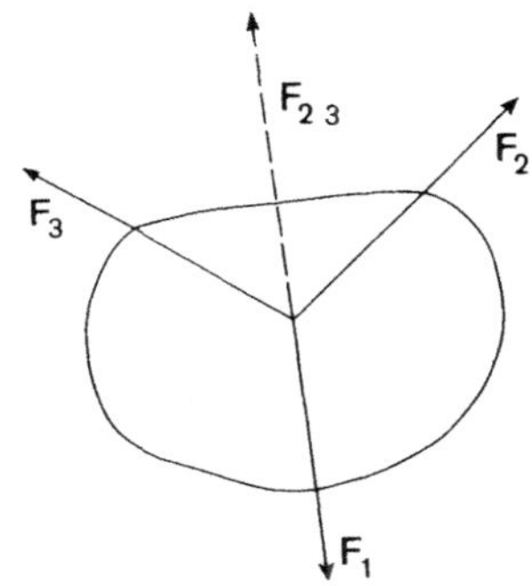

Fig. 17 Equilibrium of three forces

Theorem

A force may be displaced along its line without changing its effect.

Moments

The moment of a force about a given point is the product of the force and the perpendicular distance of that point from the line of action of the force. The moment of a force is considered to be positive, if there is the tendency to produce a counterclockwise rotation.

Theorem

The algebraic sum of the moments of any number of coplanar forces about any point in their plane is equal to the moment of their resultant about that point.

Graphical Representation of forces by means of the link polygon

The magnitude and direction of the resultant of any number of coplanar forces is found in a vector polygon as the geometric vector sum of the force vectors (Fig. 18). In order to find the location of the line of action of the resultant we choose any pole in the vector diagram and join the beginnings of the vectors at the pole. In the space diagram starting from any point on the line of action of the first force we draw lines parallel to the rays in the vector polygon. The line of action of the resultant passes through the point of intersection of the first and the last line.

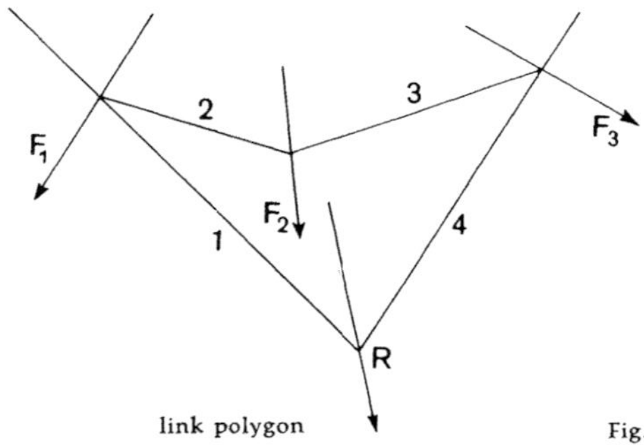

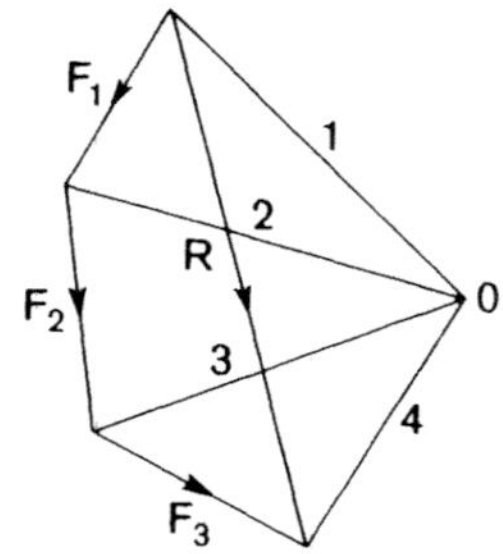

link polygon Fig. 18 force polygon

Analytical Determination of the composition of forces

The magnitude of the resultant is found by:

$$R_x = \Sigma F_x \qquad R_y = \Sigma R_y \qquad R = \sqrt{R_x^2 + R_y^2}\,.$$

In order to find the location of the resultant, any point is chosen and the perpendicular distance (a) ot the resultant from that point is calculated by

$$R \cdot a = \Sigma\,(F_i \cdot a_i) \quad \text{or} \quad a = \frac{\Sigma\,(F_i \cdot a_i)}{R} \tag{12}$$

where a_i = the perpendicular distance of a single force from the chosen point.

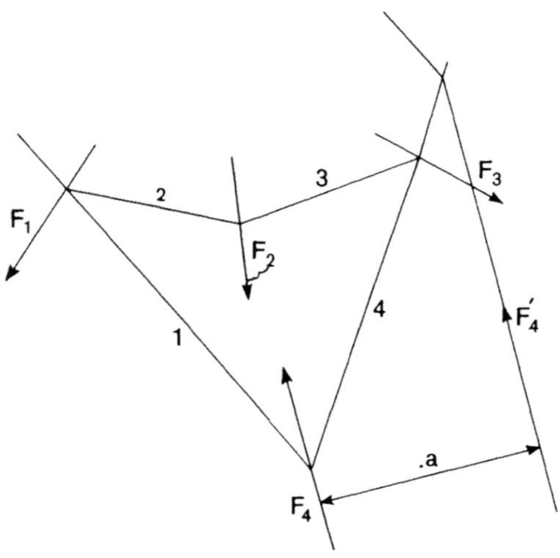

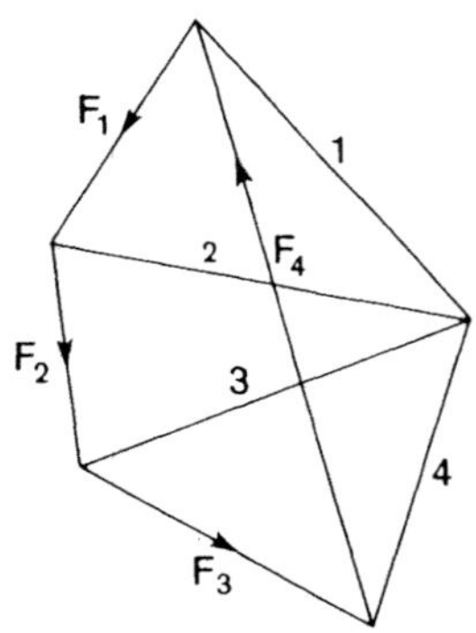

Fig. 19

Equilibrium of coplanar, non-concurrent forces

Analytical

The algebraic sum of the x and y components of the forces must be zero. The algebraic sum of the moments of the forces with respect to any point must be zero.

$$\Sigma F_x = 0 , \qquad \Sigma F_y = 0 \quad \text{and} \quad \Sigma M = 0 . \tag{13}$$

Graphical

The vector polygon and the link polygon must be closed, if the forces are in equilibrium (Fig. 19). If the line of action of the force F_4 does not pass through the point of intersection of the first and the last link in the link polygon, then a free moment of the amount $F_4\, a$ exists.

Couple

Two equal non-collinear parallel forces acting in opposite directions constitute a couple (Fig. 20). The effect of a couple is to produce a tendency to rotation. The moment of a couple is the product of one of the forces and the perpendicular distance between the lines of action of the forces. It is independent of the position of the pole. The couple can be transferred to any desired position.

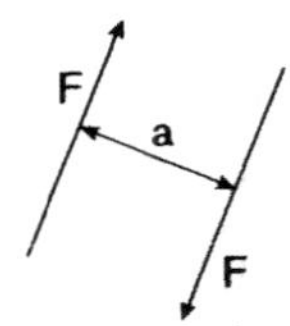

Fig. 20 Couple

A force, the line of direction of which does not pass through the centre of rotation, tends to produce a rotation as well a translation. This force can be replaced by a force of the same magnitude as the given force passing through the centre of rotation together with a moment represented by a couple (Fig. 21).

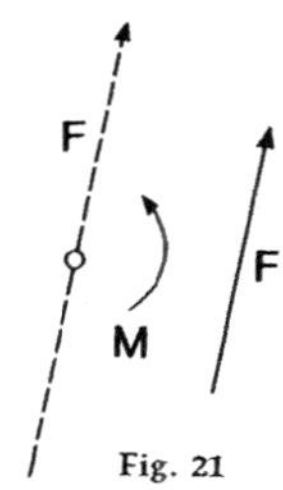

Fig. 21

2.1.3. Stability of a Rigid Body

A body is at stable equilibrium, if when subjected to a small displacement, it tends to regain its original position. It is at unstable equilibrium, if when subjected to a small displacement it tends to move to a stable position of equilibrium. It is at neutral equilibrium, if it will stay in any new position after displacement (Fig. 22).

In Fig. 23 the stability of a body, resting on a plane surface which is loaded by two forces is demonstrated. The body is stable in regard to the edge I, if the sum of the counterclockwise turning moments is smaller than the sum

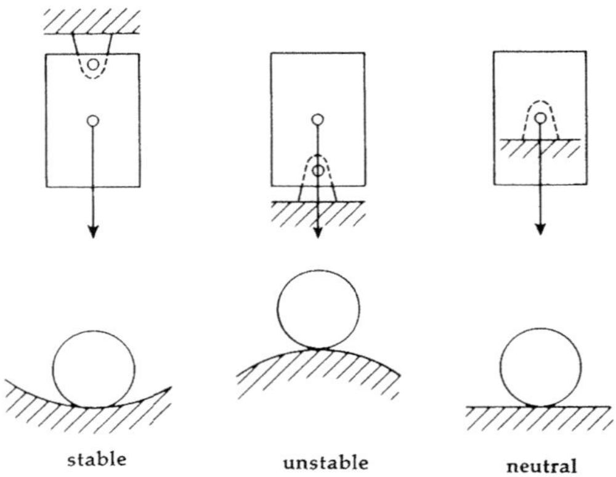

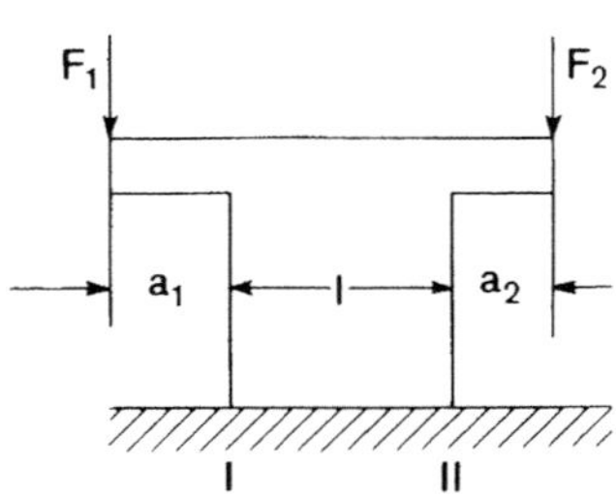

Fig. 23

Fig. 22 Stability of rigid bodies

of the clockwise turning moments and stable in regard to the edge II, if the sum of the clockwise turning moments is smaller than the sum of the counterclockwise turning moments.

$$\begin{aligned} &\text{Stability in regard to edge I:} \quad F_1\, a_1 \leqq F_2\,(l + a_2) \\ &\text{Stability in regard to edge II:} \quad F_2\, a_2 \leqq F_1\,(l + a_1) \end{aligned} \tag{14}$$

2.1.4. Structures

Fig. 24 shows some types of structures

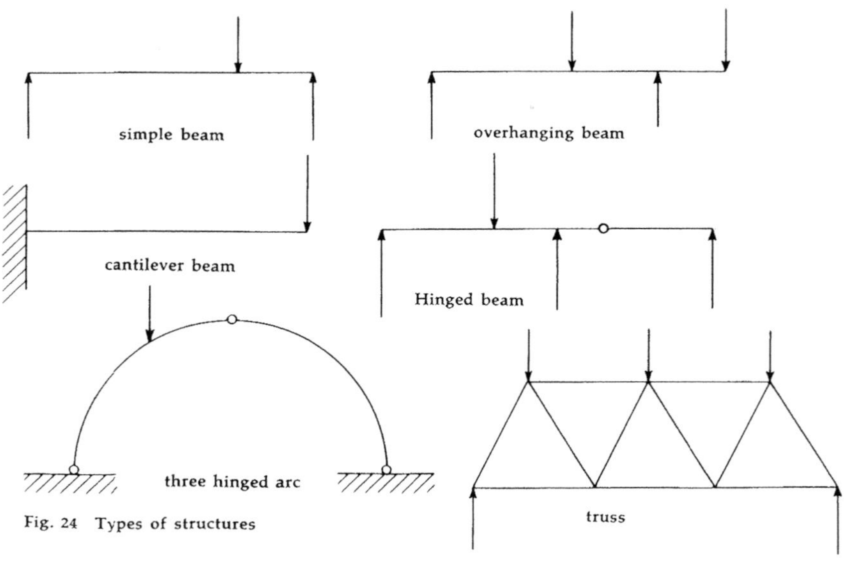

Fig. 24 Types of structures

The loads may be either concentrated or distributed. Distributed loads may be uniform, uniformly varying or non-uniformly varying. The non-uniformly distributed load may usually reduced to the triangle or to the trapezoidal form. In Fig. 25 some different kinds of loads acting on a simple beam are shown. The forces exerted on the supports are called reactions. Loads and reactions are external forces that hold the stable structure in equilibrium.

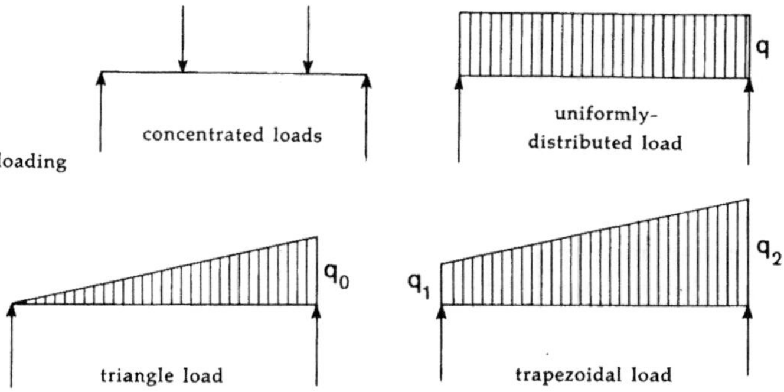

Fig. 25 Kinds of loading

There are three types of supports:

The roller support. It transmits one reaction force.

The pinned support. It transmits two reaction forces.

The built-in or encastre support. It transmits two reaction forces and one reaction moment (Fig. 26).

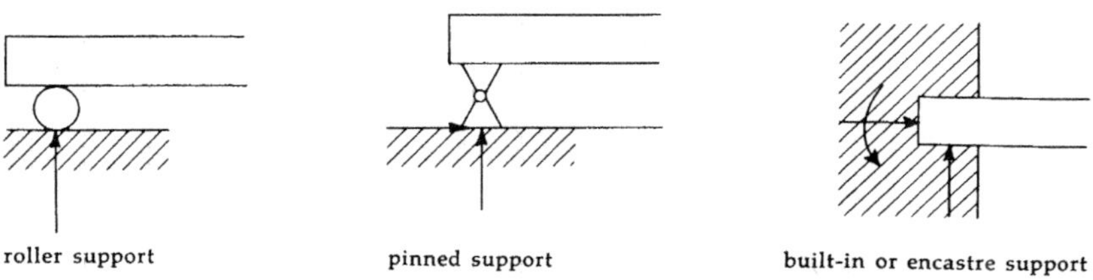

Fig. 26 Types of supports

The reactions can be found by use of the three equations of equilibrium. When the number of reactions is greater than the number of equilibrium equations, then the beam is statically indeterminate. Fig. 27 shows a statically indeterminate beam.

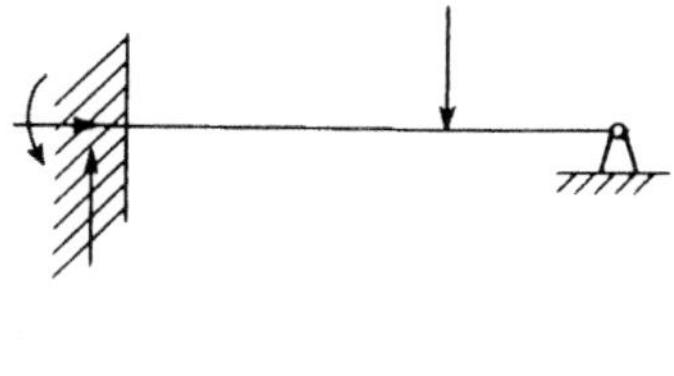
Fig. 27 Statically indeterminate beam

Shearing force and bending moment

The loads and reactions acting on a beam are assumed perpendicular at the axis of the beam.

Shearing force at any section of a beam is defined as the algebraic sum of the forces acting on either side of the section. It is considered positive, if the resultant of the forces of the left portion is directed upward.

Bending moment at any section is defined as the algebraic sum of the moments of the forces acting on either side of the section with respect to any point of the section. It is considered to be positive if the moment on the left portion is clockwise.

The shearing forces and the bending moments are represented in a shear force diagram and in a bending moment diagram,respectively (Fig. 28).

The area of shear force diagram lying above the axis is equal to the area of shear force diagram lying below the axis.

The shearing force and the bending moment at any point on a beam carrying a uniformly-distributed load is found from:

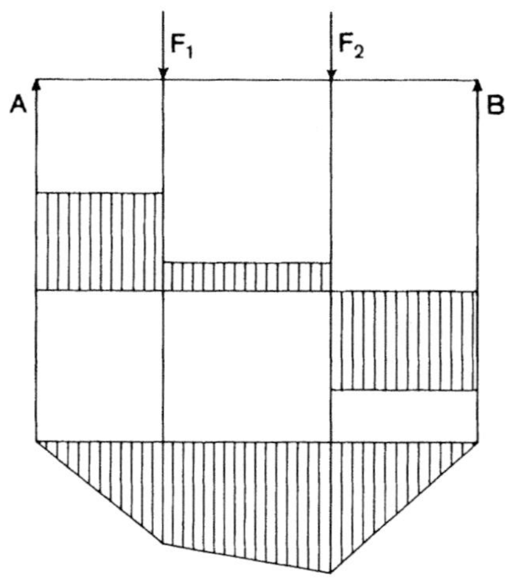

Fig. 28
Shear force diagram and bending moment diagram

Shear force

$$Q = A - q\,x = \frac{q\,l}{2} - q\,x\,.$$

Bending moment (15)

$$M = A\,x - q\,x\,\frac{x}{2} = \frac{q\,l}{2}\,x - \frac{q\,x^2}{2}$$

where q = the intensity of the load or the load per unit length of beam.

The shearing force and the bending moment at any point on a beam carrying a uniformly-varying load is found from:

Shear force

$$Q = A - \frac{q\,x}{2} = \frac{q_0\,l}{6} - \frac{q_0\,x^2}{2\,l}\,.$$

Bending moment (16)

$$M = A\,x - \frac{q\,x}{2}\,\frac{x}{3} = \frac{q_0\,l}{6}\,x - \frac{q_0}{6\,l}\,x^3$$

where q_0 = the maximum load intensity.

The reaction A and B and the bending moment M of a beam carrying concentrated loads may be found graphically, as shown in Fig. 29. In the link polygon a line is drawn between the intersection of the first ray and the line of action of the reaction A and the intersection of the last ray and the line of action of the reaction B. A line parallel to that line drawn in the force polygon through the pole delineates the reactions A and B. The vertical intercept y in the link polygon multiplied by the pole distance H represents the bending moment.

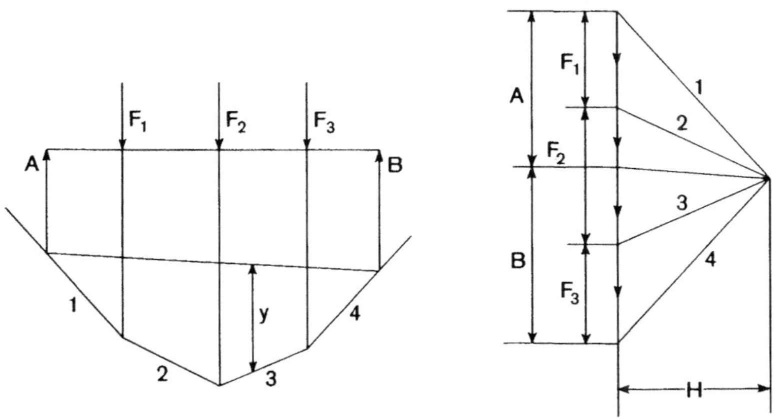

Fig. 29 Graphical determination of the reactions in a beam

The relationships between the load, the shear force and the bending moment are as follows: The first derivative of the bending moment is the shear force. The first derivative of the shear force and the second derivative of the bending moment are equal to the load intensity.

Mathematically expressed:

$$\frac{dM}{dx} = Q$$
$$\frac{dQ}{dx} = \frac{d^2M}{dx^2} = q \,. \tag{17}$$

The condition for maximum bending moment is: $\frac{dM}{dx} = Q = 0 \,.$

The bending moment has its maximum value at that section at which the shear force is zero.

Hinged beam

A hinged beam needs for each hinge an additional roller support. From the condition, that the bending moment for each hinge is zero, for every hinge there is an additional equation. Thus the beam remains statically determinate. For instance for a beam with two hinges, G_1 and G_2 the reactions can be calculated from the equations:

$$\Sigma F_x = 0\,, \quad \Sigma F_y = 0\,, \quad \Sigma M = 0\,, \quad M_{G1} = 0\,, \quad M_{G2} = 0 \tag{18}$$

where M_{G1} and M_{G2} are the bending moments at the two hinges, respectively.

Three hinged arc

The reactions A and B can be calculated from the three equations of equilibrium and the conditions for the hinge, $M_G = 0$.

Graphically the reactions may be found from the condition that the lines of action of three forces in equilibrium meet in a common point. The line of action of the reaction B on the unloaded portion must pass through the hinge, since there is no bending moment at G (Fig. 30).

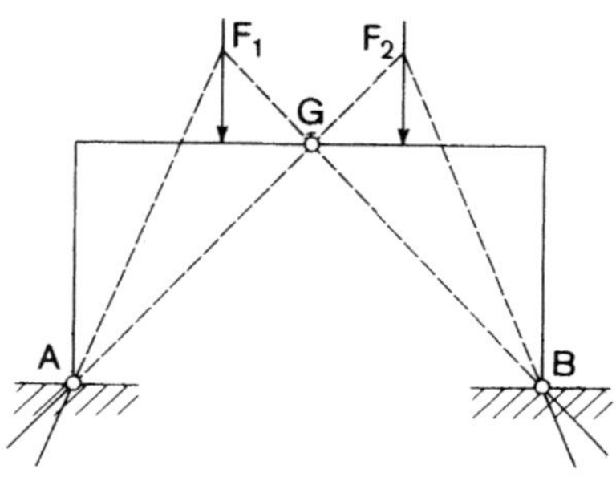

If the two sides of the three-hinged arc are loaded, the reactions may be found graphically by means of the principle of superposition. The principle of superposition states that the effect of a number of agents in a linear system acting at the same time is equal to the vector sum of the

Fig. 30 Determination of the directions of reactions on a three-hinged arc single effects produced by each agent acting separately. According to this principle the reactions A and B may be found as a vector sum of the reactions A_1 and B_1 produced by the force F_1 and the reactions A_2 and B_2 produced by the force F_2.

The force acting at the hinge is found from the condition that one portion of the three-hinged arc must be in equilibrium by itself, as demonstrated in Fig. 31.

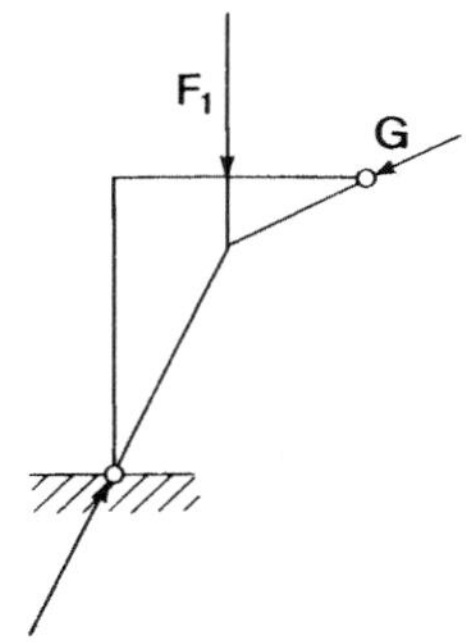

Fig. 31 Force acting of the hinge of a three-hinged arc

Trusses

A truss is a structure composed of members which are connected in such a way that they form triangles (Fig. 32).

Simplifying assumptions:

The members are connected by frictionless pins. All members of the truss and all forces acting on the truss lie in one plane.

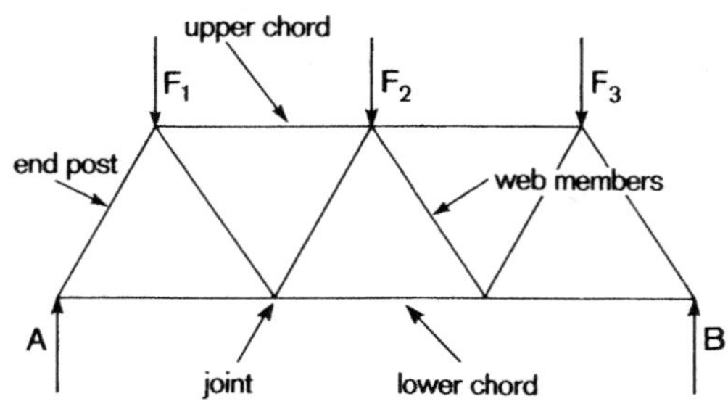

Fig. 32 Truss

The weight of members is neglible compared with the acting forces. The loads and the reactions act on the pins. Thus the internal forces in the members act along their axes. The stresses in the members may be compressive or tensile.

A truss is statically determinate, if the number of the members is

$s = 2p - 3$ where $p =$ the number of joints;
if $s < 2p - 3$ then the truss is a mechanism;
if $s > 2p - 3$ then the truss is statically indeterminate.

Reciprocal force diagram

Between the space diagram and the force diagram the following relationship exists: For each point in the space diagram there is a force polygon in the force diagram and for each point in the force diagram there is a triangle in the space diagram (Fig. 33).

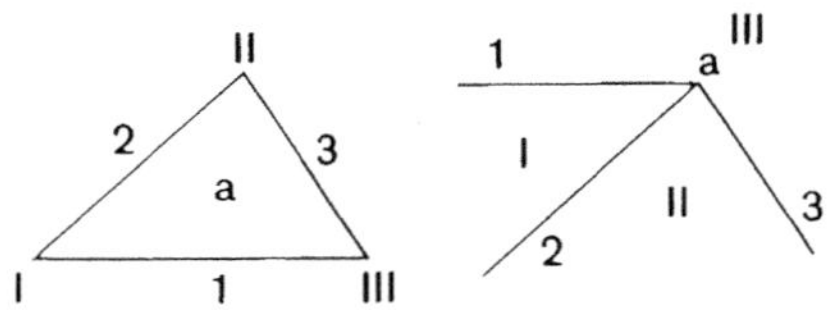

space diagram force diagram

Fig. 33

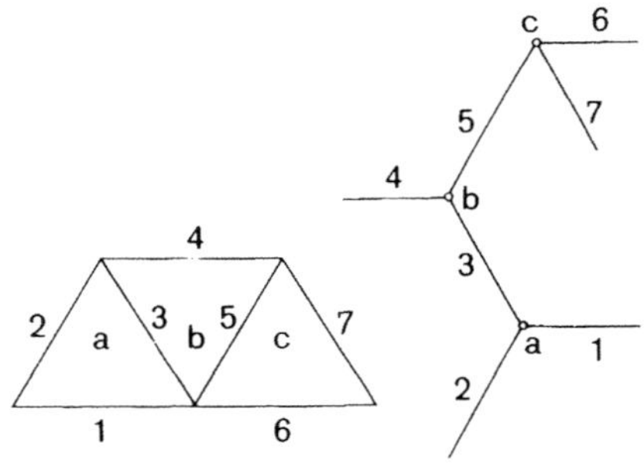

space diagram force diagram

Fig. 34

As shown in Fig. 34, two adjacent panels *a* and *b*, and *c* and *b*, in the space diagram have common lines 3 and 5. These lines correspond in the force diagram to the connecting lines between the points *a* and *b* and the points, *c* and *b*. Hence, the point *b* in the force diagram may be found as the point of intersection of the lines 3 and 5. Similarly, all points in the force diagram may be found. Fig. 35 shows a reciprocal force diagram for a truss. The arrows of the vectors are found from the vector polygons corresponding to the single joints of the truss.

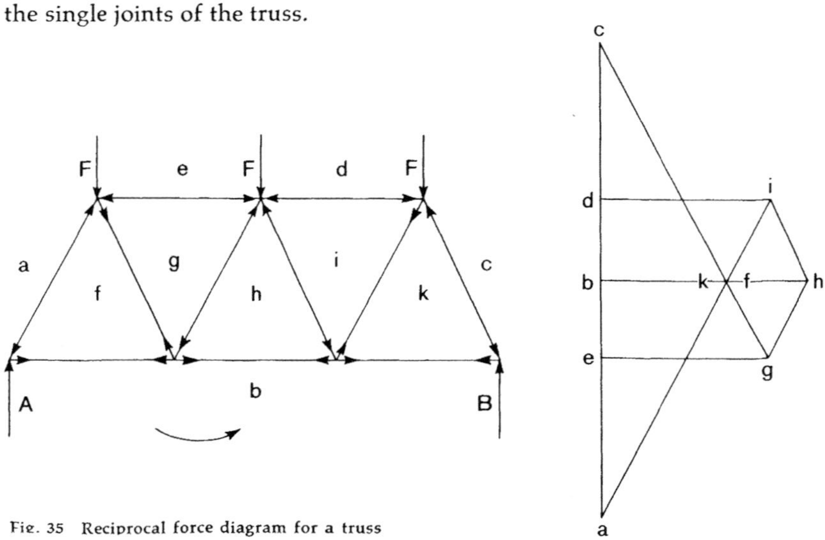

Fig. 35 Reciprocal force diagram for a truss

2.1.5. Centre of Gravity and Centroid

All particles of a body are attracted by practically parallel forces to the centre of the earth. The resultant force is called the weight of the body. That point through which, at any position of the body, the resultant force acts is called the centre of gravity of the body.

The coordinates of the centre of gravity x_s, y_s and z_s can be calculated, by taking moments about the axes x, y and z, from the equations:

$$\begin{aligned} G\,x_s &= \Sigma\,G_i\,x_i \\ G\,y_s &= \Sigma\,G_i\,y_i \\ G\,z_s &= \Sigma\,G_i\,z_i \end{aligned} \qquad (19)$$

where

$G =$ the weight of the body
$G_i =$ the weight of a particle and
x_i, y_i and $z_i =$ distances of the particles from the axes.

Centroid of areas

That point of an area, corresponding to the centre of gravity of a thin uniform flat sheet of the same contours of the area, is called the centroid of the area.

The centroid of an area can be calculated from the equation:

$$\begin{aligned} A\,x_s &= \Sigma\,A_i\,x_i \\ A\,y_i &= \Sigma\,A_i\,y_i \end{aligned} \qquad (20)$$

in which A is the total area and A_i a partial area. If an area is divided into two partial areas and the connecting line between the centroids of the areas is chosen as the axis of x (Fig. 36) it then follows from the second equation that:

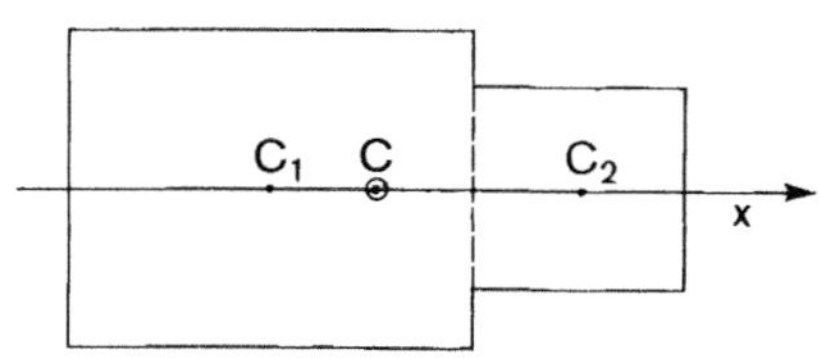

Fig. 36

$$A\,y_s = A_1\,y_1 + A_2\,y_2 = 0\,.$$

But since y_1 and y_2 are both zero, $y_s = 0$.

The centroid of an area lies on the line joining the centroids of two partial areas.

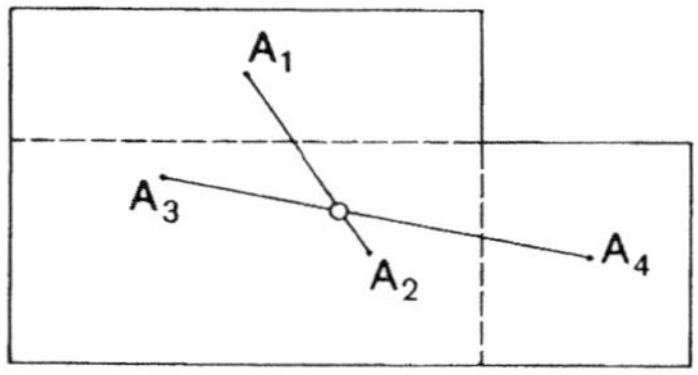

Fig. 37 Graphical determination of centroid

In Fig. 37 a graphical method of obtaining the centroid of an area is shown. The centroid of the given area is found as the point of intersection of the line joining the centroids of the partial areas 1 and 2 and the line joining the centroids of the partial areas 3 and 4.

The centre of gravity of a plate can also be found by trial. The plate is suspended at any point and a vertical line is drawn on the plate. Then the plate is suspended at some other point and another vertical line is drawn. The intersection of the two lines gives the centre of gravity of the plate.

Theorems of Guldinus

The area of a surface of revolution produced by a curve revolving about an axis which lies in the plane of the curve is equal to the product of the lengths of the curve and the path of the centroid of the curve.

The volume of a solid of revolution produced by a plane area revolving about an axis which lies in the plane of the area is equal to the product of the area and the length of the path of the centroid of the area.

2.1.6. Friction

When two bodies which are pressed together with a normal pressure N move or tend to move with respect to each other a force acts in the plane of contact which tends to resist this motion. This resisting force is called "friction". It acts on a body in a direction opposite to its motion.

As long as motion between the pressed bodies does not occur the resistance to motion is known as "static friction". Its greatest amount, when motion is about to begin, is called the limit of static friction. When two bodies move relative to each other, the resisting force is called "kinetic friction". The limit of static friction is generally greater than the kinetic friction. The ratio of the limit of static friction to the normal pressure is defined as the coefficient of friction

$$\mu = \frac{F_f}{N}.$$

The magnitude of the coefficient of friction depends upon the roughness of the rubbing surfaces and the material of which the surfaces are made. It is independent of the area of the rubbing surfaces and the magnitude of the pressure.

Angle of friction and cone of friction

If a force of any magnitude is acting on a block, which is considered, to be without weight, then the block will begin to move at a certein angle ϱ of the force with respect to the normal. As Fig. 38 demonstrates, the force R can be resolved into the vertical component N and the horizontal component F_f. The horizontal component represents the force of friction.

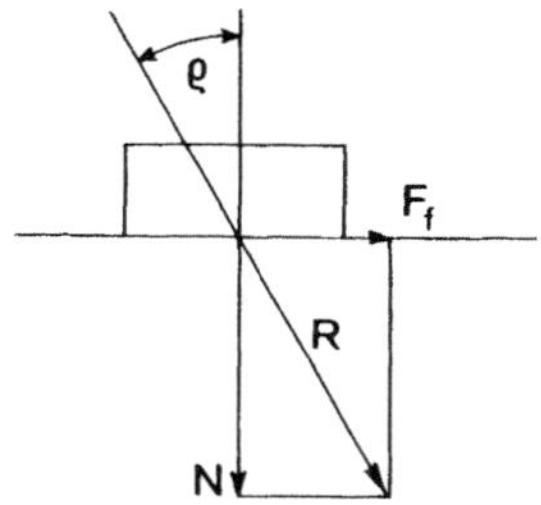

Fig. 38

Then,

$$F_f = N \tan \varrho = N \mu \,. \tag{21}$$

The angle ϱ is called the angle of friction.

The value of the least force F_f which will allow motion of a block up an inclined plane may be calculated from Fig. 39. Thus,

$$F_f = G \tan (\alpha + \varrho) \quad \text{where } G = \text{the weight of the block.} \tag{22}$$

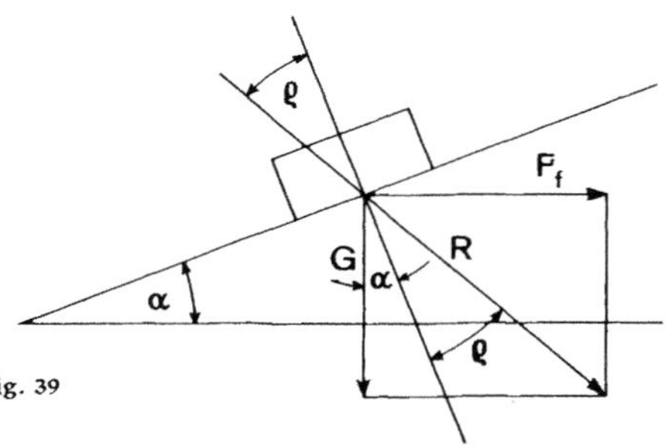

Fig. 39

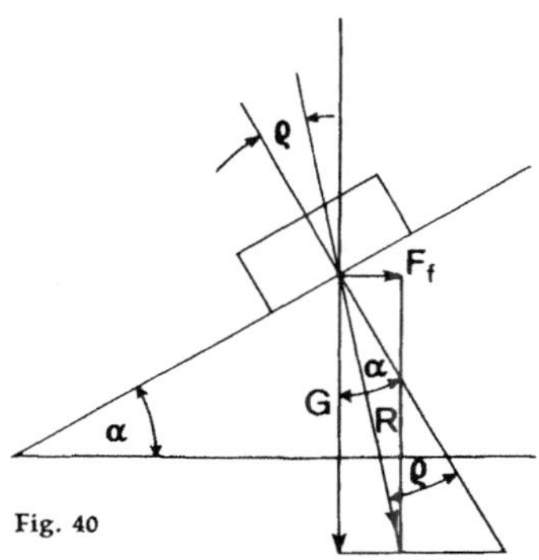

Fig. 40

If the block will move down the inclined plane (Fig. 40) the value of the horizontal least force which tends to prevent the motion may be computed from the equation

$$F_f = G \tan (\alpha - \varrho) \,. \tag{23}$$

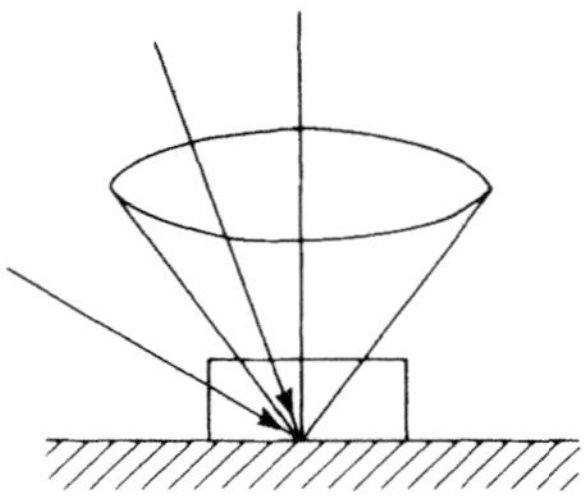

All lines which include the angle of friction with the normal line lie on the surface of a cone (Fig. 41). That cone bounded by these lines is called the "cone of friction". If a force acting on a block lies within the cone no motion occurs; if the force lies outside the cone motion occurs; if the force lies on the surface of the cone, motion is impending.

Fig. 41 Cone of friction

2.2. Dynamics

Dynamics is the study of the motion of bodies. Dynamics is divided into kinematics and kinetics. Kinematics deals with the motion of bodies without reference to the forces causing the motion. Kinetics deals with the relationship of the motion of bodies and the forces producing the motion.

2.2.1. Kinematics

There are two kinds of motion: Translation and Rotation. If a body moves from one place to another without rotating, it is said to undergo translation. It may move in a straight line or in a curve. When all particles of a body move about a fixed axis, the body is said to rotate. General motion may be regarded as composed of translation and rotation. In pure translation all particles of a body have the same velocity.

2.2.1.1. Plane Motion of a Particle

If a body only undergoes translation or, if the rotation of the body in regard to its translation may be neglected, the body can be considered as a particle and it is only necessary to study the motion of the particle.

Rectilinear motion of a particle

Velocity is defined as the time rate of change of displacement. It is a vector quantity, whereas speed only represents the magnitude of a velocity without reference to the direction and is a scalar quantity.

If displacement is represented by s and time by t, then velocity is given by

$$v = \frac{ds}{dt} .$$

When velocity is constant, the above relation becomes

$$v = \frac{s}{t}.$$

Acceleration is defined as the time rate of change of velocity. It is given by

$$a = \frac{dv}{dt} = \frac{d^2s}{dt^2}.$$

When acceleration is constant, the above relation becomes

$$a = \frac{v - v_0}{t}$$

where v = final velocity and v_0 = initial velocity.

By rearranging and integrating the above equations we obtain

$$v = \int a \, dt \quad \text{and} \quad s = \int v \, dt = \iint a \, dt^2 .$$

Special cases of rectilinear motion

Constant velocity:

$$v = \text{constant}, \qquad s = v\,t, \qquad a = 0. \tag{24}$$

Constant acceleration:

$$a = \text{constant}, \qquad v = a\,t + v_0$$

$$s = \frac{1}{2} a\,t^2 + v_0\,t. \tag{25}$$

When the velocity of a body decreases, the acceleration becomes negative and is then called "retardation".

Problem: Two cars, A and B start from rest with uniform acceleration. The acceleration of A is 1 m/s² and the acceleration of B is 1.2 m/s². B is 8 m behind A. How long will it take B to overtake A and what is the distance travelled by B?

Solution: Let t represent the time for each car and x the distance travelled by B

then for B $\quad x = \frac{1}{2}\, 1.2\, t^2$

and for A $\quad x = 8 + \frac{1}{2}\, 1\, t^2$

by equating the equations we obtain

$t = 8.95$ s and $x = 48$ m.

Freely falling body, body projected vertically downward and upward

When a body falls toward the earth, it moves with an acceleration which can be considered constant. This acceleration is called the "acceleration due to gravity" and is denoted with g. Its value varies slightly with the distance from the earth's centre and is for sea level 9.81 m/s². Since the resistance of the air may generally be neglected, the formulae 25 can be used.

For a freely falling body with

$$v_0 = 0, \qquad a = g \quad \text{and} \quad s = h$$

$$v = g\,t \quad \text{and} \quad h = \frac{1}{2}\,g\,t^2 \tag{26}$$

from

$$h = \frac{1}{2}\,g\,t^2, \quad \text{it follows that} \quad t = \sqrt{\frac{2\,h}{g}}.$$

The final velocity is then

$$v = \sqrt{2\,g\,h} \tag{27}$$

where h = the distance travelled by the falling body.

When a body is projected vertically downward with a given initial velocity, then the final velocity is

$$v = v_0 + g\,t \text{ and the distance travelled is } h = v_0\,t + \frac{1}{2}\,g\,t^2. \tag{28}$$

When a body is projected vertically upward with a given initial velocity, then the velocity at a given time is

$$v = v_0 - g\,t \quad \text{and the height} \quad h = v_0\,t - \frac{1}{2}\,g\,t^2. \tag{29}$$

The height has its maximum, when $v = 0$

$$h_{max} = v_0\,t - \frac{1}{2}\,g\,t^2$$

substituting $g\,t$ for v_0 gives

$$h_{max} = \frac{v_0^2}{2\,g}. \tag{30}$$

Plane curvilinear motion of a particle

The velocity of a body moving along a curved path is at any given point in the direction of the tangent to the path at that point (Fig. 42). The motion can be considered as rotation about the instantaneous centre of curvature with an angular velocity ω.

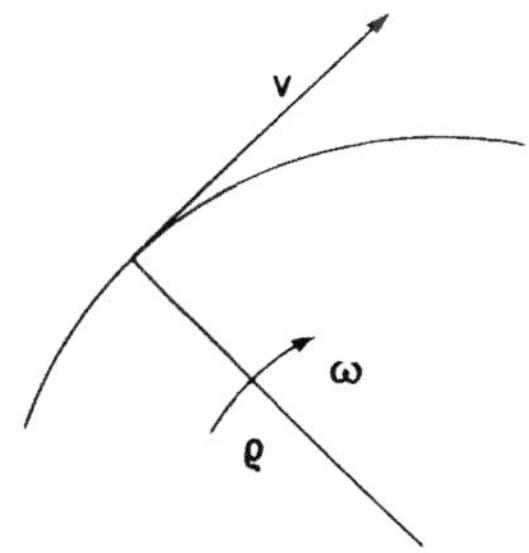

Fig. 42 Velocity of a particle moving along a curved path

The velocity of the particle may be calculated from

$$v = \varrho\, \omega$$

where ϱ = radius of curvature
and ω = angular velocity.

The velocity may change both its magnitude and its direction. From the change of the magnitude of the velocity there is a tangential acceleration. From the change of the direction of the velocity there is a normal acceleration. The total acceleration is therefore composed of two components: the tangential acceleration taken in the direction of the velocity and the normal acceleration taken in a direction perpendicular to the velocity.

$$a = a_n \widehat{+} a_t \,.$$

The normal acceleration is calculated, as in Fig. 43, in the following way

$$a_n = \frac{dv_n}{dt} = v \frac{d\varphi}{dt} = v\,\omega = \varrho\,\omega^2 = \frac{v^2}{\varrho}\,. \tag{31}$$

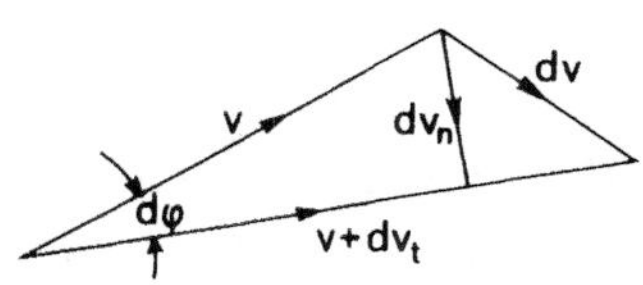

Fig. 43 Determination of the normal acceleration

The tangential acceleration is found to be

$$a_t = \frac{dv_t}{dt} = \varrho \frac{d\omega}{dt} = \varrho\,\varepsilon \tag{32}$$

in which $\varepsilon = \dfrac{d\omega}{dt}$ = angular acceleration.

Special cases of curvilinear motion

Rotary motion

If the radius of curvature is constant, the velocity is:

$$v = r\,\omega\,,$$

and the acceleration is:

$$a = r\,\omega^2 + r\varepsilon\,. \tag{33}$$

Motion of a projectile

The motion of a projectile is influenced by the initial velocity, gravity and by other factors such as air resistance, rotation of the projectile caused by rifling within the barrel. Here the motion will only considered to be influenced by the initial velocity of the projectile and by gravity. Air resistance and rotation of the projectile will be neglected.

The projectile projected with a given initial velocity the vector of which includes a given angle α with the horizontal describes a path which is called the trajectory (Fig. 44).

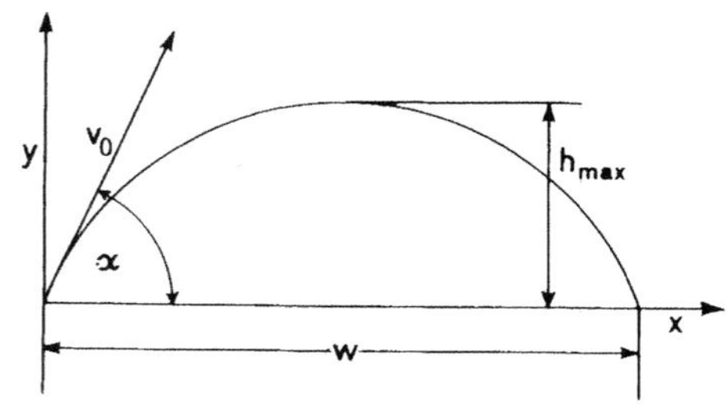

Fig. 44 Trajectory of a projectile

The equation of the trajectory is given by

$$y = x \tan \alpha - \frac{g\,x^2}{2\,v_0 \cos^2 \alpha} \tag{34}$$

where g = the acceleration due to gravity.

The horizontal and vertical components of the velocity at any point of the path are given by

$$v_x = v_0 \cos \alpha \quad \text{and} \quad v_y = v_0 \sin \alpha - g\,t\,. \tag{35}$$

The horizontal range is

$$w = \frac{v_0^2 \sin 2\,\alpha}{g}\,. \tag{36}$$

It is greatest, when the direction of the initial velocity is at an angle of 45° with the horizontal. The greatest range is given by

$$w_{\text{max}} = \frac{v^2}{g}\,.$$

The greatest elevation attained by the projectile is:

$$h_{max} = \frac{v_0^2 \sin^2 \alpha}{2g} . \tag{37}$$

The time which elapses during the projection is:

$$T = \frac{w}{v_0 \cos \alpha} . \tag{38}$$

2.2.1.2. Coplanar Motion of a Rigid Body

Rotation of a plate

The velocity and the acceleration of any point of a rotating plate is given by the equation 33

$$v_B = r_B \, \omega$$

$$a_B = r_B \, \omega^2 + r_B \, \varepsilon .$$

The angular velocity of the plate is, referring to Fig. 45,

$$\omega = \frac{v_B}{r_B} = \frac{v_C}{r_C} = \tan \alpha .$$

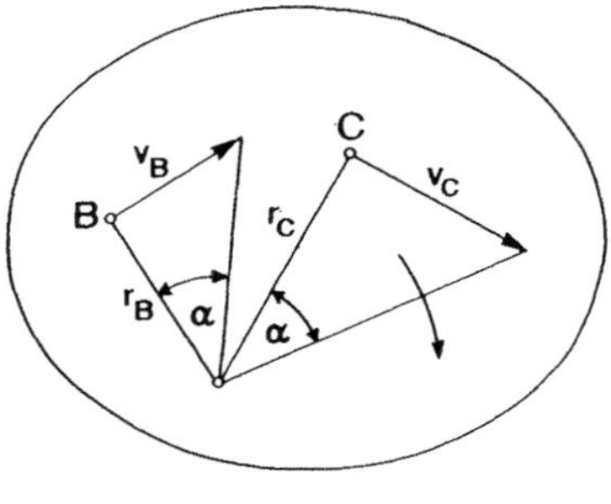

Fig. 45 Determination of the velocity of a point on a rotating plate

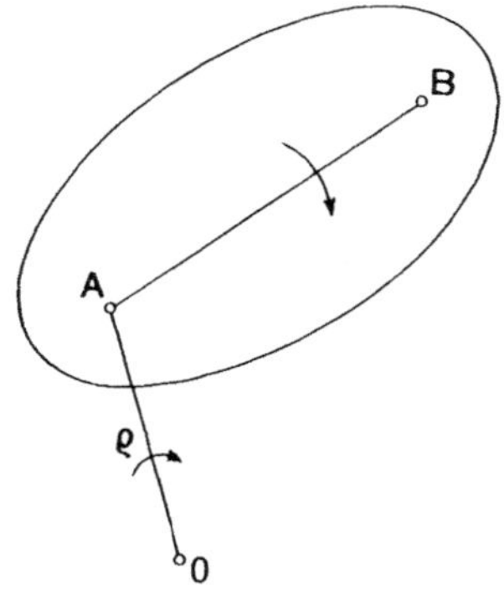

Fig. 46

Translation and rotation of a plate

The motion of any point B on a plate can be considered as the motion of another point A on the plate and a superposed rotation about the point A. The motion of the point A can be considered as a rotation about an instantaneous centre 0 (Fig. 46).

Therefore

$$v_B = v_A \widehat{+} v_{BA} \quad \text{and}$$
$$a_B = a_A \widehat{+} a_{BA} . \tag{39}$$

Velocity of a point of a four- bar linkage

The principal parts of a four- bar linkage are: the frame, the two cranks and the connecting rod (Fig. 47).

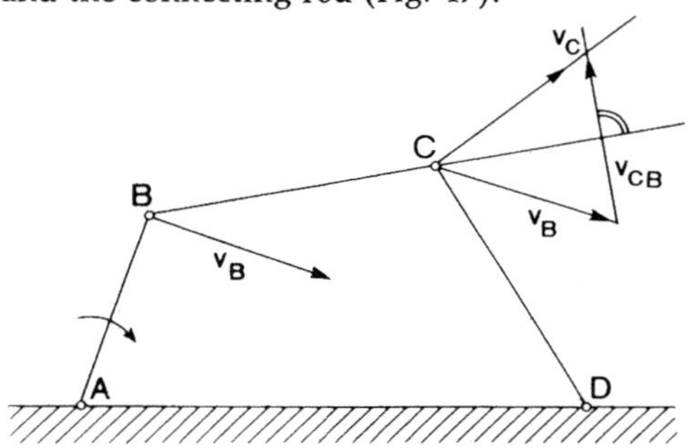

Fig. 47 Determination of the velocity of a point on a four-bar linkage

The velocity of the point C is found as the vector sum of the vectors v_B and v_{CB}.

The magnitude and the direction of the vector v_B is known. The direction of the vector v_{CB} is perpendicular to the line $B\,C$. The direction of the vector v_C is perpendicular to the line $C\,D$.

Graphical determination of velocities of points on a plate by means of the instantaneous centre

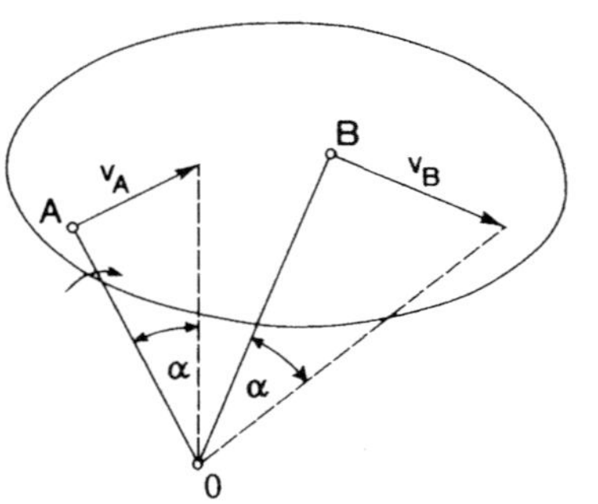

Fig. 48
Instantaneous centre of a moving plate

For a plate having any coplanar motion exept pure translation there is a point inside or outside the boundaries of the plate which at one instant remains at rest. This point is called the "instantaneous centre". For an instant the motion of the plate can be considered as rotation about this point. The instantaneous centre is found as the point of intersection of two lines through any two points A and B perpendicular to the velocities of these points as shown in Fig. 48.

In Fig. 49 it can be seen that the velocity of a point of a four-bar-linkage may be found by means of the instantaneous centre.

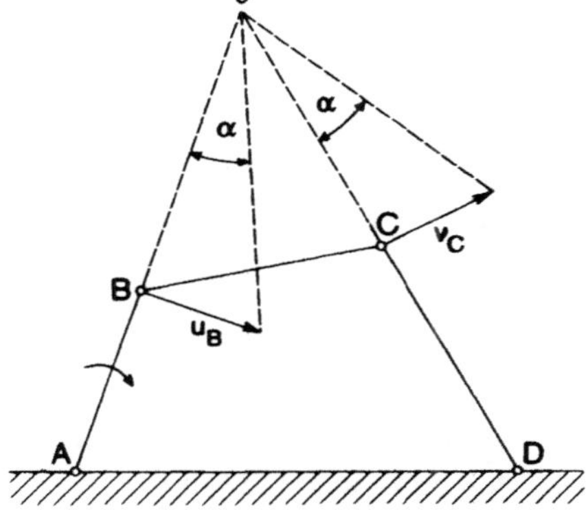

Fig. 49 Determination of the velocity of a point on a four-bar linkage by means of instantaneous centre

Relative motion

In a moving guide attached to a plate, a block is sliding. The relative velocity of the block with respect to the plate is denoted with v_r. The motion of the plate can be considered as rotation about an instantaneous centre 0 (Fig. 50). The velocity of the block is composed of the velocity v_g of the block fixed on the plate and the relative velocity v_r.

$$v_B = v_g + v_r$$

The acceleration of a point (C) on a for-bar-linkage (Fig. 47) can be found from the vector sum of the vectors a_B and a_{CB}. The acceleration of a block sliding in a moving guide is composed of the acceleration of the block fixed on the plate and the relative acceleration of the block with respect to the plate and a further acceleration called "Coriolis acceleration".

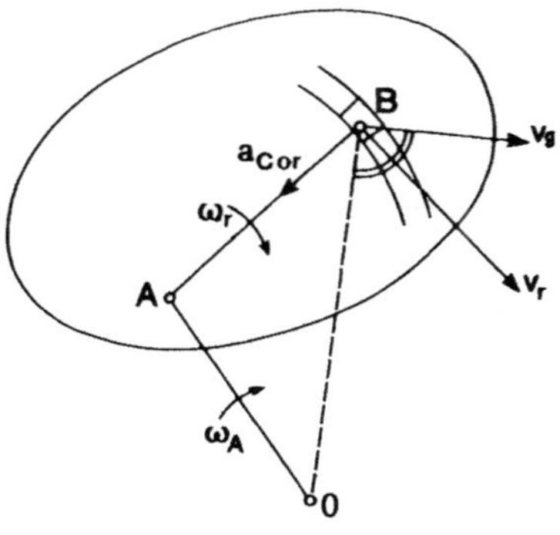

Fig. 50 Determination of the velocity of a point which moves on a moving plate

$$a_B = a_g \widehat{+} a_r \widehat{+} a_{Cor} \,. \qquad (41)$$

The Coriolis acceleration is given by $a_{Cor} = 2\, v_r\, \omega_A$ in which $\omega_A =$ the angular velocity of the plate.

2.2.2. Kinetics

2.2.2.1. Newton's Second Law of Motion for Translation

A body subjected to the action of an unbalanced force will experience an acceleration proportional to the force and inversely proportional to the mass of the body. This relation, known as „Newton's second law of motion" is expressed by the equation

$$a = \frac{F}{m}. \tag{42}$$

The acceleration is in the direction of the force. If several forces act on the body, the acceleration is produced by the resultant force and is therefore in the direction of the resultant force. If no force is acting on the body, then it will not experience an acceleration. In this case the body remains in its state of rest or uniform motion in a straight line. This is implicit in Newton's first law of motion.

If the acting force is the weight of the body, the body experiences accelerations due to gravity.

$$g = \frac{G}{m} \quad \text{or} \quad m = \frac{G}{g}. \tag{43}$$

The mass can therefore be defined as weight/acceleration due to gravity. If an additional frictional force is acting on the body, the difference between the external force and the frictional force will produce the acceleration.

$$F_a = F - F_f = m\,a\,.$$

Problem: A horizontal force $F = 30$ kp is acting on a block $G = 40$ kp which slides on a 30° inclined plane. The coefficient of friction is $\mu = 0.1$ (Fig. 51). Determine the acceleration of the block.

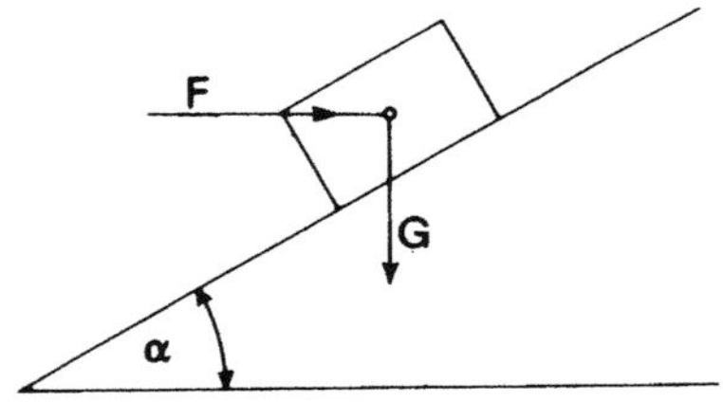

Fig. 51

Solution: The frictional force is calculated from

$$F_f = (G \cos \alpha + F \sin \alpha)\,\mu = (40 \text{ kp} \times \cos 30° + 30 \text{ kp} \times \sin 30°) \times 0.1$$

$$F_f = 4.96 \text{ kp}\,.$$

The external force in the direction of motion is

$$F' = (F \cos \alpha - G \sin \alpha) = 30 \text{ kp} \times \cos 30° - 40 \text{ kp} \times \sin 30° = 6 \text{ kp}\,.$$

The acceleration is then

$$a = \frac{F' - F_f}{m} = \frac{(6 \text{ kp} - 4.96 \text{ kp}) \times 9.81 \text{ m/s}^2}{40 \text{ kp}} = 0.255 \text{ m/s}^2\,.$$

D'Alembert's principle

A body on which an unbalanced force is acting experiences acceleration and is therefore not in equilibrium. On each particle of the body a force of inertia is acting which is in a direction opposite to the direction of the acceleration of the body. The resulting force of inertia acts on the centre of gravity of the body and has the magnitude of the actual force. If the resulting force of inertia is added to the actual force, then the system of forces is in equilibrium and the equations of static equilibrium may be applied. This statement is knwon as D'Alembert's principle and the added force is called: "D'Alembert force or inertia force".

Problem: A block $G_1 = 65$ kp is connected to a block $G_2 = 25$ kp hanging vertically by a cable running over a pulley, as shown in Fig. 52. The coefficient of friction is $\mu = 0.3$. Determine the acceleration of the two blocks and the cable tension.

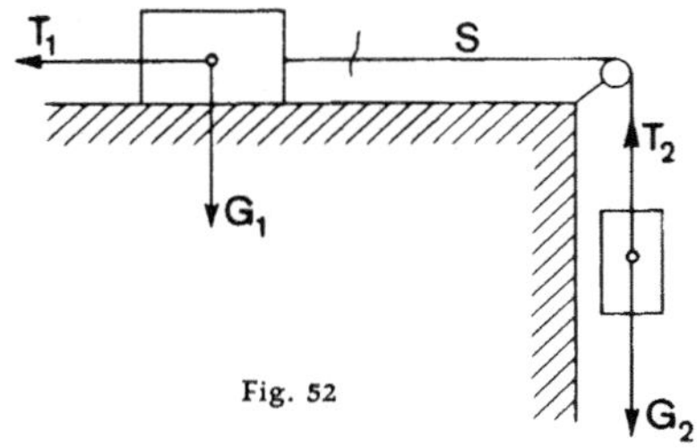

Fig. 52

Solution: The unbalanced force is

$$F_a = G_2 - \mu\, G_1 = 25\ \text{kp} - 0.3 \times 65\ \text{kp} = 5.5\ \text{kp}.$$

The acceleration is found from

$$a = \frac{F_a}{m_1 + m_2} = \frac{5.5\ \text{kp} \times 9.81\ \text{m/s}^2}{65\ \text{kp} + 25\ \text{kp}} = 0.6\ \text{m/s}^2 .$$

The cable tension is

$$S = G_2 - T_2 = 25\ \text{kp} - \frac{25\ \text{kp} \times 0.6\ \text{m/s}^2}{9.81\ \text{m/s}^2} = 23.47\ \text{kp} \quad \text{or}$$

$$S = \mu\, G + T_1 = 0.3 \times 65\ \text{kp} + \frac{65\ \text{kp} \times 0.6\ \text{m/s}^2}{9.81\ \text{m/s}^2} = 23.47\ \text{kp} .$$

The inertia force of a body rotating about an axis with uniform velocity is called "centrifugal force"

$$C = m\, a_n = m\, r\, \omega^2 \tag{44}$$

2.2.2.2. Newton's Second Law of Motion for Rotation

Consider a force, acting on a rigid body, which causes the body to rotate about an axis. According to D'Alembert's principle the body is in equilibrium, when to the actual force the forces of inertia are added.

$$F - \int dm\, a = 0 \quad \text{or} \quad F = \int dm\, a = \int dm\, r\, \varepsilon .$$

By multiplying both sides of the equation by r we obtain

$$F\, r = \int r^2\, dm\, \varepsilon \quad \text{or} \quad M = I_m\, \varepsilon . \tag{45}$$

$I_m = \int r^2 \, dm$ is called the "moment of inertia of mass". For bodies with constant thickness, referring to Fig. 53,

$$I_{mz} = \int r^2 \, dm = \int r^2 \frac{s\,\gamma}{g} \, dA = \frac{s\,\gamma}{g} I_p \quad (46)$$

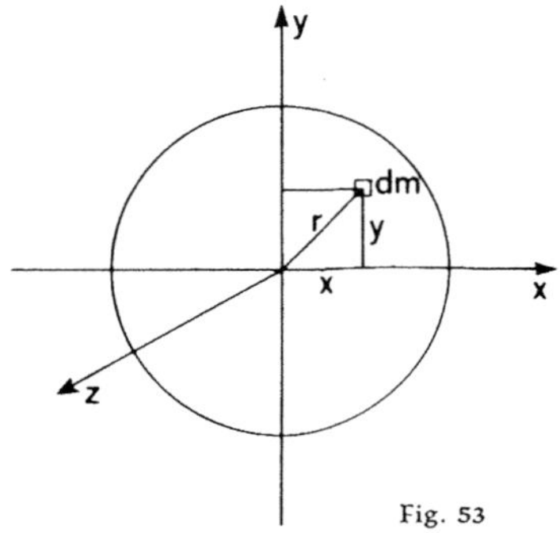

Fig. 53

where s = thickness of the body
γ = specific gravity
g = acceleration due to gravity
I_p = polar moment of inertia.

$$I_p = \int r^2 \, dA = \int y^2 \, dA + \int x^2 \, dA = I_{Ax} + I_{Ay} \quad (47)$$

where I_{Ax} and I_{Ay} are the axial moments of inertia with respect to the x-axis and the y-axis, respectively.

The polar moment of inertia is equal to the sum of its two perpendicular axial moments of inertia.

2.2.2.3. Work, Energy and Power

Work

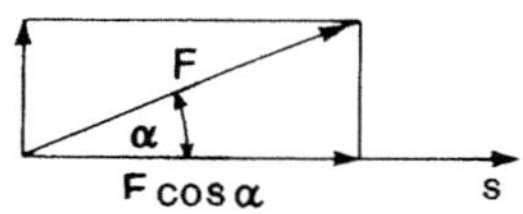

Fig. 54

When a change in the position of a body is done against a resistance, the force does work. The work is the product of the component of the force in the direction of displacement and the displacement (Fig. 54).

$$A = \int F \cos \alpha \, ds \,. \quad (48)$$

The work done by a freely falling body is

$$A = G\,h \quad (49)$$

where G = weight of body and h = height.

The work done by a helical spring is

$$A = \int F_x \, dx = c \int x \, dx = \frac{c\,f^2}{2} \quad (50)$$

where c = Spring stiffness it is that force which stretches as or compresses unit length of spring, respectively.

and f = the increase or decrease, respectively of the length of the spring.

Energy

Energy is the ability of a body to do work. There are two forms of energy in mechanics: potential and kinetic energy.

If a body of weight G is lifted to a height h, then it possesses the potential energy or the energy of position W.

$$W = G\,h\,. \tag{51}$$

A body in motion is capable of doing work and is said to possess kinetic energy. If a body is falling from a higher to a lower position, its potential energy is transformed into kinetic energy. The change in potential energy is equal to the gain in kinetic energy that the body received after falling. Expressed mathematically this is,

$$W = G\,h = m\,g\,\frac{1}{2}\,g\,t^2 = \frac{m}{2}\,(g\,t)^2 = \frac{m\,v^2}{2}\,. \tag{52}$$

Potential energy can be transformed into kinetic energy and conversely.

Kinetic energy of rotation

The kinetic energy of a differential mass is

$$\mathrm{d}W = \frac{1}{2}\,\mathrm{d}m\,v^2 = \frac{1}{2}\,\mathrm{d}m\,r^2\,\omega^2\,.$$

The kinetic energy of the entire body becomes

$$W = \frac{\omega^2}{2}\int \mathrm{d}m\,r^2 = \frac{I_m\,\omega^2}{2}\,. \tag{53}$$

Power

Power is the time rate of doing work.

Thus

$$P = \frac{\mathrm{d}A}{\mathrm{d}t} = \frac{F\,\mathrm{d}s\cos\alpha}{\mathrm{d}t} = F\,v\cos\alpha\,. \tag{54}$$

Work may be written thus,

$$A = \int P\,\mathrm{d}t\,.$$

2.2.2.4. Impulse, Momentum and Impact

Linear impulse and linear momentum

Newton's second law of motion for translation may be written thus

$$F = m\,a = m\,\frac{\mathrm{d}v}{\mathrm{d}t} = \frac{\mathrm{d}\,(m\,v)}{\mathrm{d}t} \quad \text{or} \quad \int F\,\mathrm{d}t = m\,v_1 - m\,v_2\,. \tag{55}$$

The product of mass and velocity is called "linear momentum". It is a vector quantity. The integral of a force and an increment of time is called "linear impulse". It is also a vector quantity.

The above formula (55) gives the relation between linear impulse and linear momentum. If no force is acting on a body, the momentum of a body remains constant.

Impact

If two bodies collide, their collision is called "impact". If the directions of motion of the two bodies are perpendicular to the plane of contact, the impact is direct, otherwise the impact is oblique. If the normal of the plane of contact passes through the centres of gravity of the two bodies, the impact is central, otherwise the impact is eccentric. Here only the direct, central impact will be discussed. The time from the beginning of contact until the greatest deformation is known as the "period of deformation" and the time from the greatest deformation until the end of contact as the "period of restitution". The velocity u at the end of the period of deformation can be calculated from the condition, that the momentum of the two bodies before and after the period of deformation remains constant.

Expressed mathematically,

$$m_1\,v_1 + m_2\,v_2 = (m_1 + m_2)\,u\,.$$

Whence,

$$u = \frac{m_1\,v_1 + m_2\,v_2}{m_1 + m_2}\,. \tag{56}$$

If the two bodies were inelastic, no separation would occur. The two bodies would continue their motion with the velocity that they have after the period of deformation. If, however, the two bodies were perfectly elastic, the relative velocity of separation would be equal to the relative velocity of approach.

The velocities of the two bodies after the period of restitution are therefore

$$w_1 = v_1 - 2\,(v_1 - u) = 2\,u - v_1$$
$$w_2 = v_2 - 2\,(v_2 - u) = 2\,u - v_2\,. \qquad (57)$$

Angular impulse and angular momentum

The angular momentum of a particle with respect to a given axis is the moment of its linear momentum,

$$B = mv\,r = m\,r^2\,\omega = I\,\omega\,. \qquad (58)$$

The momentum of a rigid body is obtained from

$$B = \int r\,v\,\mathrm{d}m = \int r^2\,\mathrm{d}m\,\omega = I\,\omega\,.$$

Newton's second law of motion for rotation may be written thus

$$M = I_{\mathrm{m}}\,\varepsilon = I\,\frac{\mathrm{d}\omega}{\mathrm{d}t} = \frac{\mathrm{d}\,(I\,\omega)}{\mathrm{d}t} = \frac{\mathrm{d}B}{\mathrm{d}t} \quad \text{or} \quad \int M\,\mathrm{d}t = B_1 - B_2\,. \qquad (59)$$

The expression $\int M\,\mathrm{d}t$ is called „angular impulse".

If no moment of external forces acts on a body, the angular momentum remains constant.

3. Strength of Materials

3.1. Introduction

Strength of materials is that branch of engineering science which deals with internal forces and the resulting deformations in structural and machine members. The external forces which act on a structure or a machine are called loads, the internal forces which resist the external forces are called stresses and the accompanying changes in shape and size of a member are called strains.

The loads may be divided with reference to the manner in which they act on a body in:

Static or steady loads which are nearly constant,
Continually changing loads,
Impact loads which are applied in a very short time.

The stresses and strains which occur in a member may not be greater than a permissible stress or strain which depends on factors such as the material of the member, the manner in which the load acts, the temperature of the member etc.

After the dimensions of a member have been fixed, they are normally checked with respect to their resistance to deformation.

Stresses and their respective strains are caused by normal or shearing forces and by bending or twisting moments. The normal force produces a normal stress which may be a tensile or a compressive stress. The shearing force produces a shearing stress. The bending moment produces a bending stress and the twisting moment a torsional shear stress.

In a section several stresses may exist at the same time. In a rectangular element which is in equilibrium three normal stresses and six shearing stresses may occur, as shown in Fig. 55. The strength of material in the face of an element is influenced by the stresses in two mutually perpendicular faces.

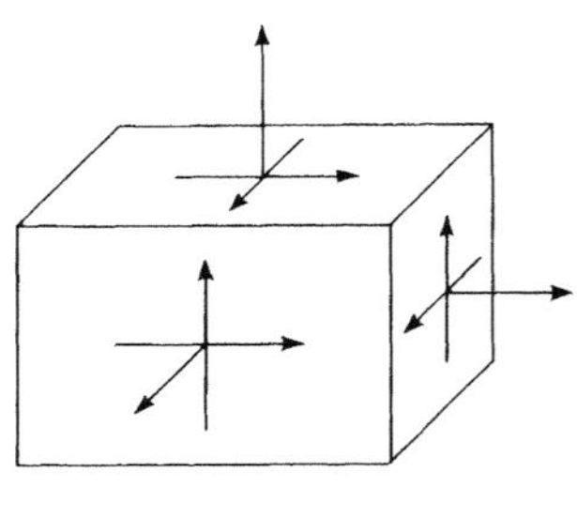

Fig. 55
Stresses on a rectangular element

The determination of Stresses

The members are replaced by simple geometrical bodies. For these idealized bodies formulae for strength can be established. It must be checked, that the error resulting from these simplifying assumptions is not too great. Stresses can only be found, if all the deformations are taken into consideration at the same time. This exact method of finding the stresses is called the "mathematical theory of elasticity". In simple members the distribution of stresses in a section may be assumed and then the magnitude of the stresses can be calculated. This method is known as the "elementary theory of the strength of materials". It must be confirmed by experiment, that the assumed distribution of stresses is correct.

Besides the theoretical methods there are some experimental methods of finding the values of the stresses. The best known methods are:

Strain measurement by extensiometer

Here the surface stresses of a specimen are determined from the strains.

Electrical resistance strain gauge measurement

At the surface of the specimen fine metal wires are attached. If the specimen is loaded, the wires are stretched and undergo a change of electrical resistance which allows the determination of the values of the stresses on the surface of the specimen to be made.

Photo-elastic stress analysis

A model of transparent material is observed in polarised light. Dark bands are produced on a screen and may be photographed. From these dark bands called "isochromatic fringes" the values of the stresses can be determined. This method also allows the stresses at any section of the specimen to be found.

3.2. Mechanical Testing of Materials

The mechanical tests give information concerning the mechanical properties of materials. The behaviour of materials in regard to resisting the external forces depends on mechanical properties such as strength, elasticity, toughness, brittleness, hardness etc.

With regard to loading the test pieces, the mechanical tests may be classified into five groupes: Tension-, compression-, shear-, torsion- and flexure tests. With respect to the rate at which the load is applied tests may be classified into the following groupes: Static tests, if a constant load is applied over a relatively short time; dynamic or impact tests, if the load is applied by striking a blow; creep tests, if a constant load is applied over a long time and fatigue tests, if the load is repeated many times.

3.2.1. Tensile Test

The most commonly-made test, from which very important information can be obtained, is the tension test, which is described now.

In a testing machine a bar, of length l, and of uniform cross-section A_0, which is usually circular, is subjected to an axial tensile load that gradually increases from zero value until the bar breaks. If F denotes any value of the load, then the stress will be $\sigma = \frac{F}{A_0}$ and the strain $\varepsilon = \frac{\Delta l}{l}$ in which Δl = elongation of the bar. The relation between the stress and the strain is represented by the stress-strain diagram (Fig. 56) which is directly obtained by an autographic attachment to the testing machine. There are some characteristic points in the stress-strain diagram of mild steel.

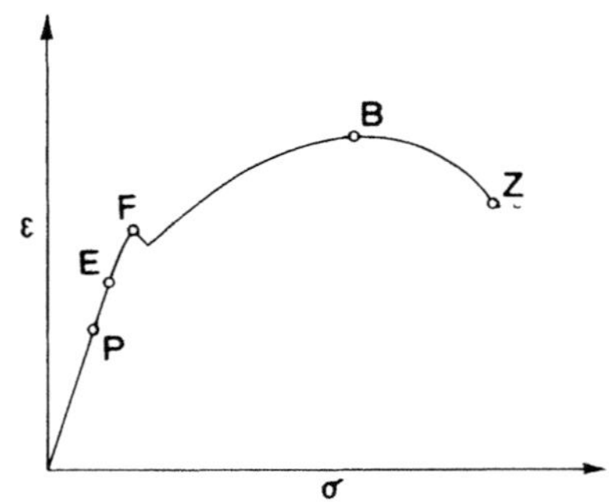

Fig. 56 Stress-strain diagram

Proportional limit (*P*).

Up to that point the stress-strain curve is nearly a straight line.

Elastic limit (*E*).

Up to that point the material has no plastic deformation upon a complete release of the load.

For steel the proportional limit coincides with the elastic limit. Normally that value of stress is taken at which the permanent deformation is 0.01 % of the gauge length.

Yield point (*F*).

From that point the material retains a permanent deformation, if the load is removed. Most of the non-ferrous metals and high-strength steels do not possess yield points. For these materials that value of stress is taken at which the permanent deformation is 0.2 % of the gauge length.

Ultimate strength (*B*).

This is the maximum stress, which a material can endure. Brittle materials break at the ultimate strength, whereas ductile materials continue to stretch.

After the ultimate strength a ductile material begins to neck. At one point of the bar the cross-section diminishes more rapidly than at other places. The true stress occuring in the neck section is much than the stress calculated by dividing the !oad by the original cross section.

Rupture strength (Z).

At this point the rupture of the bar occurs.

Hooke's law.

Within the proportional limit the ratio of the stress to the strain is constant. Expressed mathematically

$$\frac{\sigma}{\varepsilon} = E .$$

The numeral value of the constant E is called the "modulus of elasticity". When the stress and the strain are caused by axial loads the above expression may be written thus:

$$E = \frac{F\,l}{A\,\Delta l} . \tag{60}$$

Poisson's ratio.

If a bar is subjected to an axial tension or compression the bar is stretched or shortened, respectively in an axial direction and at the same time the lateral dimension of the bar decreases or increases, respectively (Fig. 57).

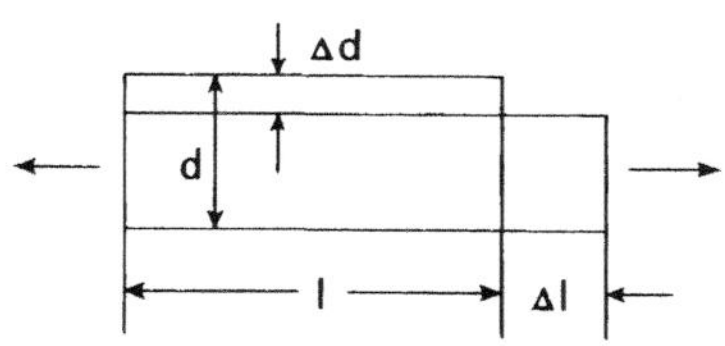

Fig. 57 Contraction of a stressed bar

The ratio of the axial strain to the lateral strain is constant and is called "Poisson's ratio"

$$\frac{\varepsilon\, l}{\varepsilon\, q} = m .$$

The numerical value of Poisson's ratio for all metals is $m = \frac{10}{3}$.

Shearing strain

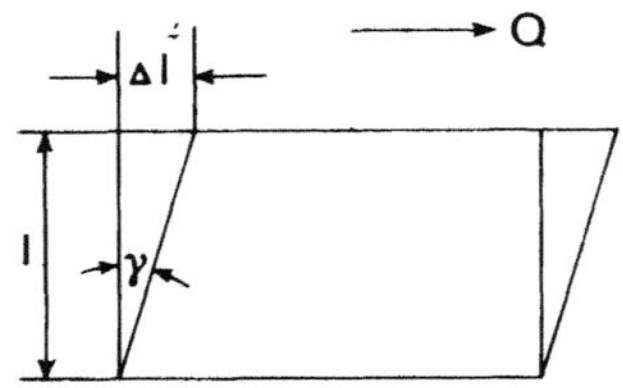

Fig. 58
Distortion caused by a shearing force

The distortion on a rectangular block caused by a shearing force is shown in Fig. 58. The shearing strain is given by $\frac{\Delta l}{l}$. As the angle γ is very small, Δl may be substituted by the arc of radius l. The shearing strain is then the angle γ expressed in radians. Within the proportional limit the ratio of the shearing stress to the shearing strain is constant.

Expressed mathematically

$$\frac{\tau}{\gamma} = G\,.$$

This is Hooke's law for shearing. The numerical value of the constant is called the "shearing modulus of elasticity" or "modulus of rigidity". The following relation exists between the modulus of elasticity and the modulus of rigidity:

$$G = \frac{m\,E}{2\,(m+1)}\,. \tag{61}$$

Strain energy

The strain energy of a body is defined as the work done by the load in straining the body.

When a tensile or a compressive load is applied to a bar and the occuring stress is below the proportional limit, the strain energy is calculated from:

$$W = \frac{1}{2}\,F\,\Delta l = \frac{1}{2}\,\sigma\,\varepsilon\,A\,l = \frac{1}{2}\,\sigma\,\varepsilon\,V \tag{62}$$

where V = the volume of the bar.

When a shearing force is applied to a rectangular block and the occuring shearing stress is below the proportional limit, the strain energy is calculated from:

$$W = \frac{1}{2}\,Q\,l = \frac{1}{2}\,\tau\,\gamma\,A\,l = \frac{1}{2}\,\tau\,\gamma\,V\,. \tag{63}$$

3.2.2. Hardness Tests

Hardness is defined as the resistance of a material to indentation. Hardness of metals may be tested by static tests, such as the Brinell test, the Vickers test, the Rockwell test or by dynamic tests such as the Shore Scleroscope test.

Brinell test

The Brinell test consists in pressing a hardened steel ball into a test specimen. In the standard test the ratio $\frac{F}{D^2}$ is 30 for steels and 10 for non-ferrous metals, where F is the applied force in kp and D is the diameter of the steel ball in mm. The full load is applied for a minimum of 10 s for steel and 30 s for non-ferrous metals. Normally a 10 mm ball and a load of 3000 kp is used for steel. Other ball diameters used are 5 mm and 2.5 mm. The diameter of the impression is measured by a micrometer microscope having a transparent engraved scale in the field of view. The Brinell hardness number (HB) is the ratio of the applied load to the spherical area of the impression. The Brinell hardness numbers obtained for steel ranges from about 100 . . . 500 kp/mm^2. It is found, that there is an approximate linear relation between the Brinell hardness number and the ultimate tensile strength. For mild steel the ratio of ultimate strength to Brinell hardness number is 0.36 and for alloy steel 0.34.

Vickers test

The Brinell test is limited to materials with Brinell hardness numbers below 500 kp/mm^2. The Vickers method may be applied for materials above and below the Brinell hardness number 500 kp/mm^2. It consists of pressing a square based diamond pyramid, in which the angle between the opposite faces is 136°, into the test specimen. The load may be varied from 5000 kp to 120 kp in increments of 5 kp. The Vickers hardness number (HV) is the ratio of the applied load to the surface area of indentation.

Rockwell test

The Rockwell test is similar to the Brinell and Vickerk test in pressing an identer into the surface of the test piece, but it differs from the previous methods in that the depth of the impression here is used as the criterion of hardness and not the area of the impression. The identer or penetrator may be either a steel ball or a conical-shaped diamond with a rounded point called a brale. Various loads are used depending on the conditions of the test. When a steel ball is applied, the load is usually 100 kp. The ball is normally 1/16 in in diameter. Larger diameters may be employed for softer materials. The Rockwell hardness number is computed from the formula

$$\mathrm{HRB} = 130 - \frac{h}{0.002} \tag{64}$$

where h = the depth of penetration in mm.

This method is used for materials with Brinell hardness numbers from 80 to 225 kp/mm². For harder materials a diamond cone witl a cone angle of 120° is applied. The Rockwell hardness number then is computed from the formula

$$\mathrm{HRC} = 100 - \frac{h}{0.002}\,. \tag{65}$$

Shore scleroscope test

The Shore scleroscope test is different from the tests discussed previously, which are static tests, and is a dynamic test. Scleroscope hardness is expressed by a number given by the height of rebound of a small pointed hammer after falling within a glass tube from a height of 10 in against the surface of the specimen. The weight of the standard hammer is 1/12 oz. When the hammer falls onto a soft surface, part of the energy of fall is absorbed and the energy available for rebound is comparatively small. But, if the hammer is dropped on a hard surface less energy is absorbed, so that the rebound is higher than before.

The relationships between the several types of hardness numbers, may be found in tables.

3.2.3. Impact Test

With the impact test the resistance of a material to an impact is determined. The impact test machine consists, as shown in Fig. 59, of a pendulum suspended from a short shaft that rotates in a ball bearing. On the end of the pendulum is a heavy disk. The specimen, a notched bar, is fixed on an anvil. The pendulum released from a fixed position strikes the specimen, which breaks, and the pendulum continues its path for some

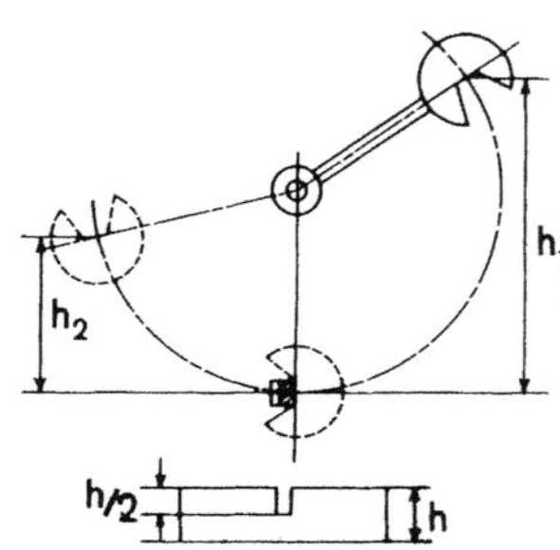

Fig. 59 Impact test machine and specimen for impact test

distance on the other side. The energy required by the pendulum for breaking the specimen is a measure of its resistance to impact. The impact resistance is computed from the formula

$$\frac{W}{A} = G\,\frac{h_1 - h_2}{A} \tag{66}$$

where h_1 = height of fall
h_2 = height of rebound
A = cross section.

3.2.4. Creep Tests

It has been found that steels at elevated temperatures and non-ferrous metals at any temperature, which are under constant stress will continue to elongate. This phenomen is called "creep". As allowable stress for a given operating temperature is taken, as that stress at which in a stated time a permissible plastic strain occurs. Since the creep test must be extended over months or years, it is usual to make short time tests and to extrapolate data obtained from these tests. The results of these tests are plotted as curves from which the allowable stress for a specified creep in a number of years may be found. In Fig. 60 such curves are plotted for a certain material for a creep in 10 years.

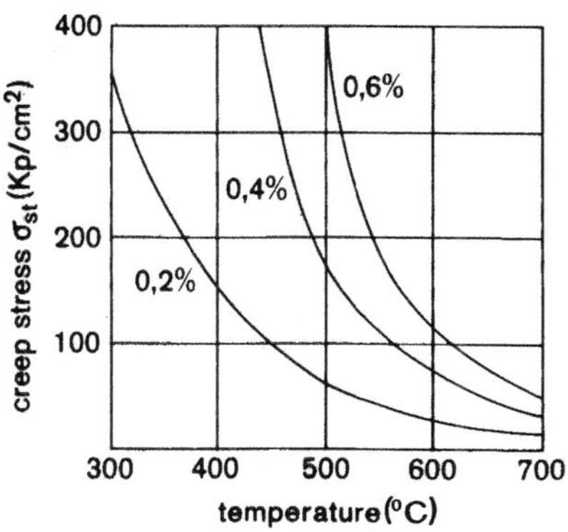

Fig. 60 Curves of a creep test

3.2.5. Fatigue Tests

Many machine parts are subjected to fluctuating stresses. The values of these stresses at which a member breaks are found by fatigue tests. The stress pulsates between two limiting values, as represented in Fig. 61. If the stress pulsates between a tensile and a compressive stress of unequal values, the stress is called "partly reversed stress".

There are two special cases of alternating stresses: the "completely reversed stress" in which the stress pulsates between a tensile and a compressive stress of equal value, the mean value of stress in this case being zero, and the stress alternating between zero and a maximum value, the mean stress in this case being one half the maximum stress.

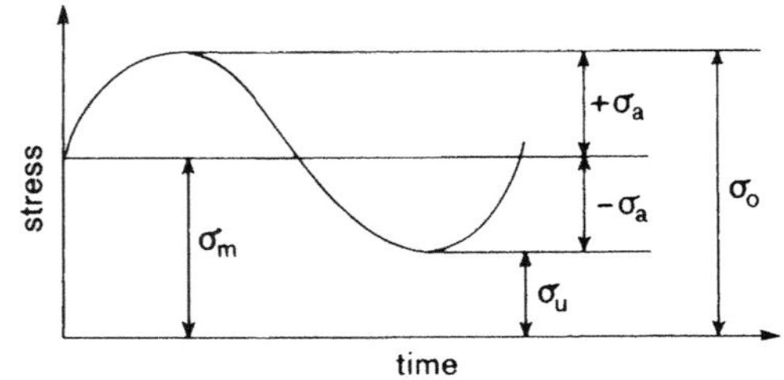

Fig. 61 Fluctuating stress

A failure produced by a fluctuating stress is called "fatigue failure". There are machines for fatigue tests of metals for cycles of axial stresses (tension and compression), for cycles of flexural stresses and for cycles of torsional shearing stresses. Machines for applying flexural stresses are the most common.

The maximum stress which a material subjected to a fluctuating load can endure is known as the "endurance limit". To determine the endurance limit of a metal, it is necessary to test a number of similar specimens. The first sepcimen as tested at a relatively high stress, so that failure will occur at a small number of applications of stress. Succeeding specimens are then tested, each one at a lower stress. The results of fatigue tests are plotted or a diagram in which values of stress are plotted as ordinates and values of numbers of cycles of stress for fracture as abscissae, as shown in Fig. 62. For all ferrous metals and for most non-ferrous metals the curve in the diagram becomes horizontal for values of numbers ranging from 10^6 to 5×10^6 cycles. The ordinate for the horizontal indicates the fatigue stress.

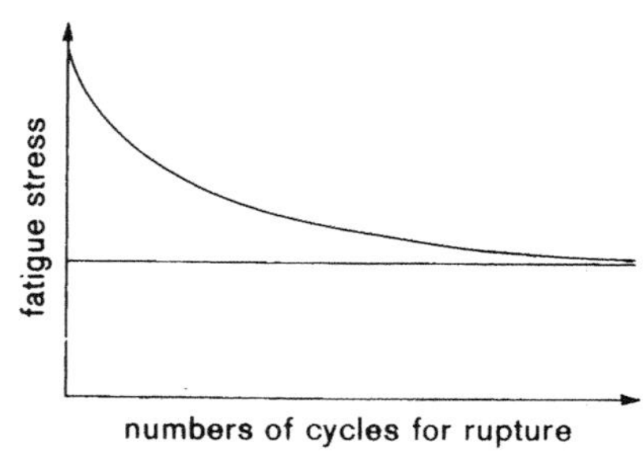

Fig. 62 Fatigue test diagram

The higher the mean stress, the smaller is the range of alternating stress. When the range of alternating stress is zero, then the mean stress is equal to the ultimate strength. The endurance limit is plotted in a diagram as a function of the mean stress. This diagram, called the fatigue stress diagram is reproduced in Fig. 63.

3.2.6. Stress Concentration

At an abrupt change of section of a member, such as occurs at a square corner, a hole in a bar or in a shaft, a keyway, a groove, a screw thread, etc. a high localized stress arises. As long as the acting force is constant, the stress-concentration has no influence on the failure of the member. If, however, a fluctuating force is acting, the allowable stress must be reduced. The ratio of the endurance limit of the material to the endurance limit of the specimen with an abrupt change of section is called the "fatigue stress concentration factor β_k".

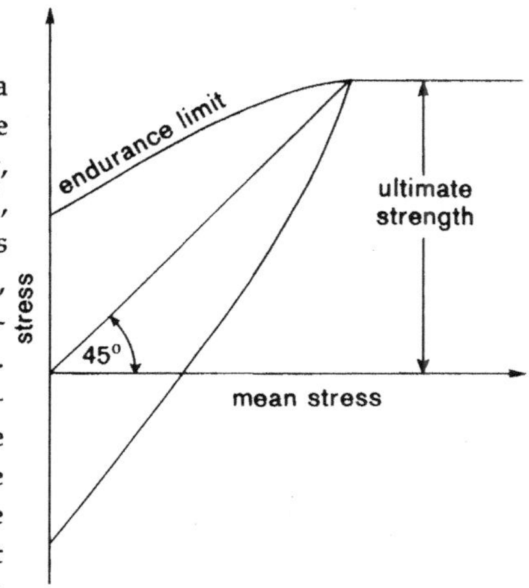

Fig. 63 Fatigue stress diagram

3.3. Tensile Stress, Compressive Stress, Shearing Stress

3.3.1. Tensile Stress and Compressive Stress

For cross sections at a sufficiently large distance from the point of application of the force, a uniform distribution of stress may be assumed. In this case the formulae for tensile and compressive stresses are given at:

$$\sigma_z = \frac{F}{A} \quad \sigma_d = \frac{F}{A} \tag{67}$$

where σ_z = tensile stress and σ_d = compressive stress.

3.3.2. Thin Spheres and Cylinders under Internal or External Forces

If the thickness of a vessel is less than about 1/20 of the internal diameter, then the vessel is said to be thin. In this case the circumferential or hoop stress in a sphere may be assumed uniformly distributed. The radial stress is small in comparison with the hoop stress and can be neglected.

The hoop stress of the sphere may be calculated from

$$\sigma = \frac{d\,p}{4\,s} \tag{68}$$

where d = internal diameter of the vessel
p = pressure of the vessel
s = thickness of the vessel.

For a thin cylinder the hoop and longitudinal stresses are also assumed constant and the radial stress may be neglected.

The longitudinal stress may be calculated from:

$$\sigma_l = \frac{d\,p}{4\,s} \tag{69}$$

and the hoop stress from

$$\sigma_t = \frac{d\,p}{2\,s}\,. \tag{70}$$

3.3.3. Shearing Stress

The shearing stress is not uniformly distributed in the cross section. Normally the average shearing stress is used

$$\tau_s = \frac{Q}{A} \tag{71}$$

where Q = shearing force.

3.4. Thermal Stresses

When the temperature of a body is raised or lowered, the material expands or contracts. If the body is restrained from changing its length, then a stress in the body occurs.

If Δt denotes the change of the temperature and α_t the thermal coefficient of linear expansion of the material, the total change in length of a bar is:

$$\Delta l = \alpha_t\, l\, \Delta t\,.$$

When the strain is prevented, the stress may be found thus:

$$\sigma = E\,\varepsilon = E\,\frac{\Delta l}{l} = E\,\alpha_t\,\Delta t\,. \tag{72}$$

3.5. Stresses and Deflexion of Beams

3.5.1. The Deflection Formula

A straight beam with a symmetrical cross section, as shown in Fig. 64, is subjected to a bending moment which acts in the plane of symmetry of the beam. The bending moment is assumed to be pure, i. e. there are no shearing stresses in any cross section. In this case the cross sections in the beam remain plane during the bending action.

In any cross section there are tensile and compressive stresses. In one plane of the beam the stress is zero. This plane is called the "neutral fibre" It is perpendicular to the plane of symmetry and passes through the centroid of the cross section. As the cross sections remain plane during the bending action, the stress on a fibre at any section is proportional to the distance of the fibre from the neutral fibre. The sum of moments of stresses on all fibres must be in equilibrium with the bending moment.

Expressed mathematically

$$M_b = \int \sigma_b \, dA \, y \, .$$

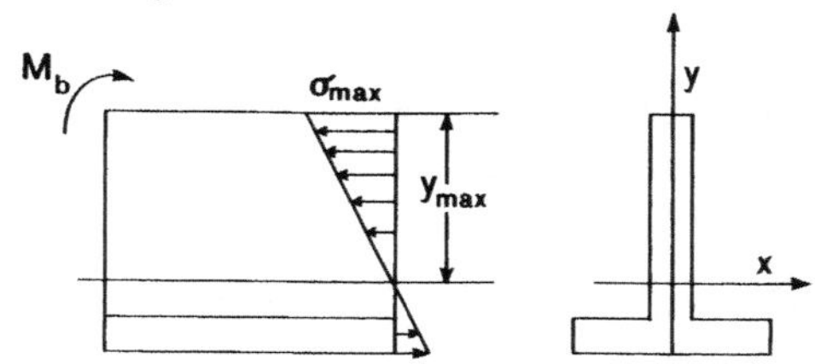

Fig. 64 Stress distribution in the cross section of a beam stressed by bending moment

$$\sigma_b = \frac{\sigma_{b\,max}\, y}{y_{max}} \, , \quad M_b = \frac{\sigma_{b\,max}}{y_{max}} \int y^2 \, dA = \frac{\sigma_{b\,max}}{y_{max}} I_x = \frac{\sigma_b}{y} I_x$$

where $\int y^2 \, dA = I_x$ = moment of inertia of the cross section

or

$$\sigma_b = \frac{M_b}{I_x} y \, . \tag{73}$$

The maximum stress is

$$\sigma_{max} = \frac{M_b}{I_x} y_{max} \quad \text{or} \quad \sigma_{max} = \frac{M_b}{W_x} \, , \tag{74}$$

where $W_x = \dfrac{I_x}{y_{max}}$ = the section modulus.

3.5.2. Unsymmetrical Bending

The deflection formula may be used, if the plane of the bending moment coincides with the plane of symmetry of the beam. The formula may be used also, if the plane of the bending moment lies perpendicular to the plane of symmetric passing through the centroid of the cross section. These perpendicular planes are called "principal planes".

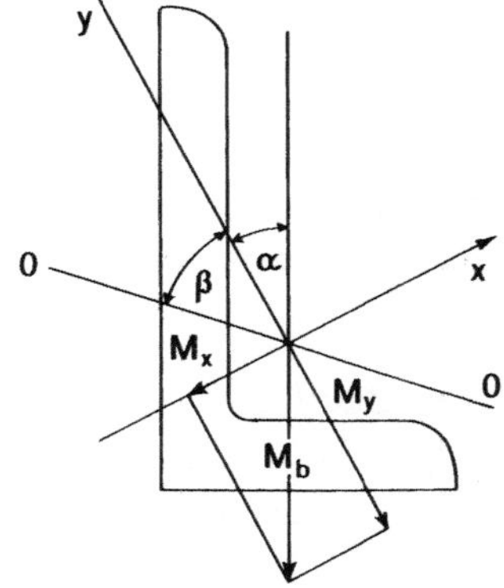

Fig. 65 Unsymmetrical bending

For unsymmetrical cross sections there are also two principal planes, for which the deflection formula may be used. If the plane of the bending moment passes through the centroid of the section without coinciding with one of the principal planes, then the bending is said to be unsymmetrical. In this case the vector of the bending moment is resolved into components along the principal planes, as shown in Fig. 65.

The bending stress is then the algebraic sum of the stresses produced by the components of the bending moment.

Expressed mathematically:

$$\sigma_b = \frac{M \sin \alpha}{I_y} x + \frac{M \cos \alpha}{I_x} y \tag{75}$$

where α = the angle between the y- plane and the plane of the bending moment.

From the condition that the stress in the neutral axis is zero, the direction of the neutral axis may be obtained.

$$\tan \beta = \frac{I_y}{I_x} \cot \alpha \,. \tag{76}$$

3.5.3. Elastic Curve Equation

Beams in structures and machines must have strength as well as stiffness, that is the maximum occuring stress and the maximum deflection may not exceed allowable values, which must be below the proportional limit.

The deflection of a beam is expressed in the elastic curve equation

$$E\, I_x \frac{d^2 y}{dx^2} = - M_b \,. \tag{77}$$

In order to determine the value of the deflection, the differential equation must be integrated twice.

3.6. Buckling

A slender member, a column or a strut, is subjected on its ends to loads acting along the axis of the member. When the strut has an initial small bending due to a slight accidental eccentricity of loading, the deflection increases for increasing values of the load. If the acting load is sufficiently high, then the strut buckles.

It may be assumed that the two ends are supported in such a way that they can freely rotate in each direction. The strut will buckle at a certain load in a plane perpendicular to the axis of the cross section about which the moment of inertia is least. The critical load or buckling load may be found for this case by the following equation derived by Euler in 1757:

$$F_E = \frac{\pi F I}{l^2} \tag{78}$$

in which

$E =$ modulus of elasticity
$I =$ minimal moment of inertia
$l =$ length of strut or column.

When both ends are fixed, the critical load is:

$$F = 4\, F_E\,.$$

When one end is fixed and the other one freely rotating, then the critical load is:

$$F = 2\, F_E\,.$$

When one end is fixed and the other one free from all restraint, the critical load is:

$$F = \frac{F_E}{2}\,.$$

The stress in a cross section of a strut or a column, whose ends are freely rotating is:

$$\sigma_k = \frac{F}{A} = \frac{\pi^2 E I}{A l^2} = \frac{\pi^2 E i^2}{l^2} = \frac{\pi^2 E}{\lambda^2} \tag{79}$$

where $i = \sqrt{\dfrac{I}{A}} =$ radius of gyration

and $\lambda = \dfrac{l}{i} =$ slenderness ratio.

The Euler formula may only employed, if the stress in a cross section is below the proportional limit i. e. if $\sigma_k \leqq \sigma_P$.

The slenderness ratio corresponding to the proportional limit may be found from equation (79) thus:

$$\lambda = \sqrt{\frac{\pi^2 E}{\sigma_P}}\,.$$

For structural steels with $E = 2.1 \times 10^6$ kp/cm² and $\sigma_P = 2000$ kp/cm², λ is 100.

For values of slenderness ratio less than λ, there are empirical formulae.

3.7. Torsion

Twisting moment

The twisting moment or torque acts in a plane normal to the axis of the shaft, whereas the bending moment acts in a plane through the axis of the shaft. The twisting moment for a section in a shaft is defined as the algebraic sum of the moments about the axis of the shaft on either side of the section.

The twisting moment may be calculated for a given power and a given speed of a shaft as follows.

The power in a shaft which rotates at a speed of n revolutions per minute is

$$P = \frac{F\,v}{75 \times 100} \quad \text{where} \quad v = \frac{2\,r\,\pi\,n}{60}.$$

Substituting for v gives,

$$P = \frac{F\,2\,r\,\pi\,n}{75 \times 60 \times 100} = \frac{M_t\,n}{71\,620}$$

whence the twisting moment is

$$M_t = 71\,620\,\frac{P}{n} \tag{80}$$

where M_t = twisting moment (cm kp)
P = power (PS)
and n = speed (r. p. m).

Shearing stress

For circular shafts i. e. shafts having solid or holow circular sections, all cross sections will remain plane during the twisting action. The shearing stress is then proportional to the distance from the centre. In a section the sum of the moments of the shearing stresses must be in equilibrium with the twisting moment (Fig. 66).

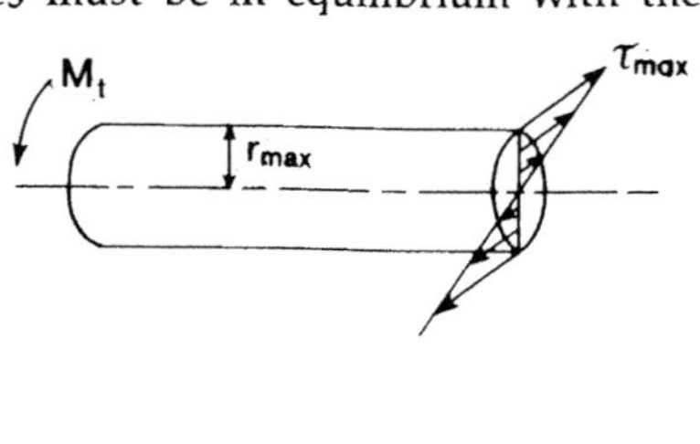

Fig. 66 Stress distribution in the cross section of a cylinder stressed by torsion

Expressed mathematically

$$M_t = \int \tau_t \, dA \, r$$

since

$$\tau_t = \frac{\tau_{t\,max}\, r}{r_{max}}, \quad M_t = \frac{\tau_{t\,max}}{r_{max}} \int r^2 \, dA = \frac{\tau_t}{r} I_p$$

or

$$\tau_t = \frac{M_t}{I_p} r \tag{81}$$

where $\int r^2 \, dA = I_p$ = the polar moment of inertia of the cross section.

The maximum shearing stress is

$$\tau_{t\,max} = \frac{M_t}{I_p} r_{max} .$$

For a solid circular section,

$$\tau_{t\,max} = \frac{16\, M_t}{\pi\, d^3} \tag{82}$$

where d = diameter of shaft.

For a hollow circular section,

$$\tau_{t\,max} = \frac{16\, M_t\, D}{\pi\, (D^4 - d^4)} \tag{83}$$

where D = outer diameter of shaft
and d = diameter of hole.

Angle of twist

The angle of twist (Fig. 67) caused by a given twisting moment may be calculated for shafts having solid or hollow circular sections from the equation

$$\varphi = \frac{M_t\, l}{G\, I_p} \tag{84}$$

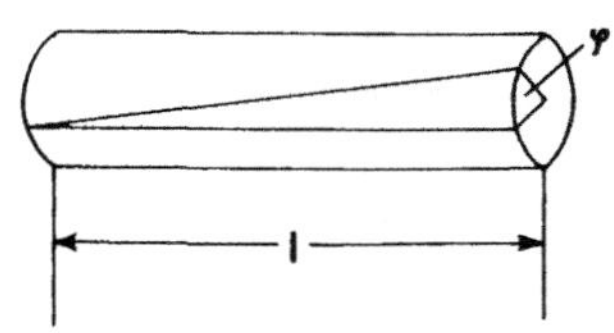

Fig. 67 Angle of twist on a beam stressed by torsion

where l = length of shaft
and G = modulus of rigidity.

For a solid circular shaft the equation becomes

$$\varphi = \frac{32\, M_t\, l}{\pi\, d^4\, G} . \tag{85}$$

For a hollow circular shaft the equation becomes

$$\varphi = \frac{32\, M_t\, l}{G\, \pi\, (D^4 - d^4)} . \tag{86}$$

3.8. Combined Stresses

At any point of a loaded member there are three planes mutually at right angles to each other in which only normal stresses occur and no shearing stresses. These planes are called "principal planes" and the corresponding stresses are called "principal normal stresses". If three principal normal stresses act at a point of the body, the state of the stresses at this point is called "triaxial", if only two principal normal stresses occur, the state of the stresses is called "biaxial" or "plane" and if only one normal stress occurs, the state of stresses is called "uniaxial". On a rectangular element three normal stresses and six shearing stresses may occur, as shown in Fig. 55, These nine stresses have the same effect as the three principal normal stresses. In the triaxial state of stresses these nine stresses or the three principal normal stresses, respectively are the cause of failure. Be means of the theories of failure all stresses occuring on a rectangular element are combined and are represented by a normal stress which has the same effect as the other stresses. There are various theories of failure. The best known theories are the "maximum shearing stress theory" and the theory of the "elastic strain energy of distorsion".

3.8.1. Maximum Shearing Stress Theory

According to this theory the failure of a body is caused by the greatest shearing stress occuring on a body. Maximum shearing stresses act in three planes at an angle of 45° to the principal planes and have the value of one half the algebraic difference of two principal normal stresses.

$$\tau_{\max 1} = \frac{1}{2}\,(\sigma_1 - \sigma_2)$$

$$\tau_{\max 2} = \frac{1}{2}\,(\sigma_1 - \sigma_3)$$

$$\tau_{\max 3} = \frac{1}{2}\,(\sigma_2 - \sigma_3) .$$

The maximum value of these three shearing stresses causes the failure. In the biaxial state of stresses becomes $\sigma_3 = 0$. That is, $\sigma_1 = \sigma_{max}$ and $\sigma_2 = \sigma_{min}$

$$\tau_{max\,1} = \frac{1}{2} (\sigma_{max} - \sigma_{min})$$

$$\tau_{max\,2} = \frac{1}{2} \sigma_{max}$$

$$\tau_{max\,3} = \frac{1}{2} \sigma_{min} .$$

If σ_{max} and σ_{min} have the same sign, then the greatest value of τ_{max} is

$$\tau_{max} = \frac{\sigma_{max}}{2} .$$

If σ_{max} and σ_{min} have different signs, then the greatest value of τ_{max} is

$$\tau_{max} = \frac{\sigma_{max} - \sigma_{min}}{2} .$$

In the uniaxial state of stresses the maximum shearing stress is converted into the combined stress $\sigma_{red} = 2\,\tau_{max}$.

Thus the combined stress for the case that σ_{max} and σ_{min} have the same sign,

$$\sigma_{red} = \sigma_{max} . \tag{87}$$

For the case, that σ_{max} and σ_{min} have different signs

$$\sigma_{red} = \sigma_{max} - \sigma_{min} . \tag{88}$$

3.8.2. Theory of Elastic Strain Energy of Distorsion

The elastic energy stored in a material is composed of the elastic energy of volume dilatation and the elastic strain energy of distortion. According to this theory the elastic strain energy of distortion in a body causes the failure, whereas the elastic strain energy of volume dilatation has no influence on the failure. The elastic strain energy of distortion in a unit volume in triaxial or biaxial state of stresses is compared with that in uniaxial state of stresses and then the combined stress is computed.

For a biaxial state of stresses the combined stress is calculated from

$$\sigma_{red} = \sqrt{\frac{1}{2} (\sigma_{max} - \sigma_{min})^2 + \sigma_{max}{}^2 + \sigma_{min}{}^2} \tag{89}$$

or, if the maximum principal stresses are replaced by the stresses σ_x, σ_y and τ_{xy}

$$\sigma_{\text{red}} = \sqrt{\sigma_x^2 + \sigma_y^2 - \sigma_x \sigma_y + 3 \tau_{xy}^2}\,. \tag{90}$$

If only one normal stress and one shearing stress occurs, the formula becomes

$$\sigma_{\text{red}} = \sqrt{\sigma^2 + 3 \tau^2}\,. \tag{90 a}$$

By means of the theories of failure a relation between the allowable normal stress and the allowable shearing stress may be found.

According to the maximum shearing stress theory,

$$\tau_{\text{all}} = 0.5\, \sigma_{\text{all}}\,. \tag{91}$$

According to the theory of elastic energy of distorsion

$$\tau_{\text{all}} = 0.578\, \sigma_{\text{all}}\,. \tag{92}$$

4. Hydraulics

That branch of mechanics which deals with fluids is called "Fluid Mechanics". Fluids are subdivided into two classes: liquids and gases. Both, liquids and gases take the shape of the vessel which contains them. The difference between liquids and gases as that a liquid only occupies a definite volume independent of the dimensions of the container and forms a free surface if the volume of the liquid is smaller than the volume of the container, whereas a gas expands and fills the container in which it is placed. Gases are compressible, liquids are normally considered incompressible since the effect of compressibility is so small that it in general may be neglected, although in some cases the compressibility may be significant.

Hydromechanics is that branch of mechanics which is concerned with the laws governing the behaviour of liquids. It is subdivided into Hydrostatics and Hydrodynamics. Hydrostatics deals with the laws of liquids at rest, hydrodynamics deals with the laws of liquids in motion. If the practical applications of the principles of hydromechanics are dealt then the term "Hydraulics" is used. In hydraulics the strict results of mathematical operations and the results of experimental researches are combined. The most important liquid that hydraulics is concerned with, is water. Another important liquid is oil.

Between the particles of a liquid in motion there is an internal friction. The frictional forces increase with the viscosity and the difference of velocity of the sliding liquid particles. If the liquid is at rest or the velocity of the liquid is very small, the internal friction may be neglected. A liquid considered incompressible and without internal friction is called a "perfect liquid".

4.1. Hydrostatics

4.1.1. Pressure of a Liquid

In hydrostatics we assume that the liquid is perfect i. e. the liquid is incompressible and without internal friction. A perfect liquid cannot withstand shear forces or tension, it can only withstand compression. On any plane in a liquid only a compression stress acts or, in other words, the pressure acting on any plane in a liquid is always normal to this plane.

The pressure of a liquid is defined as: $p = \frac{dF}{dA}$ where dA is an infinitesimal small plane and dF the force acting on this plane.

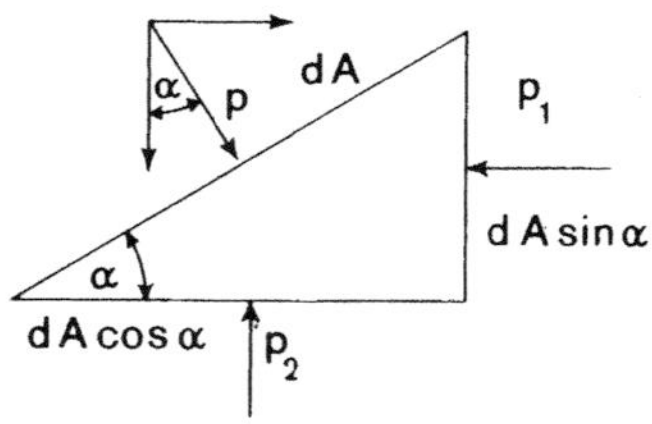

Fig. 68 Internal forces acting on a liquid prism

From the condition of equilibrium of the forces acting on an infinitesimal liquid prism (Fig. 68) we obtain, if we neglect the infinitesimal weight of the prism:

$$-p_1\, dA \sin\alpha + p \sin\alpha\, dA = 0$$
$$p_2\, dA \cos\alpha - p \cos\alpha\, dA = 0$$

whence

$$p_1 = p = p_2\,.$$

In words, the pressure at any point in a static liquid is the same in all directions.

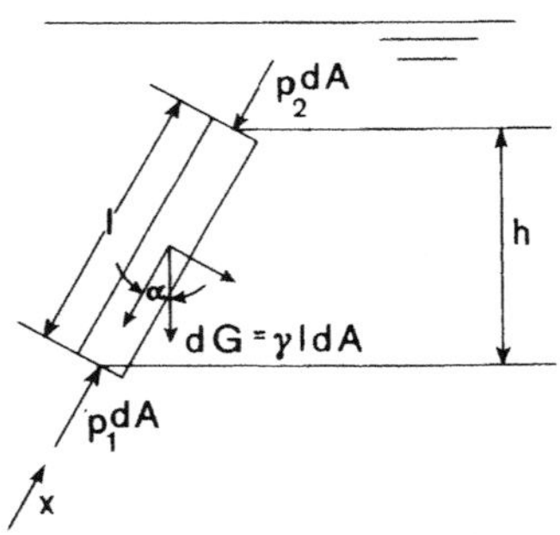

Fig. 69 Increasing pressure of a liquid in the depth

From the condition of equilibrium of the forces acting on a liquid prism in the x direction (Fig. 69) we obtain:

$$p_1\, dA - p_2\, dA - \gamma\, l\, dA \cos\alpha$$

or if h is substituted for $l \cos\alpha$

$$p_1 - p_2 = \gamma\, h \qquad (93)$$

where γ = specific weight of the liquid and h = the pressure head.

If $p_2 = 0$, then the equation becomes:

$$p = \gamma\, h\,.$$

Sometimes the liquid pressure is given by the height of a mercury column or a water column. We can write instead of 1 at = 1 kp/cm²

$$h_{Hg} = \frac{p}{\gamma} = \frac{1000\ \text{p/cm}^2}{13.596\ \text{p/cm}^3} = 73.6\ \text{cm} = 736\ \text{mm} = 736\ \text{Torr} \quad \text{and}$$

$$h_{H_2O} = \frac{p}{\gamma} = \frac{1000\ \text{p/cm}^2}{1\ \text{p/cm}^3} = 1000\ \text{cm} = 10\ \text{m}\,.$$

If $\gamma\, h$ (in the equation $p_1 - p_2 = \gamma\, h$) is very small in comparison with p_1 or p_2, as is the case in a liquid under high pressure, then the equation becomes:

$$p_1 = p_2 = p\,.$$

In other words, pressure in a liquid which is under high pressure is transmitted equally in all directions (Pascal's Law).

4.1.2. Buoyancy

The formula $p_2 - p_1 = \gamma h$ applied to an element of a body submerged in a liquid (Fig. 70) gives, if both sides are multiplied by dA

$$p_2 \, dA - p_1 \, dA = \gamma h \, dA$$

or

$$dF_2 - dF_1 = \gamma \, dV .$$

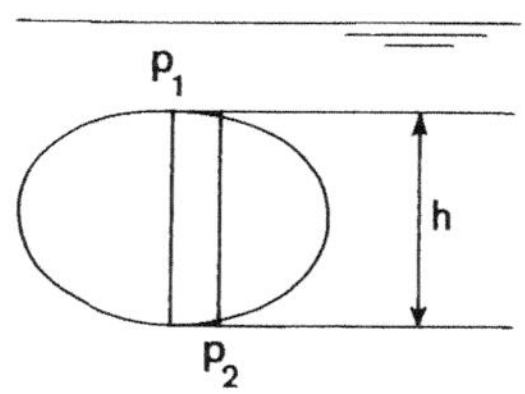

Fig. 70 Buoyancy

Integrating:

$$F_2 - F_1 = \gamma V$$

where V = the volume of the submerged body.

On a submerged body a force acts vertically upward through the centre of gravity of the body, the magnitude of which is equal to the weight of the displaced liquid (Archimedes' Law). This force is called the "force of buoyancy or the buoyant force" and its magnitude is independent of the depth of submergence.

4.1.3. Pressure Forces on Submerged Planes and Curved Surfaces

Plane surfaces

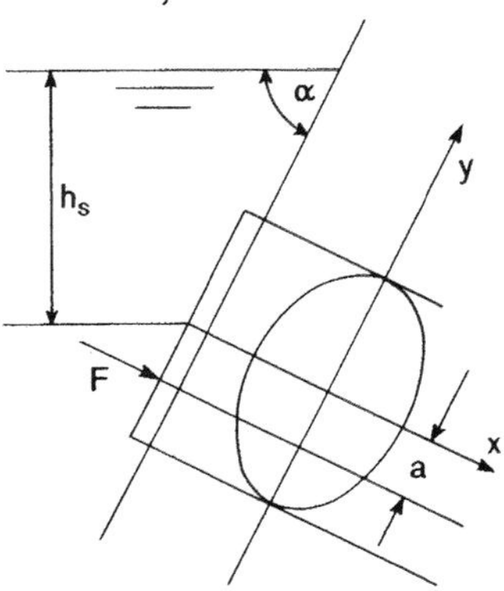

Fig. 71 Pressure force acting on a submerged plane

A reservoir filled with liquid is limited on one side by an inclined wall making an angle α with the free liquid surface. There is a submerged opening in the wall covered by a flat plate of arbitrary shape (Fig. 71). Examples of such plates submerged in liquids are lock gates and valves. The problem is to calculate the magnitude of the pressure force exerted on the plate and its location which is called "centre of pressure". The magnitude of the pressure force may be calculated by the equation:

$$F = \gamma h_s A . \tag{94}$$

where γ = specific gravity of the liquid, h_s = vertical distance of the centroid of the plate from the free surface of liquid and A = area of plate.

The equation shows that the force acting on a plane area is the product of the pressure at its centroid multiplied by the area of the plate.

The location of the pressure force may be calculated from:

$$a = \frac{I_x \sin \alpha}{h_s A} \tag{95}$$

where I_x = axial moment of inertia of the area about the horizontal axis through the centroid of the plate.

The equation shows that the centre of pressure lies always below the centroid of the plate by the distance a.

Graphical method of finding pressure force and centre of pressure

For plates having constant width the pressure force and the centre of pressure may be found graphically.

The pressure force may be calculated from the equation:

$$F = \gamma\, h_s\, A = \gamma\, h_s\, H\, b$$

where b = width of the plate.
The pressure $\gamma\, h_s$ is drawn according to a convenient scale perpendicular to the plate surface (Fig. 72). The hatched area in Fig. 72 represents the term $\gamma\, h_s\, H$. This area multiplied by the width of the plate gives the pressure force. The centre of pressure is found by projecting the centroid of the hatched area on the plate.

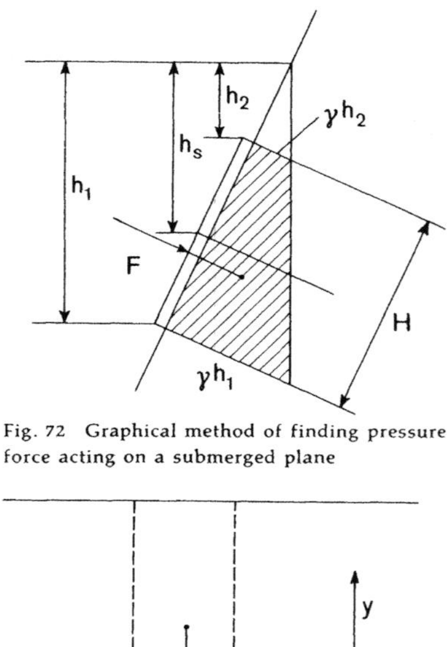

Fig. 72 Graphical method of finding pressure force acting on a submerged plane

Curved surface

In order to find the resultant force acting on a curved surface we imagine the force resolved into its vertical and

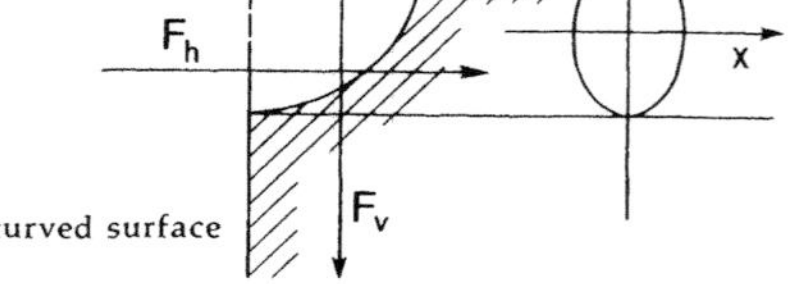

Fig. 73 Pressure force acting on a submerged curved surface

horizontal components. The vertical component F_v is the weight of the liquid column lying above the curved area and it acts through the centre of gravity of the liquid column. The horizontal component F_h is equal to the horizontal force acting on a plane which is the vertical projection of the curved area (Fig. 73). The magnitude and location of this force is the same as for a force acting on a vertical area equal in size of the projected area and it may be found from the formulae given in (94) and (95).

The pressure force acting on a curved surface outside or inside an inclined plane, shown in Fig. 74 is found by the combination of a force acting on the area projected on the inclined plane and the weight or the buoyant force of the liquid inside the curved area.

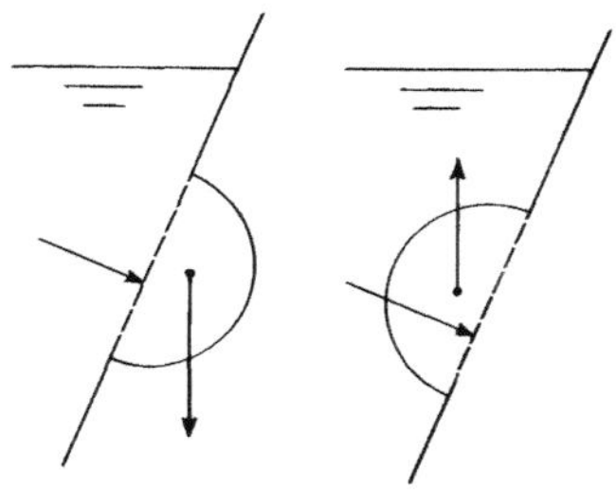

Fig. 74 Pressure force acting on a submerged curved surface outside or inside an inclined plane

4.1.4. Flotation of Bodies

On a body completely immersed in a liquid there are two forces acting, the upward force of buoyancy and the downward force of the body weight. The buoyant force acts vertically upward through the centre of gravity of the displaced liquid, whereas the weight of the body acts vertically downward through the centre of gravity of the immersed body. If the weight of the body is greater than the buoyant force then the body sinks; if the weight of the body is equal to the buoyant force, then the body rests in any position wherever placed; if the buoyant force is greater than the weight of the body then the body rises and passes through the free liquid surface. From this moment the volume of the displaced liquid and consequently the buoyant force is continually decreasing. The body comes to the state of rest or will float, if the two forces, the buoyant force and the weight of the body are equal and opposite, and the centre of buoyancy lies on a vertical line through the centre of gravity of the body. The connection line between the centre of buoyancy and the centre of gravity of the body is called the "floating axis". It must always be vertical. There are bodies which may float in several positions or which have several floating axes. A cylindrical body, the centre of gravity of which is in the middle of the body, has an infinite number of positions of floating.

Depth of flotation

The depth to which a body will sink into a liquid may be calculated from the condition that the weight of the displaced liquid equals that of the body.

The simplest case is that of a body having the shape of a right prism floating with its base horizontal and its sides vertical (Fig. 75). An example of this case is a caisson or a pontoon.

$$b\,t\,l\,\gamma = G \quad \text{whence} \quad t = G/b\,l\,\gamma \tag{96}$$

where t = immersed depth
G = body weight
b = width of body
l = length of body
γ = specific weight of body

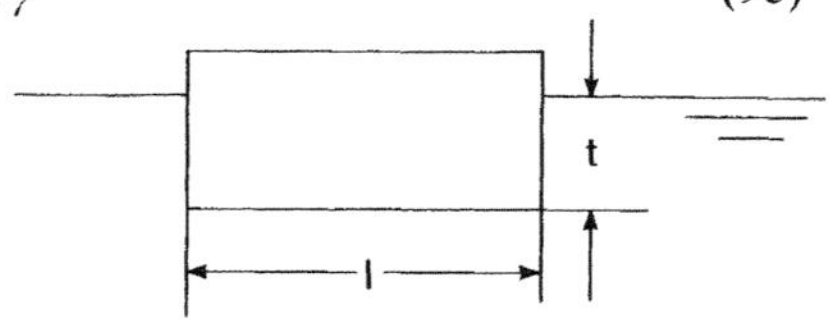

Fig. 75 Immersed depth of a floating body

Stability of floating bodies

A body supported in any way is said to be stable or in stable equilibrium, when there is a tendency to return toward its equilibrium position after beeing slightly displaced. The body is brought to its equilibrium position by a restoring moment. For a floating body this moment is caused by a couple made of the buoyant force and the weight of the body, (Fig. 76 a). The body is unstable, when there is a tendency to move away from the equilibrium position after disturbance, (Fig. 76 b). When the body has no tendency to return to or to move away from its former position after disturbance the body is said to be in neutral equilibrium (Fig. 76 c).

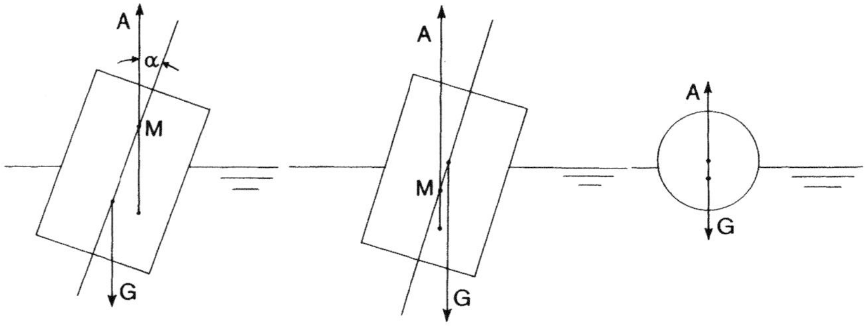

a) stable b) unstable c) neutral

Fig. 76 Stability of floating bodies

If the angle of inclination (α) is increased, the centre of buoyancy moves away from the floating axis and the point of intersection (M) between the floating axis and the line of action of the buoyant force moves away from the centre of gravity of the body. For a decreasing angle α, the point M approaches the centre of gravity. If the angle α approaches zero, the point M approaches a limiting position above the centre of gravity. The limiting position of the point M is called its "metacentre". The distance from the metacentre to the centre of gravity is called the "metacentric height". For stability the metacentre must lie above the centre of gravity. The higher the metacentre lies above the centre of gravity, the greater is the stability of the floating body. For ocean liners the metacentric height is from 0.3 to 0.7 m and for war-ships from 0.8 to 1.2 m.

4.2. Motion of Liquids

4.2.1. Flow of a Perfect Liquid

In the solution of problems of liquids in motion the friction, between the liquid particles and the friction between the liquid and the limiting walls introduces a difficulty for engineers. Only, if friction between particles is neglected the theoretical considerations can be made simple. But the results of these simplified considerations are in most cases not entirely in accordance with practical experience. Thus, the following method is used by engineers: the laws of liquids in motions are obtained from the principles of mechanics and are later improved by adding correction factors to the formulae.

4.2.1.1. Continuity Equation

Any liquid flowing is supposed to be made up of a number of stream lines. Stream lines are theoretical lines which are traced by connecting the tangents of velocity of neighbouring particles. They show the direction of flow at any point in the liquid stream at any one instant of time. Stream lines inside a tubular space bounded by an imaginary surface consisting of streamlines are stream tubes. Since no liquid crosses the walls of a stream tube, the stream tubes behave like a tube with rigid walls inside which liquid flows. The flow of any liquid stream is said to be steady, if at any point in the stream the velocity remains constant with time. The velocity may change from point to point in the stream, but never at any one point. In steady flow, all stream lines keep their shape. The flow pattern is unchangeable with time, the stream lines coincide with the paths of the liquid particles. If at any point in the stream the velocity is changing with time, the flow is said to be unsteady. The stream lines do not keep their shape, the flow pattern is changing with time.

Immagine a steady flow in a stream tube. Since no liquid is added or taken away between two specified cross-sections 1 and 2 the liquid quantity passing through section 1 must be equal to the liquid quantity passing through section 2 (Fig 77).

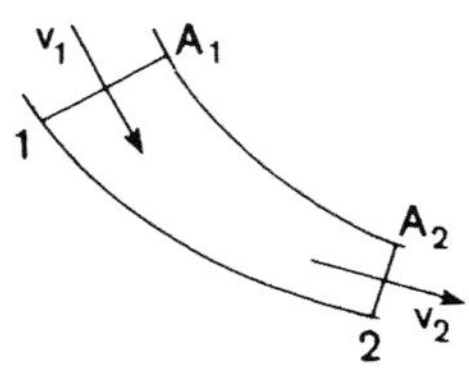

Fig. 77

Expressed mathematically

$$A_1 \, v_1 = A_2 \, v_v \quad \text{or} \quad A \, v = \text{const.} \tag{97}$$

where A_1 and A_2 = cross-sections in the stream tube
and v_1 and v_2 = mean velocities of the liquid stream in the cross-sections 1 and 2, respectively.

This is called the equation of continuity.

4.2.1.2. Flow through Orifices

The term orifice is applied to any opening, having a closed circumference, made in a wall or partition of a reservoir. The orifices differ in geometrical shape and in the manner in which their edges are formed. Circular and rectangular orifices are most commonly used in engineering structures.

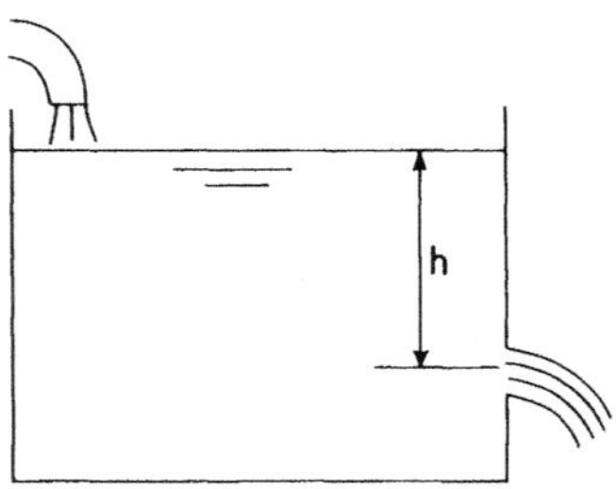

Fig. 78
Discharge at constant liquid head

First we assume the liquid level remains constant by allowing in the same liquid quantity through a tap as flows out through the orifice. In this case the flow of discharge is steady (Fig. 78).

The ideal exit velocity of the jet may be calculated from the equation stated by Torricelli

$$v_0 = \sqrt{2 \, g \, h} \tag{98}$$

where h = head of liquid
and g = acceleration due to gravity

The real exit velocity of the jet is a little less than the ideal due to the liquid's viscosity. This is represented by the coefficient of velocity, c_1.

The stream lines in the vessel are crowded in the orifice and pass through as a jet of smaller area than the area of the orifice. The ratio between the area of the jet and the area of the orifice is called the "coefficient of contraction", c_2.

The liquid quantity flowing from the orifice in unit time may be computed as the product of the contracted area $c_2 A$ and the real velocity

$$v = c_1 \sqrt{2 g h}.$$

Therefore

$$Q = c_1 c_2 A \sqrt{2 g h} \quad \text{or} \quad Q = c A \sqrt{2 g h} \tag{99}$$

where $c = c_1 c_2 =$ coefficient of discharge.

If there is no flow into a reservoir, discharge occurs due to the falling head. The time required for lowering the liquid level from the head h_1 to the head h_2 (Fig. 79) may be calculated from the condition that the volume of liquid between the falling heads in the vessel equals the volume of liquid that passes through the orifice.

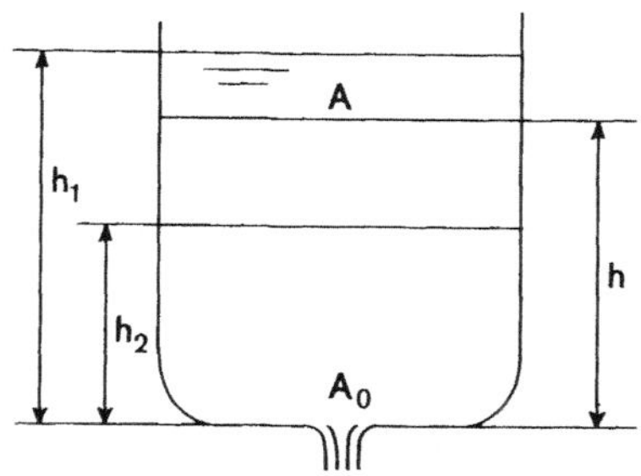

Fig. 79 Discharge at falling liquid head

Expressed mathematically

$$A\,dh = c_2 A_0 v\,dt$$

where $A =$ area of liquid surface of the height h
and $A_0 =$ area of the orifice

$c_1 \sqrt{2 g h}$ substituted for v gives

$$c A_0 \sqrt{2 g h}\,dt = A\,dh \quad \text{whence} \quad dt = \frac{A\,dh}{c A_0 \sqrt{2 g} \sqrt{h}}$$

the equation integrated gives

$$t = \frac{1}{c A_0 \sqrt{2 g}} \int_2^1 \frac{A\,dh}{h}. \tag{100}$$

4.2.1.3. Bernoulli's Theorem

In 1738 Daniel Bernoulli, a Swiss mathematician, demonstrated a general theorem of hydraulics by means of which many problems of liquids in motion may be solved. A perfect liquid in steady flow is assumed. After, friction may be considered by an additional term in the equation.

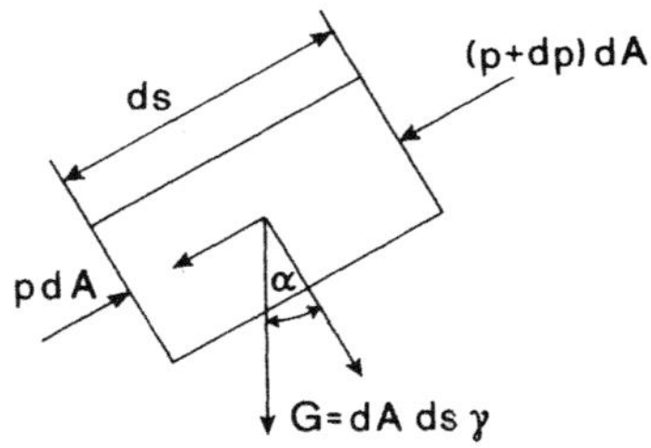

Fig. 80

We consider an infinitely small portion of liquid in a stream tube. The length of this element is, ds, the area normal to the flow, dA, the specific weight, γ and the acceleration due to gravity, g (Fig. 80).

According to the Newton's Second Law of Motion the resultant of all forces acting in the direction of motion must be equal to the product of mass and acceleration.

Expressed mathematically

$$p\,\mathrm{d}A - (p + \mathrm{d}p)\,\mathrm{d}A - \mathrm{d}A\,\mathrm{d}s\,\gamma \sin\alpha = \mathrm{d}A\,\mathrm{d}s\,\frac{\gamma}{g}\,a$$

where $a =$ acceleration of the liquid particle.

Substituting for $a = \mathrm{d}v/\mathrm{d}t = v\,\mathrm{d}v/\mathrm{d}s$ and for $\mathrm{d}s \sin\alpha = \mathrm{d}z$ the equation becomes, after simplifying,

$$-\,\mathrm{d}p - \gamma\,\mathrm{d}z - \mathrm{d}s\,\frac{\gamma}{g}\,v\,\frac{\mathrm{d}v}{\mathrm{d}s} = 0 \quad \text{or}$$

$$\frac{\mathrm{d}p}{\gamma} + z + \frac{v}{g}\,\mathrm{d}v = 0\,.$$

After integrating each separate term between the limiting values we obtain

$$\frac{p_1}{\gamma} + \frac{v_1^2}{2\,g} + h_1 = \frac{p_2}{\gamma} + \frac{v_2^2}{2\,g} + h_2 \quad \text{or}$$

$$\frac{p}{\gamma} + \frac{v^2}{2\,g} + h = \text{const.} \tag{101}$$

This is the Bernoulli equation for steady flow with friction reglected. All terms have the dimensions of a linear distance. The term p/γ is given the name pressure head, the term $v^2/2\,g$ the name velocity head and h is the height of the particle above any assumed plane and is called the elevation head. The sum of all heads is called the total head, H.

In words, the theorem of Bernoulli may be expressed as follows: In steady flow without friction, the sum of the pressure head, velocity head and elevation head is constant along a stream line.

If friction is considered the equation is modified to

$$\frac{p_1}{\gamma} + \frac{v_1^2}{2\,g} + h_1 = \frac{p_2}{\gamma} + \frac{v_2^2}{2\,g} + h_2 + h_w$$

where h_w = the lost head due to friction.

In words: In steady flow with friction present the total head at any section equals that at any subsequent section increased by the lost head occuring between the two sections.

Bernoulli's theorem has another very important interpretation. If we consider all terms in the Bernoulli equation multiplied by unit weight, then we can consider the term p/γ as pressure energy, the term $v^2/2\,g$ as kinetic energy and h as elevation energy. The theorem for flow without friction is then expressed: the sum of the pressure energy, kinetic energy and elevation energy remains constant along a stream line. This is the principle of conservation of energy. For flow with friction the theorem is expressed as: The total sum of energy at any section of a stream line equals that at any subsequent section plus the lost energy.

4.2.1.4. Flow Measurement

For measuring pressures of liquids at rest or in motion a variety of manometers are in use. The most common are the Piezometer tubes and the Bourdon gauge which are described here.

The Piezometer tube consists of a glass tube inserted in a vessel or in a pipe (Fig. 81). Since the top of the tube is open to the atmosphere the relative pressure is measured. The Piezometer is used for measuring small pressures. For measuring high pressures the Bourdon gauge is used. It consists of a bent tube one end of which is rigidly held and the other end closed and connected by a rack and pinion to a pointer (Fig. 82). When pressure is applied to the tube it tends to straighten the bent tube and the pointer moves over a graduated scale.

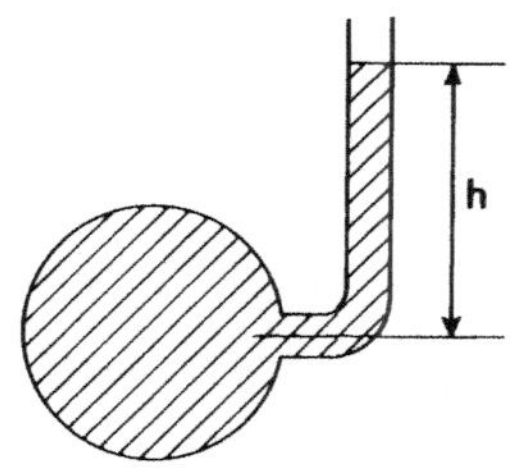

Fig. 81 Piezometer tube

Fig. 82 Bourdon gauge

If we want to measure the difference of two pressures a differential manometer is used. The U-type, as shown in Fig. 83 is the simplest type. It is made from a U-shaped glass tube. The tube is filled with mercury for measuring pressure of water and with water or alcohol for measuring pressure of gases.

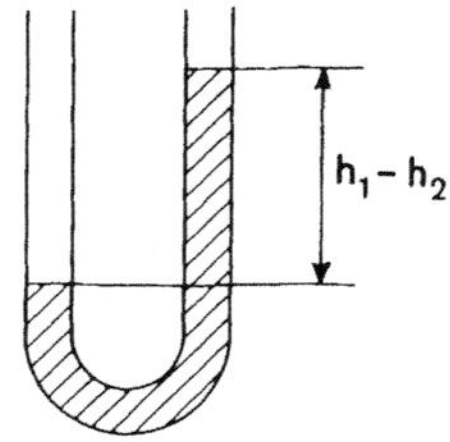

Fig. 83 Differential U tube gauge

For measuring the static pressure of liquid in motion the liquid entry into the manometer tube must occur perpendicular to the direction of flow (Fig. 84).

For measuring the stagnation pressure, this is the sum of the static pressure and the pressure corresponding to the velocity head

$$p + \frac{v^2}{2g} = p + \varrho \frac{v^2}{2}$$

where $\varrho = \frac{\gamma}{g}$ = density of liquid,

the Pitot tube is used. Here the liquid entry into the manometer tube must occur in the direction of flow (Fig. 85).

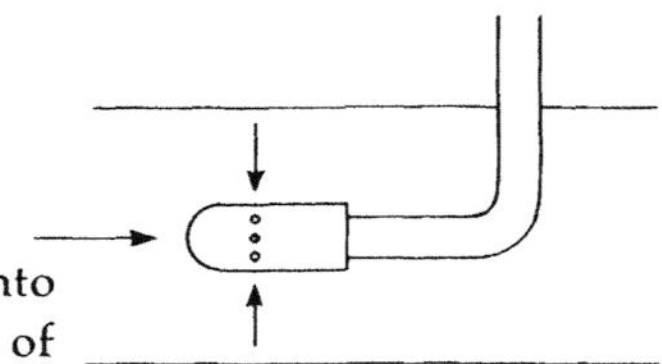

Fig. 84 Manometer tube for measuring static pressure of liquid in motion

The Pitot tube may be combined with a tube for measuring static pressure, shown in Fig. 86, and is then called the Pitot-static-tube or Prandl-tube, since it was first constructed by the German physicist Prandl. This instrument may be used for measuring velocity in a conduit. From the difference of the stagnation and static pressure we may obtain the velocity of the liquid flow.

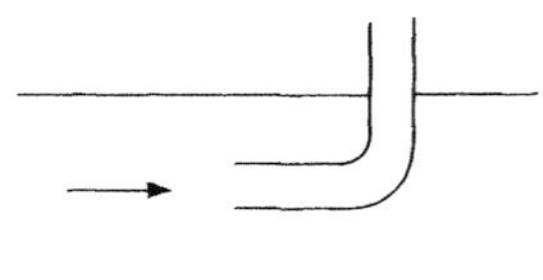

Fig. 85 Pitot tube

$$p_2 - p_1 = p_1 + \varrho \frac{v^2}{2} - p_1 = \varrho \frac{v^2}{2}.$$

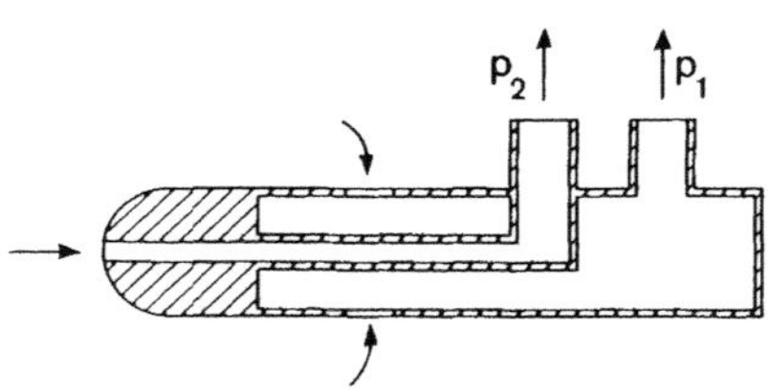

Fig. 86 Prandl-tube

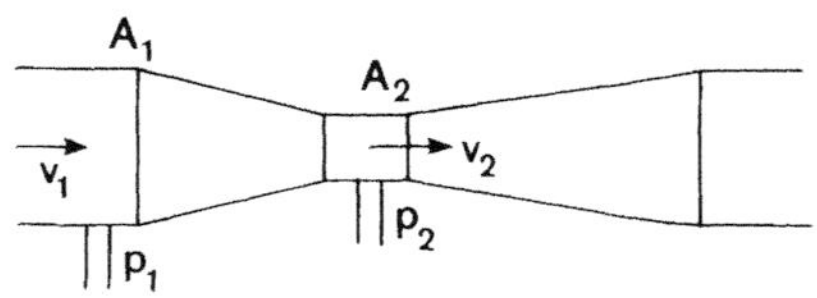

Fig. 87 Venturi Meter

To measure the rate of flow in pipelines the Venturi Meter is used, named after the Italian hydraulic engineer Venturi. It consists of a converging, a parallel and a diverging portion (Fig. 87). The function of the converging portion is to increase the velocity of the liquid and to lower the static pressure. The diverging cone or diffuser serves to change the area of the stream back at the exit area.

Applying Bernoulli's theorem we obtain:

$$\frac{p_1}{\gamma} + \frac{v_1^2}{2\,g} = \frac{p_2}{\gamma} + \frac{v_2^2}{2\,g}$$

using the continuity eqation

$$Q = A_1\,v_1 = A_2\,v_2 \quad \text{or} \quad v_1 = v_2\,\frac{A_2}{A_1}$$

with the Bernoulli equation gives

$$\frac{p_1}{\gamma} - \frac{p_2}{\gamma} = \frac{v_2^2}{2\,g}\left(1 - \left(\frac{A_2}{A_1}\right)^2\right).$$

The ideal discharge is

$$Q = A_2\,v_2 = \sqrt{\frac{2\,g}{(1-(A_2/A_1)^2)}}\sqrt{\frac{p_1 - p_2}{\gamma}} \quad \text{or}$$

$$Q = c\sqrt{\frac{p_1 - p_2}{\gamma}}\,. \tag{102}$$

4.2.1.5. Linear Momentum Equation in Hydraulics

If the velocity of a flowing liquid is changed in magnitude or direction, forces occur on the limiting walls. Such limiting walls are the walls of pipes or channels, the guide vanes and runner blades on turbines or the diffusion vanes and impeller blades on centrifugal pumps. The forces may be calculated by means of the linear momentum equation (55).

$$\int F\,dt = m\,v_1 - m\,v_2\,.$$

For steady flow of a perfect liquid the impulse for unit time is

$$\int F\,dt = F\,t = F\,.$$

The mass of liquid flowing per unit time is

$$m = Q\,\frac{\gamma}{g}$$

where Q = volume of liquid per unit time.

This value substituted in the previons equation gives

$$F = \frac{Q\gamma}{g} v_1 - \frac{Q\gamma}{g} v_2 \text{ (impulse per unit time)} \tag{103}$$

Examples

1. *Force exerted by a liquid jet upon a plate* (Fig. 88)

Since the y-components of the velocity are self cancelling only the x-component of the velocity need be considered.

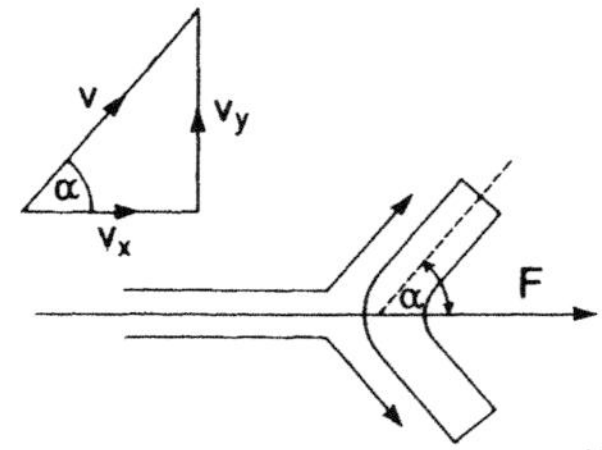

Fig. 88 Force of a liquid jet upon a wall

The equation $F = \frac{Q\gamma}{g}(v_1 - v_2)$ becomes

$$F = \frac{Q\gamma}{g}(v - v\cos\alpha) = \frac{Q\gamma v}{g}(1 - \cos\alpha)$$

For $\alpha = 90°$ the equation becomes $F = \frac{Q\gamma \cdot v}{g}$

For $\alpha = 180°$ the equation becomes $F = 2\,\frac{Q\gamma \cdot v}{g}$

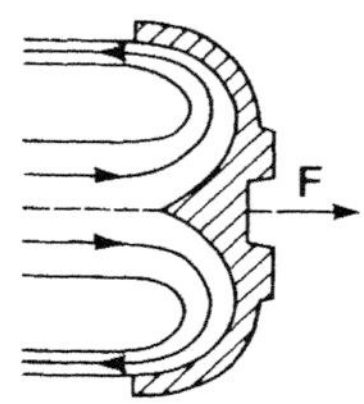

Fig. 89
Force of a liquid jet upon a vane

2. *Force exerted on a converging pipe bend* (Fig. 90)

The force exerted on a pipe bend is the resultant of the liquid weight G, of the pressure forces $p_1 A_1$ and $p_2 A_2$ on the ends of the pipe bend and the force arising from the change of momentum.

It is assumed that the velocity v_2 is horizontal. The component of the resultant force in the x-direction is

$$F_x = p_1 A_1 \cos\alpha - p_2 A_2 + \frac{Q\gamma}{g}(v_1 \cos\alpha - v_2)\,.$$

Fig. 90 Force exerted on a converging pipe

The component of the resultant force in y direction is

$$F_y = p_1 A_1 \sin\alpha + G + \frac{Q\gamma}{g} v_1 \sin\alpha\,.$$

The resultant force is

$$F = \sqrt{F_x^2 + F_y^2}\,.$$

4.2.2. Motion of Viscous Liquids

4.2.2.1. Viscosity

Whilst perfect liquids are considered to be incompressible and free from viscosity, real liquids display viscosity and compressibility. Since the effect of compressibility is very small and therefore need only be considered in some special cases, it will be ignored here and only the effect of viscosity will be considered. Viscosity only occurs in flowing liquids; in stationary liquids the effect of viscosity does not appear. Viscosity offers resistance to the relative motion between liquid and solid bodies and to the motion between the layers of a flowing liquid.

The velocity of the particles of a liquid flowing through a conduit is not uniform across the section and varies from point to point. The velocity distribution may be represented by velocity curves. These curves are generally determined experimentally. Between the adjacent layers of liquid shear stresses occur which are dependent upon the slip velocity of the layers.

Consider two layers at an infinitesimal distance dy apart. The velocity difference between the two layers is dv. The rate of change dv/dy, at a certain point P, of the velocity curve is the tangent of the angle α enclosed by the tangent drawn to the curve at the point P and the perpendicular to the flow, as in Fig. 91. According to Newton the shear stress τ is proportional to the velocity gradient $\frac{dv}{dy}$.

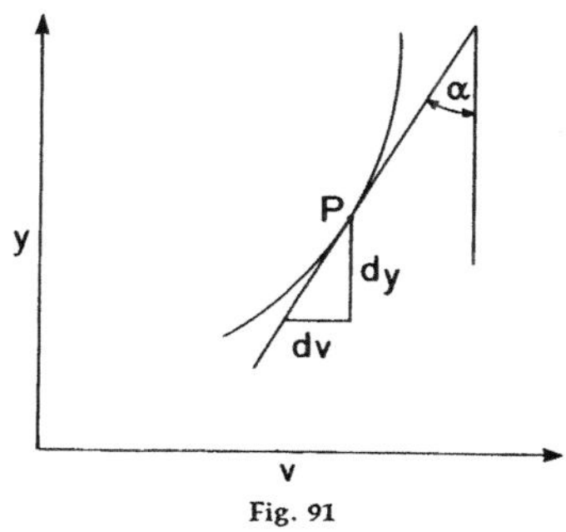

Fig. 91

$$\tau = \eta \frac{dv}{dy} \quad \text{or} \quad \eta = \frac{\tau}{dv/dy}\left(\frac{\text{kp s cm}}{\text{cm}^2\ \text{cm}} = \text{kp s/cm}^2\right). \tag{104}$$

The quantity η is independent of the liquid motion and represents a liquid property which is called the coefficient of viscosity. The shear stress disappears where the velocity gradient is zero and has its maximum value where the velocity gradient is the largest.

Consider a liquid flowing inside a closed conduit. The velocity gradient and consequently the shear stress has its maximum value at the boundery. It is

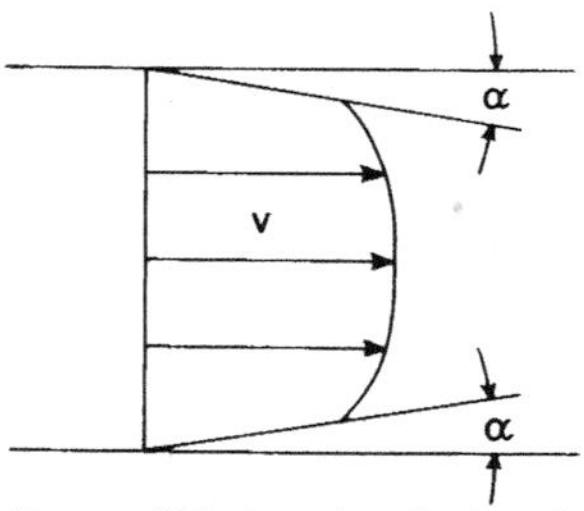

Fig. 92 Veloricity distribution of a liquid flowing in a closed conduit

not possible for the velocity curve to approach the boundary tangentially, as this would invoke an infinitely large shear stress. Thus an angle of finite magnitude must always exist between the velocity curve and the boundary surface (Fig. 92).

The coefficient of viscosity can be determined experimentally either by the efflux or by the torque method.

In the efflux method a small tank with a small orifice at the bottom is used. For determining the viscosity of any liquid the time is measured which the liquid takes for emptying the tank. The ratio of this time and the time which water at 20 °C takes for running out gives a number from which the coefficient of viscosity can be calculated by means of an empirical formula.

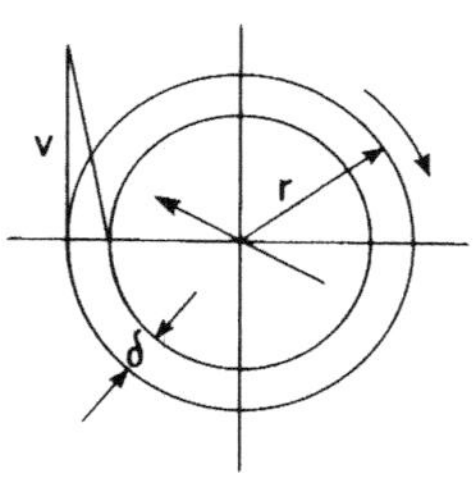

Fig. 93 Torque method for determining the viscosity of a liquid

In the torque method a thin film of liquid is enclosed between two coaxial cylinders (Fig. 93). The inner cylinder, suspended by a torsion wire or torsion spring, is stationary. The outer ring rotates at a suitable speed. Since the clearance between the two cylinders is very small the velocity distribution in the clearance may be assumed to be linear. Thus the velocity gradient is $dv/dy = v/\delta$. The torque occuring in the inner cylinder is registered by a pointer fixed on the torsion wire or torsion spring.

The coefficient of viscosity may be calculated from the formula given in (104).

$$\eta = \frac{\tau}{dv/dy}$$

substituting $M_t/r\, 2\, r\, \pi\, l$ for τ (where M_t = torque, r = radius of clearance = inner radius of the outer cylinder and l = length of the two cylinders) and v/δ for dv/dy gives

$$\eta = \frac{M_t\, \delta}{2\, r^2\, \pi\, l\, v}.$$

4.2.2.2. Laminar and Turbulent Flow

The behaviour of liquids in motion are not determined theoretically only. The results of experiments show in which way the result obtained by mathematical operations may be modified. Models are used to reproduce the liquid flow. Having obtained experimental data for a certain flow, it is desirable to compare the behaviour of flow on models with that on similar structures by using the same or another liquid. Two flows exhibit the same characteristic only when they are completely similar. They are completely similar, if the following conditions are fulfilled:

1. Model and real structure must be geometrically similar i. e. all dimensions of outlines of the surfaces bounding the liquid must be proportional.
2. The surface roughness of the bounding surfaces of the model and the structure must be geometrically similar.
3. The mechanical quantities such as velocities, accelerations, forces etc. at corresponding points in the flow must be proportional.

Forces due to gravity, liquid friction, inertia and elasticity will also be present.

A complete similarity of all forces acting on corresponding points cannot be attained. Only two forces, the most important in the special problem, can be considered, all the others must be neglected. The inertia force must be considered in all problems. For the flow in closed conduits the weight and elasticity of liquid have no effect on the characteristics of the flow. Thus, besides the inertia force only the frictional force is considered.

According to Newton's Second Law of motion, $F = m\,a$, the ratio between the frictional forces occuring on the model flow F_{fM} and the real flow F_f may be written

$$\frac{F_f}{F_{fM}} = \frac{m\,a}{m_M\,a_M}$$

where m = the mass of a liquid particle and a = acceleration of the liquid particle.

Since according to equation (104)

$$\frac{F_f}{F_{fM}} = \frac{\tau\,A}{\tau_M\,A_M} = \frac{\eta\,v\,l_M\,l^2}{l\,\eta_M\,v_M\,l_M^2} \quad \text{and}$$

$$\frac{m\,a}{m_M\,a_M} = \frac{V\,\varrho\,v\,t_M}{t\,V_M\,\varrho_M\,v_M} = \frac{l^3\,\varrho\,v\,t_M}{l_M^3\,\varrho_M\,v_M\,t}$$

we obtain

$$\frac{\eta\, v\, l}{\eta_M\, v_M\, l_M} = \frac{l^3\, \varrho\, v\, t_M}{l_M^3\, \varrho_M\, v_M\, t}$$

or after simplifying and substituting ν for η/ϱ = kinematic viscosity

$$\frac{v\, l}{\nu} = \frac{v_M\, l_M}{\nu_M} \quad \text{or} \quad \frac{v\, l}{\nu} = \text{const}\,. \tag{105}$$

The dimensionless quantity $\frac{v\, l}{\nu}$ is called "Reynolds number" and designated by *Re*.

The flows in two circular pipes are similar if $\frac{v_1\, d_1}{\nu_1} = \frac{v_2\, d_2}{\nu_2}$ where d_1 and d_2 = internal diameters of the pipes and v_1 and v_2 = mean velocities in the pipes.

There are two different types of flow: laminar and turbulent flow. In laminar flow the liquid moves in layers, in turbulent flow irregular motions are superimposed on the average flow. The two kinds of flow may be demonstrated with an apparatus consisting of a tank filled with water, a small tank containing dye and a horizontal glass tube on the end of which there is a regulating valve (Fig. 94). The water flow velocity can be varied by adjusting the valve. The dye is introduced into the flow at the mouth of the glass tube through a small diameter tube. At low velocities the coloured filament remains straight and stable.

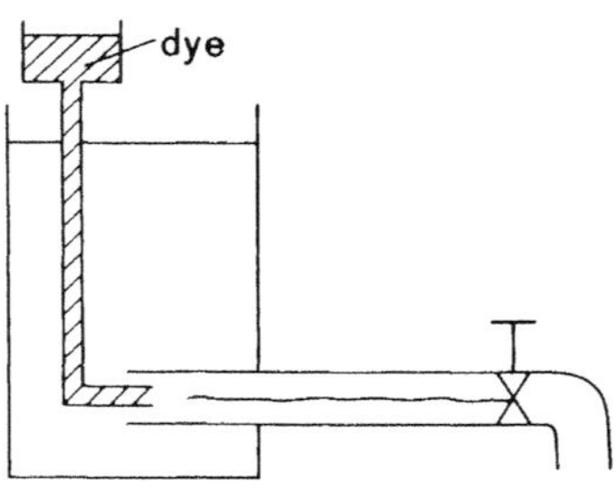

Fig. 94 Apparatus for determining the critical velocity in a pipe

At a critical velocity the filament begins to become irregular. Experiments made in tubes having different diameters and different liquids exhibit that the change from the laminar to the turbulent flow always occurs at the same Reynolds number, the critical Reynolds number. For straight circular pipes the critical Reynolds number is 2320. If there are no shocks it is possible to maintain the laminar flow also above the critical velocity, but with the smallest disturbance turbulent flow occurs. The laminar flow above the critical velocity is unstable, whereas it is stable below the critical velocity.

In water conduits and in channels there is always turbulent flow. Laminar flow occurs in the pipes of hot water systems and in oil pipes. The velocity distribution of laminar flow in pipes follows a parabolic curve (Fig. 95). The maximum velocity attained in the centre is exactly twice the value of the average flow velocity.

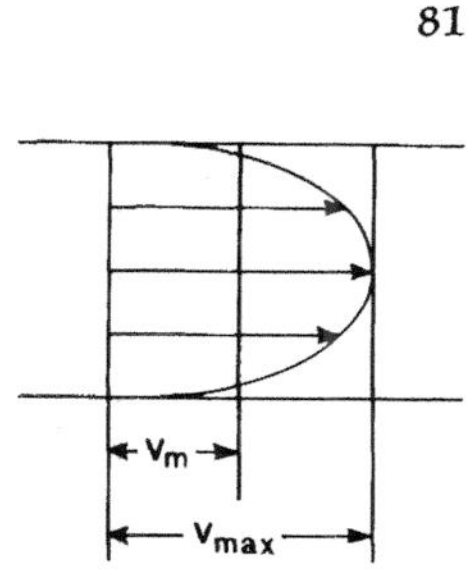

Fig. 95 Velocity distribution of laminar flow in pipes

The velocity curve in turbulent flow is much flatter than the corresponding laminar flow parabola for the same average velocity and becomes flatter with increasing Reynolds number. The maximum velocity is about 1.23 times the value of the average flow velocity (Fig. 96).

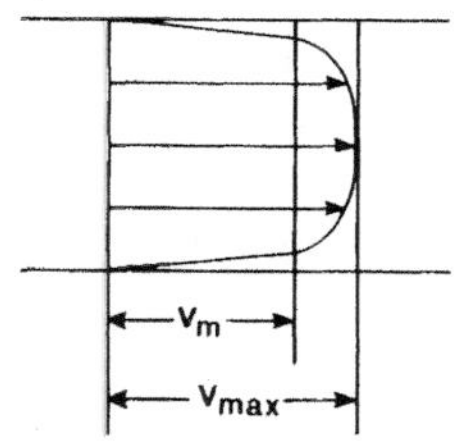

Fig. 96 Velocity distribution of turbulent flow in pipes

4.2.2.3. Flow through Closed Conduits

Closed conduits are commonly circular in section. Of all conduits having equal sections, but differing in shape of section, the circular conduit has the smallest perimeter of section and therefore the smallest inside wall area. It follows that the resistance which a circular conduit offers to the liquid flow is less than in a conduit having any another section. The resistance to flow in a closed conduit designated as "loss head" or "loss energy" is dependent on the wetted perimeter of the conduit, the length of conduit, the surface roughness of the walls of the conduit, the liquid viscosity and the kind of flow i. e. if the flow is laminar or turbulent. The surface of conduit walls may be regarded as smooth or rough depending on the geometry of the roughness of the surfaces. Asphalted sheet iron, bright-drawn brass, copper, lead and glass tubes are considered as smooth, whereas cast pipes, concrete pipes and riveted sheet iron tubes are considered to be rough. Due to corrosion smooth pipes gradually become rough.

It is customary to express the loss head in the form

$$h_{\mathrm{w}} = \psi \frac{v^2 l}{2 g r_{\mathrm{h}}} \qquad (106)$$

where ψ = friction factor, v = mean velocity of the flow, l = length of conduit and r_{h} = hydraulic radius $= \frac{\text{cross sectional area}}{\text{wetted perimeter}}$. For pipes the formula will be written $h_w = \frac{\lambda}{4} \frac{v^2 l}{2 g r_h}$ where $\lambda = 4 \Psi$ = the friction factor for pipes. For full-flowing circular pipes the hydraulic radius is $r_h = \frac{d}{4}$. For laminar flow in pipes the friction factor is $\lambda = \frac{64}{R}$. Here the friction factor is independent of surface roughness. For turbulent flow, for smooth surfaces, the friction factor is independent of the surface roughness and only dependent on the Reynolds number. There are two formulae established by Blasius and Nicuradse for calculating the friction factors for pipes for this case. In the case of turbulent flow in pipes having rough surfaces the friction depends on the surface roughness as well as on the Reynolds number, but for large Reynolds numbers the friction factor becomes independent of Reynolds number. There are several formulae stated by hydraulic engineers for this case.

A pipe line may consist of several straight circular pipes having different diameters and of transitions such as bends, elbows, tee-, cross- and branch pieces, besides different kinds of valves for regulating the flow. In all these components energy losses occur, which may be calculated from the formula

$$h_i = \zeta \frac{v^2}{2 g} \tag{107}$$

where v = flow velocity immediately behind the part mounted in the pipe line and ζ = the loss coefficient.

The values of loss coefficient are dependent on the design of the part and may be found experimentally and are given in special tables.

4.2.2.4. Flow through Open Channels

The term open channel includes any conduit in which the liquid flowing has a free surface subjected only to atmospheric pressure. The conduit may be either open like natural rivers and artificial channels or closed like sewerage tunnels. Conduits of the latter type can only be considered as open, if they are running partly full. The flow in open channels is produced by the slope of the bed. The flow is normally turbulent, steady or unsteady.

The mean velocity is found from equation 106

$$v = \sqrt{\frac{2 g r_h H}{\psi l}}$$

if c is substituted for $\sqrt{\frac{2\,g}{\psi}}$ and I for $\frac{H}{l}$ = slope of the channel the equation becomes

$$v = c\,\sqrt{I\,r_h} \tag{108}$$

where r_h = the hydraulic radius.

The maximum velocity and hence the maximum discharge will occur, when the sectional area is so shaped that the hydraulic radius is so large as possible or since $r_h = \frac{\text{area}}{\text{wetted perimeter}}$ if the wetted perimeter has its minimum.

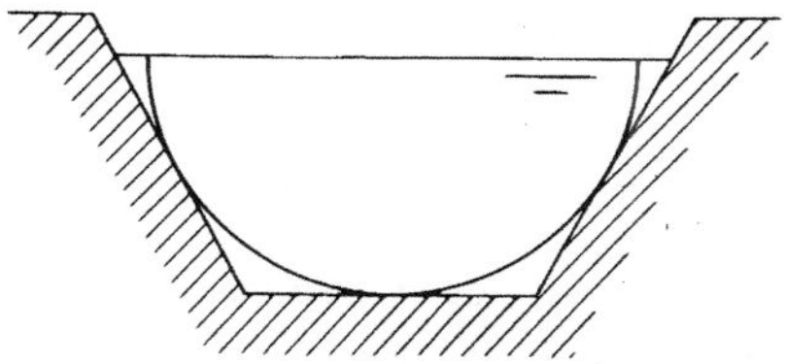

Fig. 97 Flow through on open channel

For trapezoidal cross sections is this the case if the sides are tangential to a semicircle described on the liquid line or if the hydraulic radius is one half the width of the liquid surface (Fig. 97).

5. Thermodynamics

5.1. Fundamental Principles

Engineering thermodynamics deals with the utilization of heat energy. There are three fundamental problems in thermodynamics: transition, storage and transformation of heat.

A heat transfer take place from a body at higher temperature to another of lower temperature. If the two bodies are in contact, the heat transfer occurs by means of conduction, if however the bodies do not touch each other, the heat transfer occurs by means of radiation.

Energy may be stored internally within matter. Such stored energy is identified by internal energy. Examples of stored mechanical energy are the energy of a stressed spring or a compressed gas. Heat is stored in a high temperature body e. g. in a vapour which has high temperature and pressure. The principal problem of thermodynamics is the transformation of heat into mechanical work. The mechanical energy may be used directly in driving a machine or may be transformed into electrical energy for the purpose of energy transport.

Units of energy

The unit of mechanical energy is 1 m kp. Other units are 1 PSh = 270 000 mkp and 1 kWh = 1.36 PSh = 367 000 mkp. The unit of heat energy is 1 kcal = 427 m kp = 1/632 PSh = 1/860 kWh. The calorie is defined as the energy required for heating 1 kg of water from 14.5 °C to 15.5 °C.

Units of pressure

Pressure is defined as force per unit area. Standard Atmospheric Pressure represents the average pressure at sea level and is equal to 1.033 kp/cm^2 = 760 Torr.

Units of temperature

In engineering thermodynamics the Centigrade Scale and the Absolute Scale are used. On the Centigrade Scale the zero point corresponds to the temperature of melting ice under a pressure of 1 standard atmosphere and the 100° point to the temperature of boiling water under 1 standard atmosphere. The absolute zero of temperature on the Centigrade Scale is − 273.16 °C. The temperature measured from absolute zero in centigrade degrees are signified as degrees Kelvin (°K).

Specific heat

Specific heat is defined as the amount of heat required to increase the temperature of the unit mass of a substance by 1 degree. For gases there are two specific heats, the first one for gases whose volume is constant during heating (c_v), the other one for gases whose pressure remains constant during heating (c_p).

The relation between the two properties is

$$\varkappa = \frac{c_p}{c_v}.$$

Heat of fusion

The heat of fusion is the amount of heat required to convert unit mass of a solid substance into a liquid at its melting point.

5.2. The Perfect Gas

5.2.1. The Ideal Gas Equation of State

Definition of gas and vapour

There are two kinds of substance; gas and vapour. Both vapour and gas can be reduced to a liquid state, if the temperature of the substance is below the critical temperature and the pressure above the critical pressure. Vapours whose pressure is low relative to the critical pressure and high relative to the critical temperature, are called gases. A gas during the thermal cycle is never be changed into the liquid phase and will always remain in the form of gas, whereas vapour will be changed into liquid during the process.

It is convenient to assume a hypothetical substance which has a simple equation of state representing the relations between the three properties: pressure, volume and temperature. This hypothetical gas is called the "perfect or ideal gas" and the equation dealing with the three terms is called the "equation of state of the ideal gas".

Boyle's law

If the temperature of a gas is constant during expansion or compression, its pressure varies inversely as the volume.

In mathematical terms,

$$\frac{p_1}{p_2} = \frac{v_2}{v_1} \quad \text{or} \quad p\,v = \text{const.} \quad (\text{if } T = \text{const.})$$

where p = absolute pressure
T = absolute temperature
and v = specific volume = volume of unit mass

Gay-Lussac's law

If the pressure of a gas is constant, its volume will vary directly with the absolute temperature.

Mathematically, this can be stated as

$$\frac{v_1}{v_2} = \frac{T_1}{T_2} \quad \text{or} \quad \frac{v}{T} = \text{const.} \quad (\text{if } p = \text{const.})$$

If the volume of a gas is constant, its pressure will vary directly with the absolute temperature.

Mathematically, this can be stated as

$$\frac{p_1}{p_2} = \frac{T_1}{T_2} \quad \text{or} \quad \frac{p}{T} = \text{const.} \quad (\text{if } v = \text{const.})$$

Equation of state of the ideal gas

If the previous laws are combined, the following equation, called the "equation of state of the ideal gas" results which represents the relationship between pressure, volume and temperature.

$$\frac{p\,v}{T} = \text{const.} \quad \text{or} \quad p\,v = R\,T \tag{109}$$

in which R is called the "gas constant".

Determination of the gas constant

According to Avogadro's law, at identical temperature and pressure equal volumes of all gases contain the same number of molecules. Therefore the densities of two gases vary directly with their molecular masses.

Expressed mathematically,

$$\frac{\varrho_1}{\varrho_2} = \frac{M_1}{M_2}$$

where ϱ = density
and M = molecular mass.

Since $p\,v = \frac{p}{\varrho} = R\,T$ or $\varrho = \frac{p}{R\,T}$,

$$\frac{\varrho_1}{\varrho_2} = \frac{R_2}{R_1} = \frac{M_1}{M_2} \quad \text{or} \quad M\,R = \text{const} = 848$$

$$\text{therefore} \quad R = \frac{848}{M}\left(\frac{\text{m kp}}{\text{kg °K}}\right). \tag{110}$$

Mole and molevolume

The mole is defined as the amount of a substance that has a mass equal to its molecular mass.

For Helium $M = 4$ therefore 1 mole = 4 kg.

Molecular masses of some gases:

Air 28.97, Oxygen (O_2) 32, Hydrogen (H_2) 2, Nitrogen (N_2) 28

Molevolume

$$\Phi = \frac{V}{m/M} = v\,M = \frac{R\,T\,M}{p} = 848\,\frac{T}{p}\,(\text{m}^3)\,. \tag{111}$$

Theorem

All gases occupy the same mole volume, if their pressures and temperatures are identical.

5.2.2. The First Law of Thermodynamics

Heat is a form of energy. Heat may be changed into mechanical work and conversely mechanical work may be changed in heat. According to the law of conservation of energy, stored energy can neither be created nor destroyed, therefore the amount of energy in a system must be the same during a cyle.

The relation between heat energy and mechanical work is:

$$1\ \text{kcal} = 427\ \text{m kp},\ 1\ \text{m kp} = \frac{1}{427}\,\text{kcal} = A\,.$$

A is called the "mechanical equivalent of heat"

Other relations of energy are:

$$1 \text{ PSh} = 75 \times 3600 \text{ m kp} = 270\,000 \text{ m kp} = \frac{270\,000}{427} = 632 \text{ kcal}$$

$$1 \text{ KWh} = 1.36 \times 270\,000 \text{ m kp} = 367\,000 \text{ m kp} = \frac{367\,000}{427} = 860 \text{ kcal.}$$

Internal energy

Each substance possesses a stored heat energy, called the "internal energy" which depends on its temperature, pressure and volume.

The internal energy of unit mass is: $u = \frac{U}{m}$.

If heat is added to a gas at constant volume, the heat energy is completely used for raising its internal energy.

In mathematical terms:

$$q = u_2 - u_1 = c_v (T_2 - T_1)$$

where q = heat added to the unit mass
and c_v = specific heat at constant volume.

If the internal energy is taken from absolute zero, then

$$u = c_v T . \tag{112}$$

If heat is added to a gas during a simultaneous increase of the volume of the gas, the added heat is used both for raising the internal energy and for doing external work.

Expressed mathematically

$$q = u_2 - u_1 + A L_{12} .$$

At constant pressure, the work done is: $A L_{12} = A p V$.

Enthalpy

Through many thermodynamic devices fluids pass continously in a steady flow. Examples are: A steadily operating steam boiler, a steam turbine or a centrifugal compressor.

The energy in a steady flow is composed of internal energy (u), the static energy or work flow (L) and the kinetic energy $\left(\frac{w^2}{2 g}\right)$. The total energy in a steady flow must remain constant.

Mathematically this can be stated as

$$E = m\left(u + A\,p\,v + \frac{A\,w^2}{2\,g}\right) = \text{const.}$$

The combination property $u + A\,p\,v$ is written by a single symbol and is signified as enthalpy (i).

Thus the equation becomes:

$$E = m\left(i + \frac{A\,w^2}{2\,g}\right). \tag{113}$$

If the speed of the substance is less than $w = 40$ m/s the kinetic energy may be neglected.

With $u = c_v\,T$ and $p\,v = R\,T$,

$$i = c_v\,T + A\,R\,T = (c_v + A\,R)\,T = c\,.$$

Since $p = \text{const}$,

$$i = c_p\,T \tag{114}$$

therefore

$$c_v + A\,R = c_p \quad \text{or} \quad c_p - c_v = A\,R\,. \tag{115}$$

Work of expansion or compression

The curne 1–2 in Fig. 98 represents a typical variation of pressure with specific volume of a compressible fluid. If it is assumed that the process is reversible, the area under the curve 1–2 represents the work output at expansion or the work input at compression of the unit mass of gas,

$$L_{12} = \int p\,\mathrm{d}v\,.$$

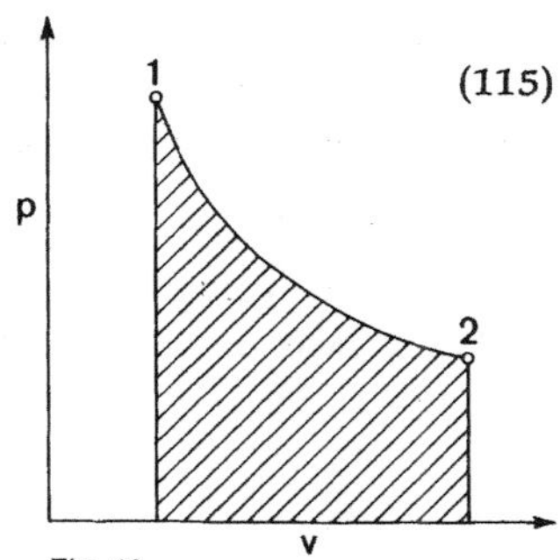

Fig. 98
Work of expansion or compression

Useful work of expansion or compression

The input energy to a machine is i_1, the output energy from the machine i_2, q_s and $A\,L_t$, as shown in Fig. 99, where q_s is the radiated energy and $A\,L_t$ the useful work output of the

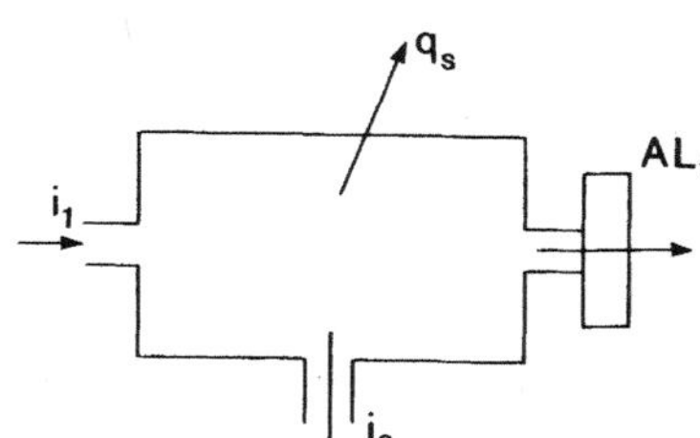

Fig. 99
Input and output energy of a machine

machine. The useful work of a machine of the unit mass of gas is determined from

$$A\, L_t = (i_1 - i_2) - q_s\,.$$

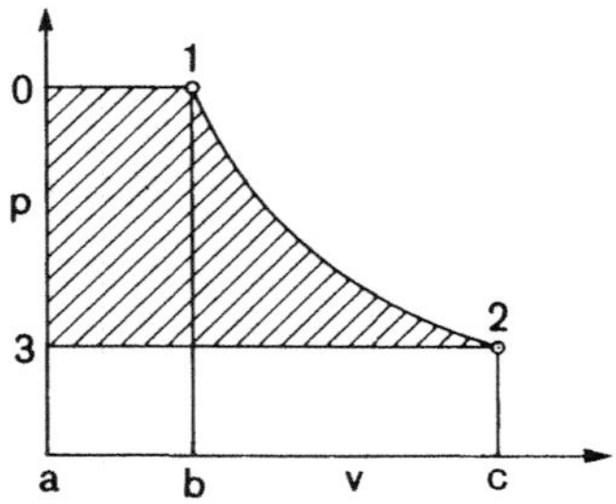

Fig. 100 Useful work of expansion or compression

Fig. 100 shows that the useful work output at expansion is composed of the work of delivery (area 0 1 b a) and the work of expansion (area 1 2 c b) reduced by the work of exhaust (area 2 c a 3). It shows that the useful work input at compression is comprosed of the work of compression (area 1 2 c b) and the work of exhaust (area 0 1 b a) reduced by the work of delivery (area 3 2 c a).

5.2.3. Reversible Processes

Reversible processes are processes in which the initial state can be reached, when the process is converted. In this case the path of expansion coincides with the path of compression.

The constant volume process

The constant volume process occurs if a substance is retained in a closed tank and heat is conducted through the tank walls. No mechanical energy reception or rejection occurs.

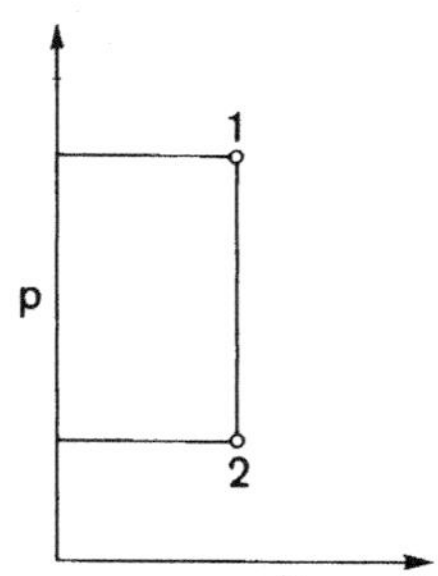

Fig. 101 constant volume process

The process is represented in Fig. 101

$$v = \text{const.} \quad \text{or} \quad p\, v^{\infty} = \text{const.}$$

The work done by this process is zero. The useful work of the process is:

$$L_t = v_1\,(p_1 - p_2) =$$
$$= R\,(T_1 - T_2)\,.$$

The heat received or reyected, respectively is found from

$$q_{12} = c_v\,(T_1 - T_2).$$

The heat received or rejected, respectively is only used for raising or lowering, respectively the internal energy of the substance.

The constant pressure or isobaric process

The constant pressure process is represented in Fig. 102.

$$p = \text{const.} \quad \text{or} \quad p\, v^\circ = \text{const.}$$

The work of the isobaric process is:

$$L_{12} = p\,(v_2 - v_1) = R\,(T_2 - T_1)\,. \tag{116}$$

The useful work of this process is zero.

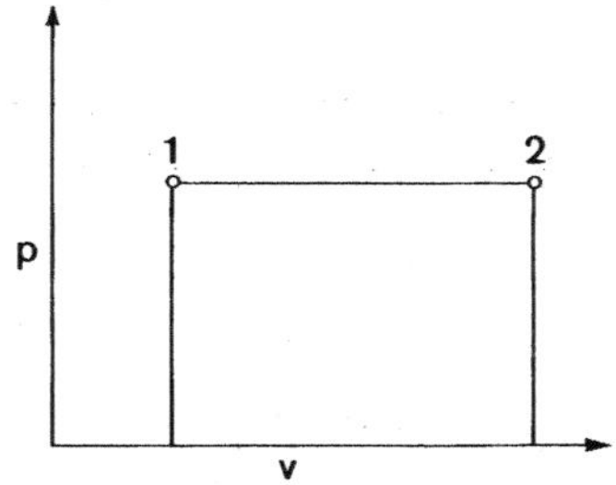

Fig. 102 constant pressure process

The heat added or rejected, respectively is found from:

$$q_{12} = u_2 - u_1 + A\,L_{12} = c_v\,(T_2 - T_1) + A\,R\,(T_2 - T_1)\,,$$
$$q_{12} = (T_2 - T_1)\,(c_v + A\,R) = c_p\,(T_2 - T_1) = i_2 - i_1\,. \tag{117}$$

The ratio between the work of the process and the added or rejected heat may be calculated thus:

$$\frac{A\,L_{12}}{q_{12}} = \frac{A\,R}{c_p} = \frac{c_p - c_v}{c_p} = \frac{\varkappa - 1}{\varkappa}\,. \tag{118}$$

For two atomic gases

$$\frac{A\,L_{12}}{q_{12}} = \frac{1.4 - 1}{1.4} = 0.286\,.$$

The isothermal process

The isothermal process is realized, if the working substance is in an effectively water jacketed cylinder with a piston. According to Boyle's law the product of pressure and volume is constant, if the temperature remains constant during the process.

$$p\,v = \text{const.}$$

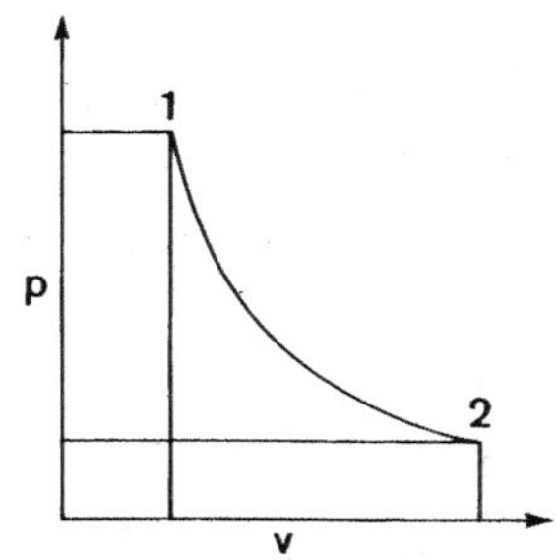

Fig. 103 Isothermal process

The process is represented in Fig. 103.

The work of the isothermal process is:

$$L_{12} = \int_1^2 p\,\mathrm{d}v = R\,T \ln \frac{p_1}{p_2}\,. \tag{119}$$

The useful work of the isothermal process is:

$$L_t = \int_1^2 v\, dp = \int_1^2 p\, dv = L_{12}\,.$$

The heat rejected at expansion and the heat added at compression may be calculated by means of the first law of thermodynamics as follows:

$$q_{12} = u_2 - u_1 + A\, L_{12} = c_v\,(T_2 - T_1) + A\, L_{12}\,.$$

Since $T_1 = T_2$, $q_{12} = A\, L_{12} = A\, L_t$.

The adiabatic process

The adiabatic process is defined as the process at which neither reception nor rejection of heat takes place.

$$q_{12} = 0\,.$$

The equations which represent the relations between pressure, volume and temperature are:

$$p\, v^{\varkappa} = \text{const.} \quad \text{or} \quad T\, v^{\varkappa-1} = \text{const.} \tag{120}$$

If p_1, v_1 and T_1 are the pressure, the specific volume and the temperature respectively at the initial state of the process, the pressure and the temperature at the end of the process are:

$$p_2 = p_1 \left(\frac{v_1}{v_2}\right)^{\varkappa} \qquad T_2 = T_1 \left(\frac{v_1}{v_2}\right)^{\varkappa - 1}. \tag{121}$$

The work of the adiabatic process is found from the condition that heat is neither added nor rejected during the process.

$$q_{12} = u_2 - u_1 + A\, L_{12} = 0$$

whence

$$A\, L_{12} = u_1 - u_2 = \frac{c_v}{A}\,(T_1 - T_2)\,. \tag{122}$$

The work done in the adiabatic process is equal to the decrease of internal energy.

The useful work in the adiabatic process is:

$$L_t = \frac{1}{A}\,(i_1 - i_2) = L_{12}\,. \tag{123}$$

As shown in Fig. 104 the work done in the isothermal process is for the area 1 T S smaller than the work done in the adiabatic process.

For compressor cylinders one tries to reach the isothermal state path by intercooling in order to save energy.

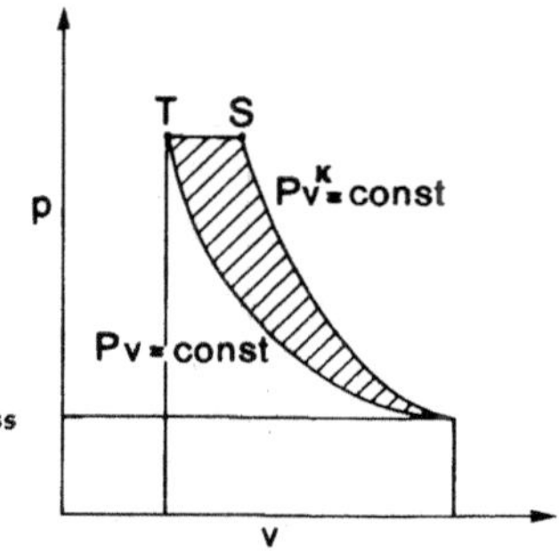

Fig. 104 Adiabatic process

Polytropic process

When a compression or expansion process in a cylinder of a reciprocating piston machine occurs, a heat transfer of varying rates and directions takes place between the gas and the surface of the cylinder. If the initial and the end state of the expansion or compression is known, the process may be represented by an equation of the form:

$$p\, v^n = \text{const.} \tag{124}$$

This is called the „polytropic process". The exponent n may have any value whatever and remains constant during the process.

The pressure and the temperature at the end of the process may be found from the equations:

$$p_2 = p_1 \left(\frac{v_1}{v_2}\right)^n \quad \text{and} \quad T_2 = T_1 \left(\frac{v_1}{v_2}\right)^{n-1}. \tag{125}$$

The specific heat remains constant during the process and may be computed thus:

$$c = \frac{n - \varkappa}{n - 1}\, c_v\,. \tag{126}$$

The work done in the polytropic process may be calculated from:

$$L_{12} = \frac{c_v}{A}\,(T_1 - T_2)\,\frac{\varkappa - 1}{n - 1}\,. \tag{127}$$

The useful work is $L_t = n\, L_{12}$.

In polytropic expansion heat is received, if the state path lies to the right of the adiabatic state path, i. e. the exponent $n < \varkappa$; heat is rejected, if the

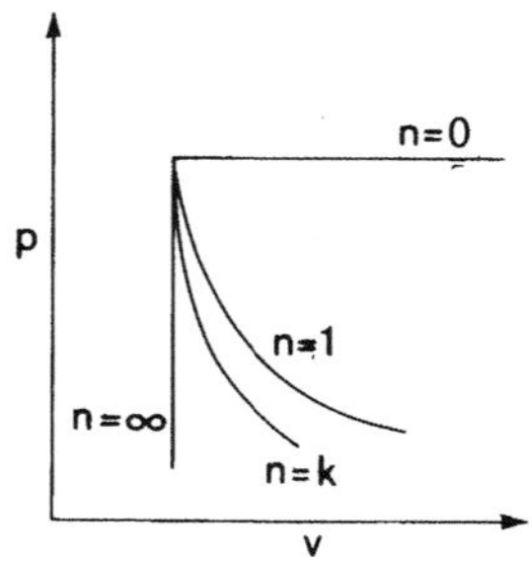

Fig. 105 Polytropic process

state path lies to the left of the adiabatic state path, i. e. $n > \varkappa$. The temperature increases, if the state path lies to the right of the isothermal state path, i. e. $n < 1$; the temperature decreases, if the state path lies to the left of the isothermal state path, i. e. $n > 1$ (Fig. 105).

In the range between the adiabatic and isothermal state paths, at expansion the temperature decreases, through heat is added.

5.2.4. Irreversible Processes

Throttling

Throttling is an irreversible process in which no external work is done. According to the law of conservation of energy

$$i_1 + \frac{A\,w_1^2}{2\,g} = i_2 + \frac{A\,w_2^2}{2\,g}$$

For *a* speed smaller than 40 m/s, $i_1 = i_2$ or $T_1 = T_2$.

For an ideal gas the temperature during throttling remains constant.

Mixture of gases

The mixture of gases is also an irreversible process. According to the principle of Dalton, any gas, in a mixture of gases acts, as if it were in a vacuum and the other gases were not present.

We assume that in all partial volumes of the gases (V_i) the same pressure (p) and the same temperature (T) exists (Fig. 106). The total volume is the sum of the partial volumes and the total mass is the sum of the partial masses.

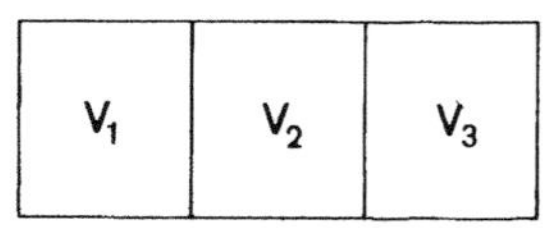

Fig. 106

During the process of mixture of the gases no external work is done and so the internal energy remains constant.

If m_i = mass of a gas
R_i = gas constant for a gas
and p_i = partial pressure of a gas i. e. that pressure which the gas exerts, if it were alone in the total volume,

then before mixture: $p\,V_i = m_i\,R_i\,T$
and after mixture: $p_i\,V = m_i\,R_i\,T\,.$

Hence

$$p_i = \frac{V_i}{V}\,p = r_i\,p \quad \text{if} \quad r_i = \frac{V_i}{V}\,. \tag{128}$$

The gas constant of the mixture may be computed as follows:

The equation of state of a gas before mixture is:

$$p\,V_i = m_i\,R_i\,T \quad \text{or} \quad m_i = \frac{p\,V_i\,V}{R_i\,T\,V} = \frac{p\,V\,r_i}{T\,R_i}\,.$$

Since the mass of the mixture is the sum of the partial masses,

$$m = \sum m_i = \frac{p\,V}{T}\sum \frac{r_i}{R_i} \quad \text{or} \quad p\,V = m\,T\,\frac{1}{\sum r_i/R_i} = m\,T\,R'\,.$$

The gas constant of the mixture is therefore:

$$R' = \frac{1}{\sum r_i/R_i}\,. \tag{129}$$

The specific heat of the mixture may be calculated as follows:

$$U = \sum U_i = \sum m_i\,c_{vi}\,T = p\,V \sum \frac{r_i\,c_{vi}}{R_i} = m\,c_v'\,T$$

whence

$$c_v' = \frac{p\,V}{m\,T}\sum \frac{r_i\,c_{vi}}{R_i} = R' \sum \frac{r_i\,c_{vi}}{R_i}\,. \tag{130}$$

Analogously,

$$c_p' = R' \sum \frac{r_i\,c_{pi}}{R_i}\,, \tag{131}$$

hence $\varkappa' = \dfrac{c_p'}{c_v'}\,.$

5.2.5. Cycles

A series of processes that may be performed repeatly in the same order in thermodynamic devices is called a cycle. There are two kinds of thermodynamics cycles: the closed and the open cycle.

In the closed cycle the working fluid passes through a series of devices, repeats the same cycle and returns to the same initial state. Examples of this cycle are: the condensing steam engine, the closed cycle type of gas turbine and the refrigerating plant.

In the open cycle a new mass of fluid is taken in each cycle within the machine. Examples of these cycles are: the steam engine which works without condensation, the internal combustion engine and the air compressor.

A cycle is reversible, if the single processes of the cycle are reversible. If one process of the cycle is irreversible, the entire cycle is irreversible. The hatched areas of the diagrams in Fig. 107 represent the work which is supplied by a heat engine and the work which must be put in to piston pumps or compressors. The first law of thermodynamics applied to the cycle of a heat engine states: the difference between the input and output heat is transformed into mechanical work (Fig. 108).

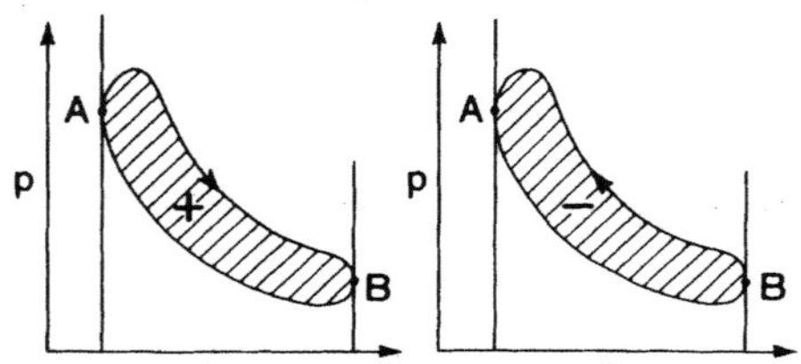

Cycle of a heat engine Cycle of a piston pump or compressor

Fig. 107

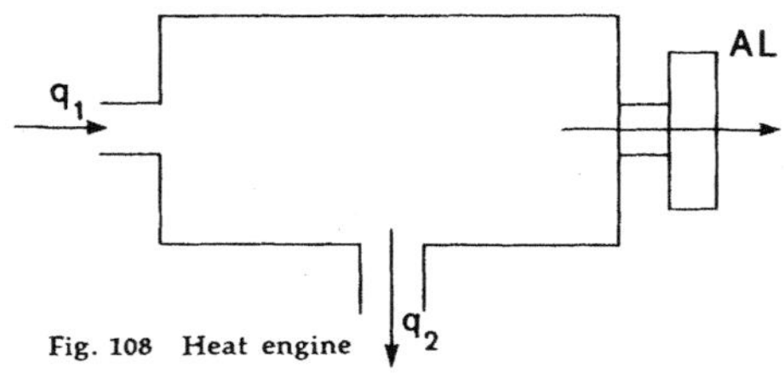

Fig. 108 Heat engine

The ratio of net work output to heat energy input is called the "thermal efficiency".

$$\eta_{th} = \frac{m\,A\,L}{Q_1} = \frac{Q_1 - Q_2}{Q_1}. \quad (132)$$

The inverted cycle for the heat engine is the cycle of the heat pump (Fig. 109).

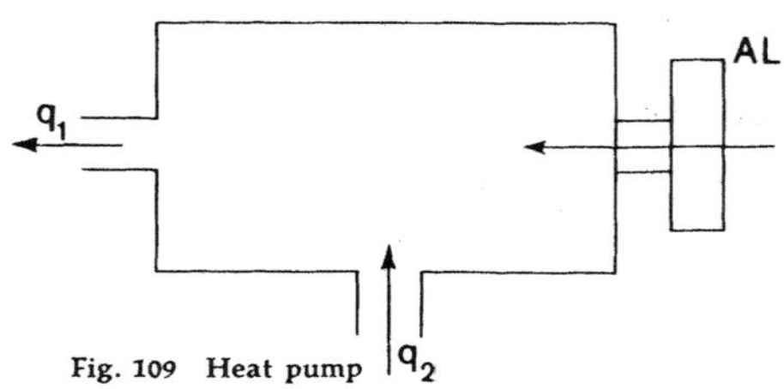

Fig. 109 Heat pump

The ratio between the heat energy extracted to the net work input is called "coefficient of performance".

$$\eta = \frac{Q_2}{m\,A\,L} = \frac{Q_2}{Q_1 - Q_2}. \quad (133)$$

Idealized cycles of heat engines

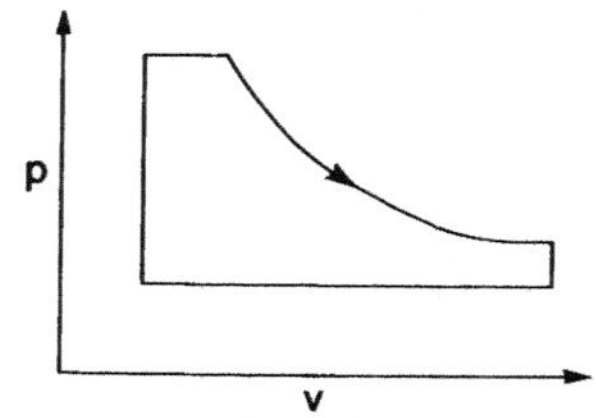

Fig. 110 Cycle of the reciprocating steam engine

The cycle of the reciprocating steam engine is composed of two constant volume processes, two constant pressure processes and an adiabatic process (Fig. 110).

The cycle of the Otto engine is composed of two adiabatic processes (expansion and compression) and two constant volume processes (combustion of the charge and escape of the burned gases) (Fig. 111).

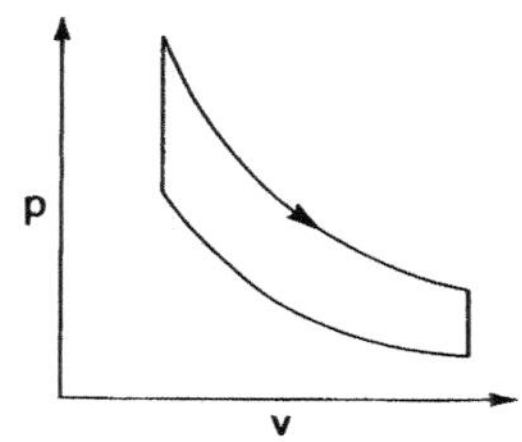

Fig. 111 Cycle of the Otto engine

The cycle of the Diesel engine is composed of two adiabatic processes (expansion and compression), a constant pressure process (combustion) and a constant volume process (exhaust of burned gases) (Fig. 112).

Idealized cycle of reciprocating compressor

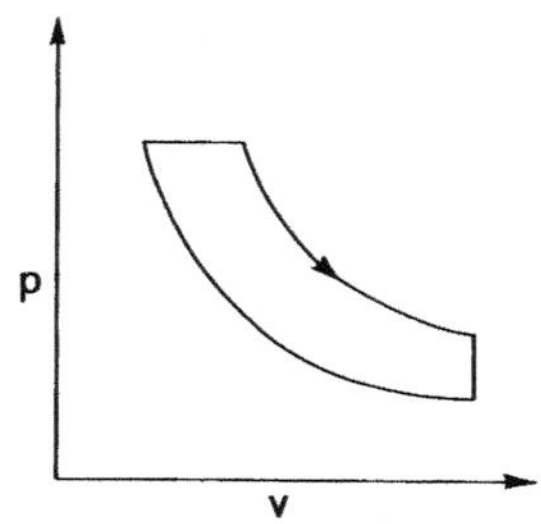

Fig. 112 Cycle of the Diesel engine

The cycle of the reciprocating compressor is composed of two adiabatic processes (compression and expansion) and a constant pressure process (exhaust of the air) (Fig. 113).

5.2.6. The Second Law of Thermodynamics

The first law of thermodynamics states that heat is a form of energy, but it does not establish the conditions under which energy transformation takes place and how the extent of the conversion is limited. These conditions and limits belong to the territory of the second law of thermodynamics.

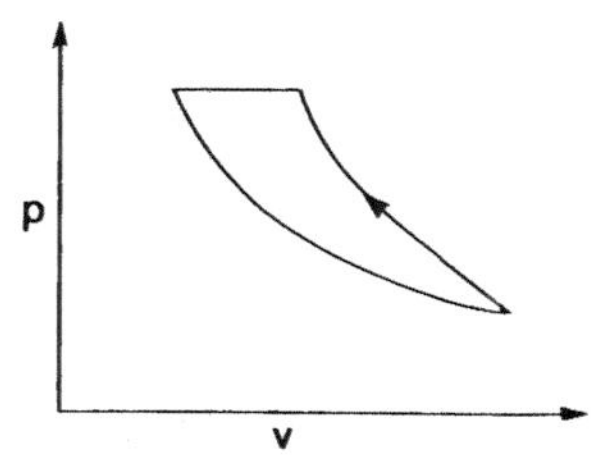

Fig. 113 Cycle of a piston compressor

Statements of the second law of thermodynamics are made which are based on observation of operations. Since the observations that are made are various, many statements are proposed to express the meaning of that law. One of the earliest and simplest statements was offered by Clausius in 1850. It states, heat cannot, of itself, pass from a colder to a hotter substance.

The work of a reversible cycle can be represented in a diagram the coordinates of which are the pressure and the volume. Pressure may be regarded as the property which furnishes the incentive. The pressure difference determines the direction of the work flow. The property with respect to heat flow which is analogous to pressure in the work diagram is temperature. Analogous to work, heat can be represented in a diagram the coordinates of which are the temperature and a property called "entropy" which corresponds to the volume in the work diagram.

When work is written in differential form as $d_L = p\,dv$

heat may be written in the form as $dq = T\,ds$.

A cycle may be represented in a diagram the coordinates of which are the absolute temperature (T) and the entropy per unit mass (s). The hatched area of the diagram in Fig. 114 represents the work expressed in heat units in a reversible cycle.

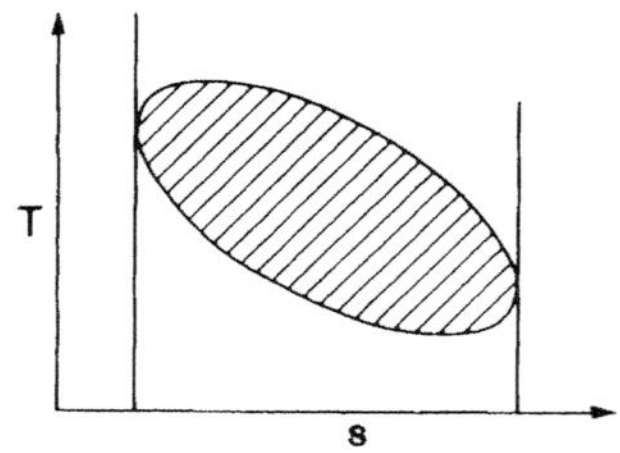

Fig. 114 Cycle in temperature – entropy diagram

Determination of entropy differences

For solids and liquids

$$ds = \frac{dq}{T} = \frac{du}{T} + \frac{A\,P\,dv}{T}.$$

Since the extension of solid and liquid bodies is very small, the second term may be neglected. Therefore,

$$ds = \frac{du}{T} = \frac{c\,dT}{T} \quad \text{and}$$

$$\Delta s = c \ln T_2 - c \ln T_1 = c \ln \frac{T_2}{T_1} \tag{134}$$

where c = specific heat of the solid or liquid body
and $T_2 > T_1$.

For an ideal gas

$$ds = \frac{du}{T} + \frac{A\,p\,dv}{T} = c_v \frac{dT}{T} + A\,R \frac{dv}{v} \quad \text{and}$$

$$\Delta s = c_v \ln \frac{T_2}{T_1} + A\,R \ln \frac{v_2}{v_1}\,. \tag{135 a}$$

Other forms of the equation are:

$$\Delta s = c_p \ln \frac{T_2}{T_1} + A\,R \ln \frac{p_1}{p_2} \quad \text{and} \tag{135 b}$$

$$\Delta s = c_v \ln \frac{p_2}{p_1} + c_p \ln \frac{v_2}{v_1}\,. \tag{135 c}$$

Reversible processes in T-s diagram

Fig. 115 shows the principal state paths in T-s diagram

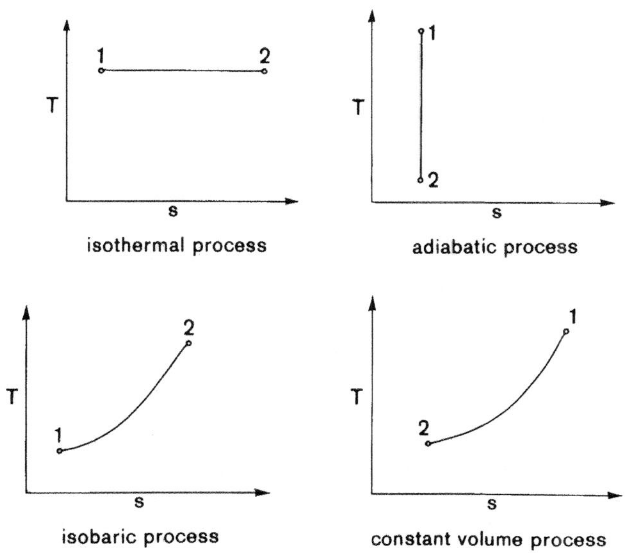

Fig. 115

The state path of the isothermal process appears as a horizontal line.

The state path of the adiabatic process appears as vertical line.

$$\text{As} \quad ds = \frac{dq}{T} = 0\,, \quad s = \text{constant.}$$

Entropy change in throttling process of an ideal gas.

$$\Delta s = c_p \ln \frac{T_2}{T_1} + A R \ln \frac{p_1}{p_2}. \qquad \text{(equation 135 b)}$$

Since $T_1 = T_2$ this becomes

$$\Delta s = A R \ln \frac{p_1}{p_2}. \tag{136}$$

Entropy change at mixture of ideal gases

Since during mixture the internal energy remains constant, the temperature remains constant.

Before mixture,

$$s_1 = \Sigma m_i (c_p \ln T - A R_i \ln p + s_0).$$

After mixture,

$$s_2 = \Sigma m_i (c_p \ln T - A R_i \ln p_i + s_0).$$

The entropy difference is:

$$\Delta s = \Sigma m_i A R_i \ln \frac{p}{p_i} \tag{137}$$

where p = pressure before mixture
and $p_i = r_i p$ = partial pressure after mixture.

The Carnot cycle

The Carnot cycle is the ideal cycle for a heat engine or a refrigerator plant. It has the highest thermal efficiency or the highest efficiency of performance that a machine can attain. It is composed of two isothermal processes and two adiabatic processes (Fig. 116).

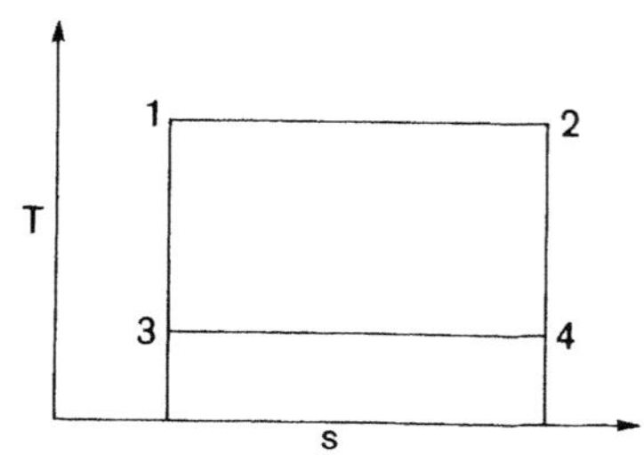

Fig. 116 Carnot cycle in temperature-entropy diagram

The thermal efficiency is:

$$\eta_{\text{th}} = \frac{Q_1 - Q_2}{Q_1} = \frac{T_1 - T_2}{T_1}. \tag{138}$$

The efficiency of performance is:

$$\eta = \frac{Q_2}{Q_1 - Q_2} = \frac{T_2}{T_1 - T_2}. \tag{139}$$

Forms of the second law of thermodynamics

First form: Heat can only be changed into mechanical work, if a gradient of temperature exists. Analogously mechanical potential energy can only be changed into kinetic energy, if a gradient of pressure exists.

Second form: For the reversible cycle the change of entropy in expansion is equal to the change of entropy in compression. The total change of entropy in the cycle is therefore zero. For the irreversible cycle losses occur as well in expansion as in compression which results in an increase of entropy. The increase of entropy in a cycle is a measure of the heat not used for work. For a reversible cycle a maximum of work can be produced from the heat.

5.3. Vapours

As water vapour is the most important vapour, only the study of this vapour will be dealt with here.

There are three physical states in which the substance water occurs. The solid, called ice, the liquid, called water and the gaseous, called vapour or steam.

When the substance "water" is heated at a particular temperature a change of state occurs which depends upon the pressure. During the transition from one state to another at constant pressure, the temperature remains constant.

At atmospheric pressure the change from the solid to the liquid state takes place at a temperature of 0 °C and the transition from the liquid to the gaseous state at a temperature of 100 °C. The first change is called the fusion or melting process, the second change the vaporisation process. The heat required to transfer unit mass of ice into water at atmospheric pressure is called the "latent heat of fusion" and the heat required to convert unit mass of water into vapour at atmospheric pressure is called the "latent heat of vaporisation". The latent heat of fusion of water at standard pressure is 80 kcal/kg and the latent heat of vaporisation at standard pressure is 540 kcal/kg.

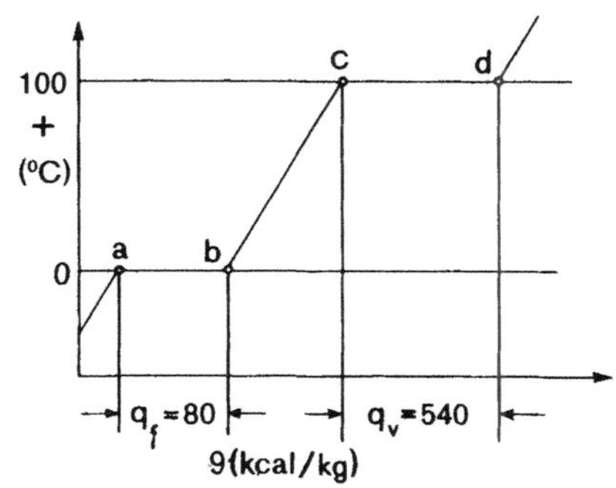

Fig. 117 Process of fusion and vaporisation of water

Fig. 117 shows graphically the process of fusion and vaporisation. The line $a - b$ indicates the fusion stage, the line $c - d$ represents the vaporisation stage. At the point c the water is in a saturated state. Between c and d wet vapour i. e. water and vapour in the saturated state exists. At the point d only saturated vapour exists. If the saturated vapour is further heated, superheated steam is produced.

The state of wet steam is defined by the dryness fraction of the steam, x.

$x =$ mass of saturated steam in 1 kg of wet steam.

$x = 0$ indicates water in the saturated state
$x = 1$ indicates steam in the saturated state.

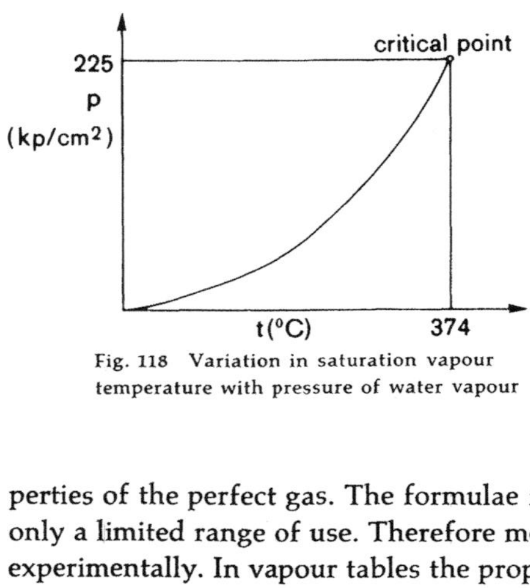

Fig. 118 Variation in saturation vapour temperature with pressure of water vapour

The vaporisation takes place at a particular temperature which depends on the pressure. The relation between the vaporisation temperature and the pressure is illustrated in Fig. 118. Above a particular temperature, called the "critical temperature" a liquefaction cannot be produced. The corresponding pressure is called the "critical pressure". The end of the curve in Fig. 118 is called the "critical point". The properties of steam cannot be found as easily as the properties of the perfect gas. The formulae for steam are complicated and have only a limited range of use. Therefore most of the properties must be found experimentally. In vapour tables the properties are tabulated.

Properties of wet vapour

1 kg wet steam contains x kg saturated steam and $(1 - x)$ kg saturated water.

If v' is the specific volume of saturated water
v'' the specific volume of saturated steam
and x the dryness fraction
then the specific volume of wet vapour is:

$$v = x\,v'' + (1 - x)\,v' . \qquad (140)$$

If u' = internal energy of 1 kg of saturated water
and u'' = internal energy of 1 kg of saturated steam,
the internal energy of wet steam is calculated from:

$$u = x\,u'' + (1 - x)\,u' . \tag{141}$$

Similary the enthalpy of wet steam is found from:

$$i = x\,i'' + (1 - x)\,i' \tag{142}$$

where i' = enthalpy of 1 kg of saturated water
and i'' = enthalpy of 1 kg of saturated steam.

The entropy of wet steam is:

$$s = x\,s'' + (1 - x)\,s' \tag{143}$$

where s' = entropy of 1 kg of saturated water
and s'' = entropy of 1 kg of saturated steam.

Wet vapour processes

The isobaric and isothermal process

As temperature and pressure are constant during vaporisation, the isobaric state path and the isothermal state path appear both in the $p\,v$ and in the $T\,s$ diagram as the same horizontal line (Fig. 119).

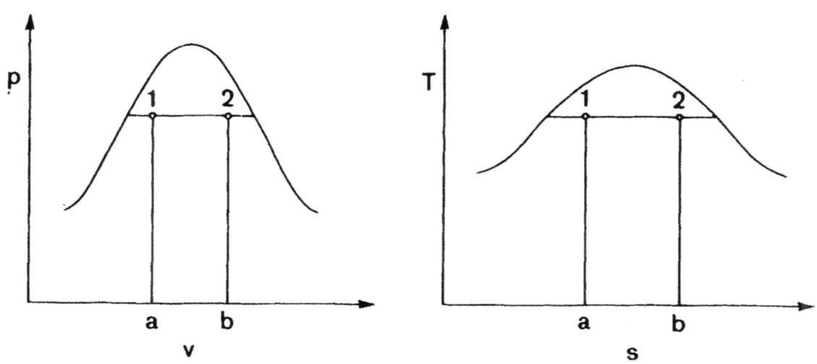

Fig. 119 Isobaric and isothermal process of water vapour

The work done by this process is: $L_{12} = p_1\,(v_2 - v_1)$ and is represented by the area $a\,1\,2\,b$ in the $p\,v$ diagram of Fig. 119.

The useful work is zero.

The heat transferred in this process is: $q_{12} = i_2 - i_1$ (equation 117) and is represented by the area $a\,1\,2\,b$ in the $T\,s$ diagram of Fig. 119.

The constant volume process

The constant volume process is represented by a vertical line in the $p\,v$ diagram of Fig. 120.

The work done by this process is zero.

The useful work is: $L_t = v_1\,(p_1 - p_2)$ and is represented by the area a 1 2 b in the $p\,v$ diagram of Fig. 120.

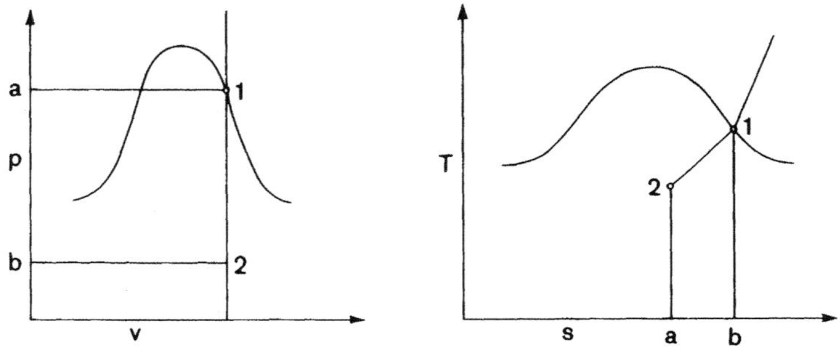

Fig. 120 Constant volume process of water vapour

The heat transferred in this process is: $q_{12} = u_2 - u_1$ and is represented by the area 2 1 b a in the $T\,s$ diagram of Fig. 120.

The adiabatic process

The adiabatic process is represented by a vertical line in the $T\,s$ diagram of Fig. 121.

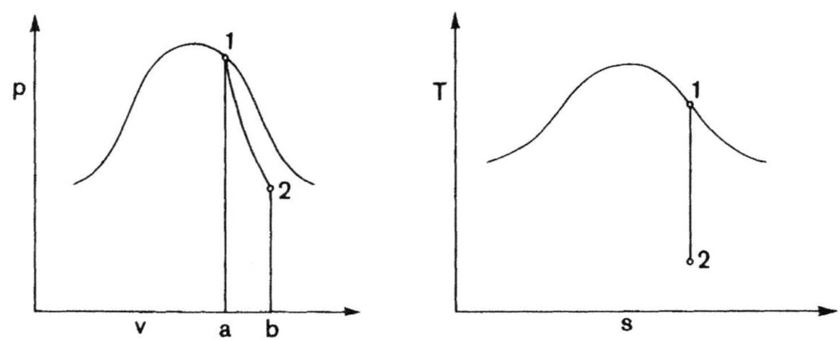

Fig. 121 Adiabatic process of water vapour

The work done by this process is: $L_{12} = \frac{1}{A}(u_1 - u_2)$ (equation 122) and is represented by the area 1 2 b a in the $p\,v$-diagram of Fig. 121.

The useful work is $L_t = \frac{1}{A}(i_1 - i_2)$. (123)

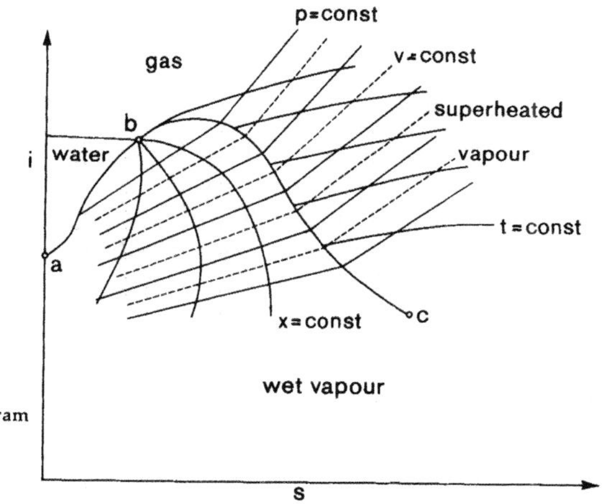

Fig. 122 Enthalpy – entropy diagram

The heat transferred in this process is zero.

The enthalpy – entropy diagram

The enthalpy – entropy diagram or, as it is often called, the "Mollier chart" is very useful to engineers for the graphical representation of the processes of steam (Fig. 122).

The point b in the diagram is the critical point, the curve $a - b$ indicates the liquid locus, the curve $b - c$ the saturation locus. There are four regions in the diagram: the water, the wet vapour, the superheated vapour and the gas region. The adiabatic process is represented by a vertical line. The lines of constant pressure, of constant temperature, of constant volume and of constant quality ($x =$ const.) are plotted in the diagram.

The throttling process

During throttling the enthalpy remains constant. The process is represented by a horizontal line in the $i - s$ diagram. As the temperature decreases, the dryness fraction becomes higher and wet steam can be changed into superheated steam, as may be seen from the $i - s$ diagram.

Mixture of steams

The enthalpy of a mixture is equal to the enthalpy of the steam parts.

$$i \sum m_i = \sum m_i \, i_i$$

where m_i = mass of a steam part and i_i = enthalpy of a steam part.

5.4. Combustion

Heat produced in a combustion process is the source of energy used in a heat engine. Combustion is a chemical process in which the combustible components of the fuels unite with oxygen of the air and form oxides releasing heat at the same time. The amount of heat liberated in a combustion process depends on the chemical composition of the fuels and the amount of oxidation of the combustible substances in the fuels.

5.4.1. Fuels

Fuels may be solid, liquid or gaseous. The most important of the solid fuels is coal. There are various types of coal such as anthracite, pit coal and lignite. Liquid fuels are: light oils, called petrols, and heavier oils such as Diesel oil and fuel oil. Gaseous fuels are: natural gas, town gas and producer gas. The combustible elements of a fuel are: carbon, hydrogen, hydrocarbons and sulphur. Other elements found in fuels are: nitrogen, oxygen, moisture and impurities. The percentage of the constituent substances is given for solid and liquid fuels normally on a mass basis, and for gaseous fuels on a volume basis at a given pressure and temperature.

The combustion of carbon forms carbon monoxide or carbon dioxide, hydrogen produces water and sulphur produces sulphur dioxide. The analysis of the products of combustion gives information about the combustion in boiler furnaces and internal combustion engines. The best known method of obtaining the analysis of products of combustion is by means of the Orsat apparaturs.

5.4.2. The Chemical Equations of Combustion Processes

Solid and liquid fuels

The molecular masses of the combustible substances of solid and liquid fuels are:

carbon 12, carbon monoxide 28, hydrogen 2, sulphur 32, oxygen 32.

Carbon

Carbon may completely burn to carbon dioxide or incompletely to carbon monoxide.

The chemical equation for complete cumbustion is:

$$C + O_2 = CO_2$$

or expressed in mole units:

$$1 \text{ mole of } C + 1 \text{ mole of } O_2 = 1 \text{ mole of } CO_2 .$$

The equation written in mass units gives:

$$12 \text{ kg of } C + 32 \text{ kg of } O_2 = 44 \text{ kg of } CO_2$$

hence 1 kg of C requires $\frac{8}{3}$ kg of O_2 and produces $\frac{11}{3}$ kg of CO_2.

The chemical equation for incomplete combustion is:

$$2\,C + O_2 = 2\,CO \text{ or } 2 \text{ moles of } C + 1 \text{ mole of } O_2 = 2 \text{ moles of } CO$$

written in mass units:

$$2 \times 12 \text{ kg of } C + 32 \text{ kg of } O_2 = 56 \text{ kg of } CO$$

hence 1 kg of C requires $\frac{4}{3}$ kg of O_2 and produces $\frac{7}{3}$ kg of CO.

Hydrogen

The chemical equation is:

$$2\,H_2 + O_2 = 2\,H_2O \quad \text{or}$$
$$2 \text{ moles of } H_2 + 1 \text{ mole of } O_2 = 2 \text{ moles of } H_2O \text{ or}$$
$$2 \times 2 \text{ kg of } H_2 + 32 \text{ kg of } O_2 = 36 \text{ kg of } H_2O$$

hence 1 kg of H_2 requires 8 kg of O_2 and produces 9 kg of twater.

Sulphur

The chemical equation is:

$$S + O_2 = SO_2 \quad \text{or} \quad 1 \text{ mole of } S + 1 \text{ mole of } O_2 = 1 \text{ mole of } SO_2$$
$$\text{or} \quad 32 \text{ kg of } S + 32 \text{ kg of } O_2 = 64 \text{ kg of } SO_2$$

hence 1 kg of S requires 1 kg of O_2 and produces 2 kg of SO_2.

The amount of oxygen by mass required for the combustion of a solid or liquid fuel may be calculated from:

$$o_{min} = \frac{8}{3} c + 8\,h + s - O$$

where c, h, s and o are the mass amounts of carbon, hydrogen, sulphur and oxygen in 1 kg of fuel.

Since air contains 23.1 % O_2 by mass, the mass of air required for 1 kg of fuel is:

$$L_{min} = \frac{100}{23.1}\left(\frac{8}{3}c + 8\,h + s - o\right).$$

Gaseous fuels

Hydrogen

$2\,H_2 + O_2 = 2\,H_2O$ or
2 moles of H_2 + 1 mole of O_2 = 2 moles of H_2O.

Since the mole volumes of all gases at the same pressure and temperature are identical,

2 m^3 of H_2 + 1 m^3 of O_2 = 2 m^3 of H_2O

hence 1 m^3 H_2 requires 0.5 m^3 O_2 and produces 1 m^3 H_2O.

Carbon monoxide

$2\,CO + O_2 = 2\,CO_2$ or
2 moles of CO + 1 mole of O_2 = 2 moles of CO_2

hence 1 m^3 CO requires 0.5 m^3 O_2 and produces 1 m^3 CO_2.

Similarly

1 m^3 methane (CH_4) requires 2 m^3 O_2 and produces 1 m^3 CO_2 + 2 m^3 H_2O
1 m^3 ethylene (C_2H_4) requires 3 m^3 O_2 and produces 2 m^3 CO_2 + 2 m^3 H_2O
1 m^3 acetylene (C_2H_2) requires 2.5 m^3 O_2 and produces 2 m^3 CO_2 + 1 m^3 H_2O
1 m^3 propane (C_3H_8) requires 5 m^3 O_2 and produces 3 m^3 CO_2 + 4 m^3 H_2O
1 m^3 butane (C_4H_{10}) requires 6.5 m^3 O_2 and produces 4 m^3 CO_2 + 5 m^3 H_2O

The amount of oxygen by volume required for the combustion of a gaseous fuel is calculated from:

$$O_{min} = 0.5\,H_2 + 0.5\,CO + 2\,CH_4 + 3\,C_2H_4 + 2.5\,C_2H_2 + \ldots - O_2$$

where H_2, CO_2, CH_4 etc. are the volume amounts in 1 Nm^3 of a gaseous fuel.

The volume of air required for 1 Nm^3 of gaseous fuel is:

$$L_{min} = \frac{100}{21}\,O_{min}$$

5.4.3. Calorific Values

Two calorific values are quoted the higher or gross calorific value, abreviated H_0 and the lower or net calorific value, abreviated H_u. The higher calorific value is the heat liberated in 1 kg of solid or liquid fuel or in 1 Nm^3 of gaseous fuel, when the products of combustion are cooled back to 0 °C. The lower calorific value is equal to the higher calorific value reduced by the heat necessary for producing water vapour. Water vapour is produced by the vaporisation of water contained in the fuel or by oxidation of the hydrogen containing in the fuel. Since the amount of heat required for producing water vapour in technical processes may not be utilized, in heat caculations normally the lower calorific value is taken. Between the two calorific values the following relations exist:

$$H_u = H_o - (w + 9\,h)\;597\;\text{kcal}/kg$$

where w = mass of water per kg of fuel
and h = mass of hydrogen per kg of fuel

$$H_u = H_0 - \left[H_2O + H_2 + \sum\left(\frac{n}{2}\,(C_m\,H_n)\right)\right]\;480\;\text{kcal/Nm}^3$$

where H_2O = volume of water per Nm^3 of fuel
H_2 = volume of hydrogen per Nm^3 of fuel
and $C_m\,H_n$ = volumes of hydrocarbons per Nm^3 of fuel.

6. Engineering Materials

The materials used in the building of structures and machines are called "engineering materials". Some of the most commonly used materials, the process by which they are obtained and their applications as machine members or members of a structure are described here.

The materials employed by engineers may be classified as metallic or non-metallic. The metallic materials may be further classified as ferrous and non-ferrous. The ferrous metals are iron and steel, the non-ferrous metals include copper, tin, zink, nickel, aluminium and their alloys. The non-metallic materials include wood, stone, brick, concrete, glass, rubber, leather, plastics, etc.

The efficient use of engineering materials requires considerable knowledge of their mechanical properties. The most important mechanical properties are: strength, stiffness, ductility, malleability, hardness, toughness and machinability. The methods for determining the properties of materials have been evolved over a long period of years and have now been standardized.

6.1. Iron and Steel

6.1.1. The Manufacture of Pig Iron

Iron is found in nature in the form of iron ores. Most iron ores are oxides which contain impurities such as sulphur, phosphorus, silicon and earthy matter. The principal iron ores are: hematite (Fe_2O_3), limonite or hydrated hematite ($2\,Fe_2O_3 + 3\,H_2O$), siderite or spatic ore ($FeCO_3$) and magnetite (Fe_3O_4). The value of an iron ore is determined by the percentage of contained iron. Ores carrying 50% or more of iron are called high grade ores and ores carrying less than 50% of iron low grade ores. Most ores of iron lie near the surface of the ground and are mined by open surface mining. In order to remove the impurities from the iron fluxes are added. These fluxes will unite with the gangue and form a fusible mixture. The fluxes have a chemical nature opposite to that of the gangue. If the ore is acid, then limestone is added, if the ore is basic, silicon is added. The combination of flux and impurities is called slag and floats on top of the molton iron and may be removed easily from the pig iron. For fuel, coke is normally used. Iron ore is smelted to pig iron in a blast furnace whose essential features are shown in Fig. 123. The blast furnace consists of a cylindrical steel shell 5 to 8 m in diameter and 40 to 50 m high lined inside with fire bricks which can withstand the high temperature inside the furnace. The ore, the fuel and the flux are carried to the top of the furnace by means of an inclined skip hoist. The proportion of the flux to the whole charge is about one sixth. The fuel is about one third and the ore about one half of the whole charge.

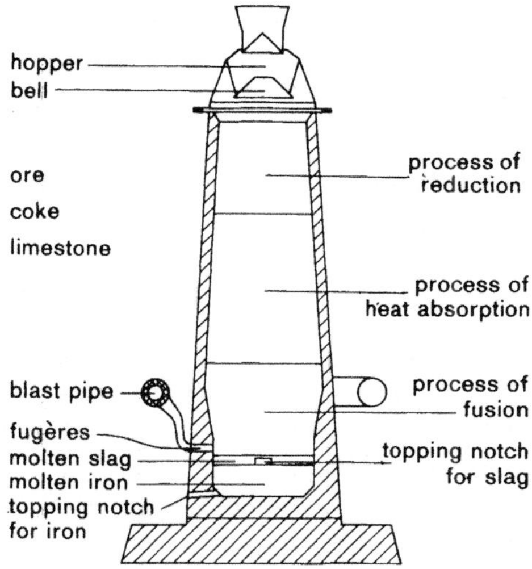

Fig. 123 Blast furnace

The combustion of the fuel takes place principally at the bottom of the furnace in a blast of preheated air with a temperature of 600 to 800 °C and a pressure of 0.5 to 1.3 atm blown into the furnace through a series of nozzles called tuyeres. The air unites with the carbon of the coke and forms carbon monoxide at a temperature exeeding 1600 °C. The carbon monoxide combines with the oxygen of the iron oxide and escapes at the top of the furnace as carbon dioxide. The bottom of the furnace has the highest temperature and here smelting takes place. The iron combines with about 3 to 5 % of the carbon from the coke which effects a reduction of the smelting temperature of the iron. The molten iron is drawn off at the bottom of the furnace and cast into small bars called pigs. The molten slag is tapped off at intervals through the cinder notch which is at 90° from the iron notch. The gas at the top of the furnace is led through washers to remove solid particles and is used to operate the blowing engines and to heat the boilers of the plant. The gas contains about 60 % nitrogen, 10 to 17 % carbon dioxide and the rest carbon monoxide.

The carbon in the iron may be present as graphite which is finely dispersed throughout the metal or as a chemical combination of carbon called iron carbide, (Fe_3C). The fracture of iron with graphite is grey and therefore this kind of iron is called grey pig iron. The fracture of iron with carbide is whiter and so this iron is called white pig iron. Slow cooling and a high silicon content allows free graphite to form. Quick cooling and a high manganese content produces iron carbide. Grey pig iron which is softer than white pig iron is used for producing cast iron, but for producing steel grey and white pig iron are used.

6.1.2. Production of Steel

6.1.2.1. The Puddling Process

All methods for producing steel are based on removing carbon together with several impurities such as silicon, manganese, sulphur and phosphorus by means of oxidation. The oldest method of refining pig iron is the puddling process. The product from this process is called "wrought iron". Wrought iron is the nearest approach to pure iron. It contains less than 0.12% of carbon.

Until the latter part of the nineteenth century wrought iron was the most important of the iron products, but with the introduction of steel production by means of the Bessemer the Thomas or the open hearth processes, the use of wrought iron has greatly decreased.

The wrought iron is produced in a reverberatory furnace which consists of a fire box, a bridge wall, a hearth with a sloping-covered top and a stack. The charge is not directly in contact with the fire, but receives its heat by reflexion from the roof of the furnace. The furnace hearth is lined with iron oxide. A special pig iron containing a high percentage of silicon is fed in to the furnace. During the melting stage nearly all the silicon, manganese, sulphur and phosphorus unites with the iron oxide to form a slag which floats on the top of the bath. From time to time the operator stirs the slag with an iron hoe to prevent particles of pasty iron from separating and sticking to the bottom. Mill scale is added to the slag to bring about the oxidation of the carbon and the other elements present in the iron. As the iron is purified, it can no longer stay molten. Lumps of pasty iron intermixed with slag is formed. These balls are removed from the furnace and are squeezed into blooms in a press. Then the blomms are rolled into smaller bars. Due to the presence of the slag, which extends through the iron in the direction of rolling, a fracture has a fibrous appearance. Thus wrought iron may be distinguished from the Bessemer (Thomas) and open hearth steels.

6.1.2.2. The Bessemer and Thomas Processes

The most common processes for making steel are the Bessemer, the Thomas, the open hearth and the electric furnace processes. In all these processes steel is drawn off in a molten state instead of being removed in pasty balls, as is the case in the puddling process.

The Bessemer process, so named after its inventor Sir Henry Bessemer who invented this process in 1885 in England, the Thomas process invented by Sidney Gilchrist Thomas and the open-hearth process are the most widely used processes for refining pig iron. They are the most rapid of all steel processes. For Bessemer and the Thomas processes a pear shaped vessel, called a "converter", is used which is lined with fire bricks. The vessel is open at the top and the bottom is pierced with tuyeres. It is so constructed that it can be revolved horizontally and tipped vertically. In the horizontal position it is charged with between 15 to 60 tons of molten pig iron often brought directly from a blast furnace. It is tipped vertically and cold air at a pressure of about 2 atm is blown through the tuyeres in the bottom of the vessel into the molten iron for about 20 minutes. During the blow period all impurities in the pig iron burn out. No outside fuel is used in this process, since the impurities in the pig iron furnish the fuel. When the flame of the converter drops, the blast air is shut off and the molten iron is poured into a ladle. The melt consists of an almost carbonless iron, containing small percentages of silicon, sulphur and phosphorus. As the carbon content of the molten steel would be too low for most purposes and as it would be impossible to stop the blowing so as to leave a definite amount of carbon present, the requisite amount of carbon is now added in the form of a ferromanganese alloy which is a variety of pig iron rich in manganese. The ladle is now lifted over a row of ingot moulds which are filled with the molten steel.

The difference between the Bessemer and the Thomas processes is that in the Bessemer process which is used for pig iron containing no phosphorus or at most small quantities of it, the lining of the furnace is acid (ground quartz or sandstone with a small amount of clay), whereas in the Thomas process which is used for pig iron containing phosphorus in higher percentage, the lining of the furnace is calcined dolomite. The basic material for the Bessemer process is grey pig iron containing silicon and for the Thomas process white pig iron containing phosphorus. The oxidation in the Thomas process is more severe than in the Bessemer process and requires more time.

6.1.2.3. The Open-Hearth Process

The Bessemer and the Thomas processes can only be used for molten pig iron. When pig iron in the molten or solid state together with steel scrap is employed for making steel normally the open- hearth process is used.

The furnace for the open hearth process is shown in Fig. 124. A shallow hearth lined with refractory bricks contains the charge. The lining may be acid or basic according to the kind of ore used in this process. The ingredients, pig iron and steel scrap are fed through a charging door. The charge ranges according to the size of the furnace from 50 to 100 tons. The sufficiently high temperature for smelting the pig iron or steel scrap is obtained by using the regenerative system. The regenerative system consists of four heating chambers called "generators" which are below the hearth. Two chambers are on each side. One chamber on each side serves for heating air, the other one for heating the fuel gas. The chambers contain a checkerwork of refractory bricks. Through the two chambers on one side exhaust gases pass from the furnace and give up heat to the brickwork and pass out through the stack. The generators on the other side which were previously heated give up the heat in one chamber to the entering air and in the other chamber to the entering fuel gas. About every 20 minutes the flow of

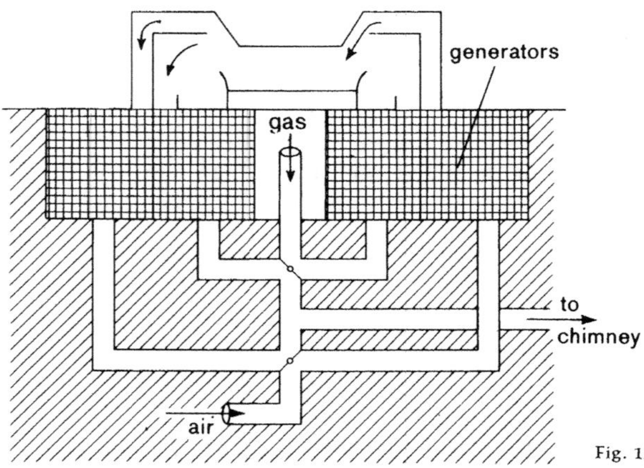

Fig. 124 Open hearth furnace

air and gas is reversed, thus the two generators on each side are heated alternately. The fuel commonly used in the open hearth steel furnace is producer gas which is generated in an apparatus called a "producer". In the producer air mixed with steam passes through incandescent coal. Producer gas is a very cheap gas of low heating power. The temperature of the air and the gas after leaving the heating chambers is about 1250 °C and the temperature of the burning gases above the hearth about 1800 °C.

The open-hearth process is much slower than the Bessemer and the Thomas processes and requires about 10 hours for each charge. But the slowness of the process makes it possible to control the quality of the steel.

6.1.2.4. The Crucible Process

In order to produce a high grade steel or an alloyed steel the crucible or the electric process is employed. In the crucible process the charge consists of powdered charcoal and bars of very pure iron and other ingredients containing elements desired in the finished steel such as manganese, nickel, chromium, tungsten, etc. The charge is placed in crucibles made of fire clay mixed with graphite and they are covered to keep out air. Each crucible holds about 20 kg of metal and is generally used for three melts before beeing discarded. Crucible furnaces are regenerative similar to those employed in the open-heath process.

6.1.2.5. The Electric Refinig Process

The electric process is normally only used for refining steel produced by the Bessemer, the Thomas or the open-hearth processes. In the electric process a very high grade plain carbon or alloy steel is produced. The great advantage of this process is the absence of gases, fumes and impurities such as are present in the other processes which are fuel fed and at which impurities may be introduced into the melt or a required constituent may be oxidized out of the melt. The temperatures in the electric process are very high and may be controlled exactly. At these high temperatures the slag containing all the impurities is in a complete liquid state. There are two types of electric furnace in use: the electric arc furnace and the electric induction furnace.

In the electric arc furnace the heat required is produced by an electric arc struck between carbon electrodes and the metal bath. The impurities are oxidized from the charge and then absorbed by the slag which lies over the charge. By means of this process alloy steels such as stainless steel, high speed steels, etc. are made.

In the electric induction furnace the bath is within a coil of wire in which an alternating current is flowing. The alternating magnetic field produced by the current induces an electric current in the steel bath which heats up the steel. The frequencies in use vary from 500 to 2000 Hertz. The induction process is used for making high purity steels such as are necessary for tool and die steel, ball bearing steel, etc.

6.1.3. Classification of Iron and Steel

6.1.3.1. Cast Iron

Cast iron is made by melting grey pig iron and iron scrap in a foundry cupola. This furnace resembles a blast furnace, but it is much smaller. To start the furnace a coke fire is lit at the bottom, then charging commences by adding alternate layers of coke and pig iron together with a little limestone. The pig iron and cast iron scrap is mixed with steel scrap, the proportion depending on the desired quality of the melt. The melting iron sinks to the bottom of the furnace and, when sufficient has accumulated, it is tapped into a ladle or directly into the moulds.
Cast iron contains from 2.5 to 4 % carbon which is in the form of graphite. Besides carbon, cast iron contains small amounts of silicon (up to about 3 %), phosphorus (up to 1.5 %), sulphur (about 0.1 %) and manganese (up to 1.5 %).

The moulds for cast iron are normally made of green sand, i. e. moist refractory sand. Sometimes the moulds are made of dry sand, i. e. a moist refractory sand which is dried before the iron is poured into it. Permanent moulds made of metal are used over and over again. For special purposes e. g. for making pipes a centrifugal cast is used. Here a centrifugal mould (i. e. a permanent mould) is employed which rotates while the iron is beeing poured into it.

The strength of cast iron is greater in compression than in tension. The ultimate tensile strength of cast iron varies between 12 and 30 kp/mm^2. In compression cast iron will withstand 60 to 70 kp/mm^2.

6.1.3.2. Malleable Cast Iron

Malleable cast iron is produced from white pig iron by an annealing process. Two methods are used for malleabilizing the castings: the blackheart and the whiteheart method. The names refer to the colour of the fracture of the castings.

In the blackheart process the pig iron is poured into moulds and then cooled rapidly. The castings are then heated to about 900 °C. At this temperature the carbon changes to graphite in the form of small grains embedded in the iron. This form of graphite is known as temper carbon.

In the whiteheart process the castings which have low silicon content are packed after rapid cooling in boxes and surounded by an oxidizing material such as powdered iron oxide. The boxes are heated to a temperature of about 900 °C and mantained at that temperature for several days, during which time a part of the carbon is oxidized out of the castings.

The whiteheart process is used for thin-walled castings up to 12 mm, and the blackheart process for castings having thicknesses from 10 to 30 mm. The carbon content of malleable cast iron is from 0.3 to 1 %. The tensile strength of malleable cast iron ranges up to 40 kp/mm^2. Malleable cast iron is used for fittings, spanners, spring hangers, differential and steering gear housings etc.

6.1.3.3. Cast Steel

Steel may be cast in the same manner as cast iron into moulds. Melting of steel occurs in small Bessemer or Thomas converters or in small open-hearth furnaces. Open-hearth steel is used more for cast steel than Bessemer or Thomas steel. The shrinkage in the steel castings is higher, about 2 %, than in the iron castings which is about 1 %. The rate of cooling is more rapid in cast steel than in cast iron. Thus cavities and thermal stresses in the castings may occur more easily than in cast iron pieces. In order to diminish the thermal stresses, the finished steel castings are annealed. The grain structure is thus refined and the internal stresses reduced. Cast steel is not as strong or as tough as forged steel of the same chemical composition. The ultimate tensile strength ranges for non-alloyed cast steel from 40 to 60 kp/cm^2.

Steel castings are used for many purposes today in place of steel forgings, since castings of complicated shape may be made easier than forgings. Steel castings are also displacing grey iron castings, because they have more strength and toughness. Alloyed steel is also used for making steel castings.

6.1.3.4. Steel

Among the materials used by engineers for structures and machines, rolled and forged steel occupies first place. When comming from the furnace or the converter the molten steel is first poured into large moulds. Each mould contains from 3 to 10 tons of steel. After solidification the ingot moulds are stripped from the steel. The ingots are then transferred to a soaking furnace where they are kept at red heat until they are uniform in temperature throughout. Then they are carried to the rolls or the steam hammer for final shaping.

The common rolled sections of steel bars are: round, hexagon, square, flat, tee, equal and unequal angles, channel, I- Z- and H-section.

Steel is an alloy of iron and carbon. The plain carbon steel is classified according to its carbon content. Dead mild steel is a steel whose carbon content is from 0.07 to 0.15 %. The carbon content of mild steel ranges from 0.15 to 0.3 %. Medium carbon steel has a carbon content ranging from 0.3 to 0.8 % and high carbon steel a carbon content ranging from 0.8 to 1.5 %. Dead mild steel is used for wire rods, thin sheets, solid drawn tubes etc. Mild steel is employed for boiler plates, structural sections, drop forgings etc. Medium carbon steel is applied for axles, shafts, high tensile tubes, wires, springs etc. and high carbon steel for springs, shear blades, chisels, punches, razors, gauges, metal cutting tools etc.

In many cases engineers need steel with high strength, with hardness and other properties which a plain carbon steel does not possess. Higher strength can be obtained by adding elements such as nickel, chromium, manganese, tungsten, molybdenum, vanadium and silicon to the steel. A steel containing such elements is called an alloy steel. Steel with a nickel content from 12 to 21 % and about 0.1 % of carbon has good strength and toughness and a good resistance to corrosion. A stainless steel contains 36 % of nickel. The common chrome steel contains 0.5 to 2 % of chromium and 0.1 to 1.5 % of carbon. The stainless chrome steel contains from 11 to 17 % of chromium. Manganese steel usually contains from 11 to 14 % of manganese and from 0.8 to 1.5 % of carbon. It possesses good hardness and ductility. Tungsten, molybdenum, vanadium and silicon are usually employed in conjunction with other elements.

6.1.4. Heat-treatment of Steel

The physical properties of steel can be changed by application of heat treatment. In order to obtain certain properties, it is necessary to control the manner in which steel is heated and cooled. The manner of heating and cooling is shown in the equilibrium diagram which represents the solidifying process of any alloy. In Fig. 125 the equilibrium diagram of iron and iron carbide is represented. Iron carbide is taken in place of carbon, since in steels carbon always occurs in the form of iron carbide. The structure of pure iron is called "ferrite", the structure of iron with a carbon content of 0.9 % is called "perlite" and the structure of pure iron carbide is called "cementite". The ferrite is the softest and cementite the hardest constituent of steel. Below 700 °C all are solid. Between 700 °C and the critical temperature which depends on the carbon content a change in the structure of iron occurs. The principal kinds of heat treatment for steel are: annealing, normalizing, hardening and case-hardening.

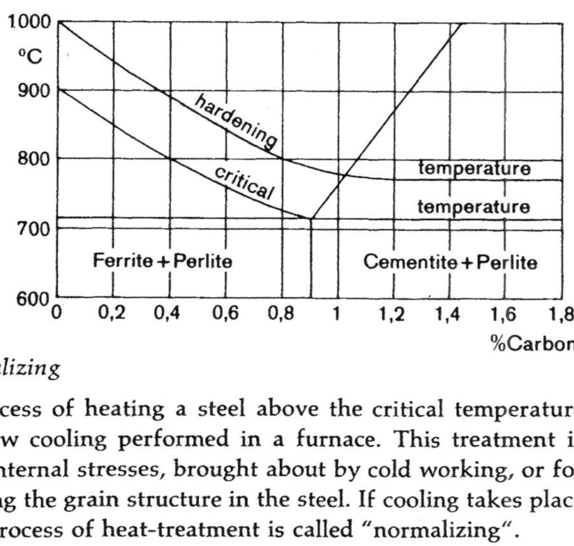

Fig. 125
Equilibrium diagram of iron and iron carbide

Annealing and normalizing

Annealing is the process of heating a steel above the critical temperature followed by very slow cooling performed in a furnace. This treatment is employed to relieve internal stresses, brought about by cold working, or for the purpose of refining the grain structure in the steel. If cooling takes place in still air, then the process of heat-treatment is called "normalizing".

Hardening

When heated above the critical temperature and cooled slowly, as Fig. 125 shows, at a carbon content below 0.9 % steel gives a mixture of ferrite and perlite. At a carbon content above 0.9 % a mixture of perlite and cementite is given. If cooling occurs very rapidly, then in the steel a transformation into a very hard structure called "martensite" take place. Martensite has a needle like structure and is a very hard substance capable of resisting wear and able to cut other metals. The hardening temperatures are about 50 to

100 °C above the critical temperatures. The lower the hardening temperature, the finer is the structure of the martensite. In order to obtain martensite, cooling must be performed very rapidly. This is made by plunging the steel in oil or water. This kind of cooling is called "quenching". Steel with a carbon content below 0.3 % needs a very high hardening temperature and has after quenching, insufficient hardness. Therefore such steels are not suitable for hardening. Since the steel after quenching is too hard for practical use, it is heated up again to a temperature below the critical temperature and cooled slowly. This heat treatment is called "tempering" and causes a reduction of the hardness and an increase of the toughness in the steel.

Case hardening

Case hardening is a process whereby only the surface of the steel to a depth up to 1.5 mm is hardened. It is used for steel having low carbon content at which the normal method of hardening has no appreciable effect. The steel is packed in a steel pot together with carbonaceous materials such as charcoal, hoof clippings, bone dust etc. and heated up to about 900° C for a sufficient length of time depending on the desired depth of hardness. The carbon diffuses into the surface of the steel and increases the carbon content on the surface. After beeing carburized the steel in the pot is slowly cooled, then the steel is removed from the pot and heated, quenched and tempered. The core of the case hardened steel contains about 0.1 to 0.15 % carbon and the case 0.7 to 1.0 % carbon. The case hardening process is applied to gauges, cams, wristpins on gudgeon pins, etc.

Another method of hardening is known as nitriding. Here the steel is packed into an air-tight box and heated to about 550 °C. Ammonia gas is passed into the heated box and it dissociates into nitrogen and hydrogen. The nitrogen combines with the iron to form a nitride. When the case is of sufficient depth, the steel is removed from the box and is ready for use. Nitriding has the advantage of producing an extremely hard surface, harder than can be obtained with other methods.

6.2. Non-ferrous Metals and their Alloys

6.2.1. Pure Metals

Non-ferrous metals and their alloys have an important place in enigneering construction. They have properties which iron alloys do not possess e. g. superior resistance to corrosion, good heat transfer, etc. and are used for such members of structures and machines for which iron or steel or iron alloys are not suitable. The most important non-ferrous metals are: copper, tin, zinc, nickel, lead and aluminium. These will now considered.

Copper

The most important copper ores are sulphides of which the principal ones are: chalcocite ($Cu_2 S$) and chalcopyrite ($Cu Fe S_2$). The ores are first roasted to burn out the sulphur leaving oxides in place of sulphides. After the roasting process the ore is smelted in a blast furnace or in a reverberatory furnace. Then the metal is purified in a Bessemer converter. Thus blister copper is obtained. After various refining processes pure copper is produced which finds applications in condensors, evaporators, cooking vats and other articles. In a electrolytic refining process electrolytic copper, a very pure copper is produced which is used for electric conductors.

Tin

The chief tin ore is cassiterite ($Sn O_2$) commonly called tin stone. Tin ore is, after roasting, smelted in a reverberatory or in a blast furnace together with coal as a reducing agent and limestone and fluorspar as flux. Crude tin is refined by liquidation or electrolytically. Tin is chiefly used for coating thin steel sheets.

Zinc

The chief zinc cores are zinc blende and zinc carbonate. The zinc ores are transformed into oxides by roasting and then mixed with charcoal or other form of carbon and heated in a special type of furnace. The carbon combines with the oxygen of the oxide. Pure zinc is separated in the form of a vapour and must pass a condenser to become liquid. Then zinc is refined by redistillation or by electrolysis. Zinc is used in galvanizing, in rolled form and in die castings.

Nickel

The copper-nickel-iron sulphide ores are smelted and refined. Refined nickel with a slight addition of magnesium or manganese is malleable either hot or cold. Nickel is used for rods, sheets, wires, tubings etc.

Lead

The most important lead ore is galena ($Pb S$). Galena is roasted, smelted in a blast furnace and refined. Lead is the heaviest of the common metals. Its specific gravity is 11.3. Lead is very soft and ductile and hence it can be rolled or extruded into pipes which are very weak, but have high resistance to corrosion. Lead is used for water pipes, as sheathing for underground electrical cables, for plates of lead-acid accumulators and for sheets in the chemical industry.

Aluminium

Aluminium is produced from a clayey mineral called „bauxite". Bauxite is roasted and then dissolved in molten cryolite at about 980 °C. The tank is lined with carbon which acts as the cathode. The anodes are also of carbon and plunged into the bath. The molten aluminium is deposited on the cathode, drained out and cast into bars. Aluminium is a very light metal. Its specific gravity is only 2.7. The tensile strength of aluminium is about 12 kp/mm^2. Since aluminium has a high resistance to corrosion. This resistance is due to a thin film of oxide covering its surface. It is generally used for cooking utensils, as a surfacing on duralumin sheets and for electrical conductors.

6.2.2. Non-ferrous Alloys

A large variety of non-ferrous metal combinations is produced by melting together various non-ferrous metals. The physical properties of these alloys differ from those of the constituent metal. They depend not only on their chemical composition, but also on the method of producing them and of the mechanical treatment. Some of the principal non-ferrous alloys which are discussed here are: bronze, brass, solder and aluminium alloys.

Bronze

The term bronze originally means alloys of copper and tin. Nowadays the term bronze is also used for some copper alloys e. g. for copper-lead alloy, copper-aluminium alloy and other copper alloys. The bronzes are then denoted as tin bronze, lead bronze etc. Bronze can be cast into moulds or rolled into wire, rods or sheets. For castings, sometimes zinc is added to the tin bronze to make it resistant to corrosion by water and the atmosphere. Phosphor bronze is a tin bronze which contains a small amount of phosphorus which is added to remove copper oxide during the alloying process.

Brass

Copper and zinc alloyed produces brass, one of the commonest non-ferrous alloys. The zinc content extends up to 45 %. The addition of 45 % zinc to copper increases the tensile strength to 35 kp/mm^2. Brass can be cast into moulds or can be rolled or drawn into sheets, tubes, rods or wire. Brass has a good resistance to corosion and is employed for fittings, pump linings and members which are exposed to moisture. Brass is also used for bearings.

Solder

For joining metals by soldering two types of alloy are employed, soft and hard solder. Soft solders are composed of tin and lead and, in some alloys, also of a little antimony. The alloy contains from 30 to 60 % tin and the remainder lead. The melting temperature is very low, depends upon its composition and ranges from 180 °C to 250 °C. Soft solders are used for joining sheet iron, in tin smithing, in plumbing and in electrical work. Hard solders include brazing solder and silver solder. The two solders are used for joints which require high strength. Brazing solder contains between 50 and 55 % copper and the remainder zinc. It is used in joining members of brass, copper, steel and cast iron. Siver solder contains between 10 and 80 % silver, between 16 and 52 % copper and between 4 and 38 % zinc. The melting point of hard solders can be varied according to its composition from about 700 °C to 870 °C.

Aluminium Alloys

The strength of aluminium may be raised, if it is alloyed with small amounts of other metals. There are alloys for castings and alloys for rolling and forging. For castings aluminium is alloyed with varying percentages of silicon, copper, iron, zinc, manganese or magnesium. The tensile strength of these alloys varies from 11 to 28 kp/mm^2. The most important aluminium alloy for forging and rolling is duralumin which contains about 4 % copper, 0.5 % manganese and 0.5 % magnesium. When annealed it has a tensile strengt of about 20 kp/mm^2. If it is heated to about 890 °C and then quenched in water, the tensile strength becomes about 35 kp/mm^2. Duralumin is used for parts which require strength combined with lightness, principally for parts of airplanes.

6.3. Non-metallic Materials

There are a large variety of non-metallic materials used in engineering constructions of which the principal ores are: wood, concrete, glass, rubber, leather and plastics.

Wood

Wood has been used as a structural materials since the earliest times. It is a cheap material, cheaper than the metallic materials. A disadvantage of wood is that it can easily be destroyed by fire and is liable to decay. Structural wood is divided into two classes: softwood and hardwood. Softwood is the wood of coniferous trees e. g. spruce, fir, pine, larch, cedar, etc. Hardwood is the wood of deciduos trees e. g. hornbeam, red beech, oak, elm, ash, poplar, maple, etc. The mechanical properties of wood especially strength and stiffness vary considerably. The tensile strength of timber in a direction parallel to the grain is much higher than in a direction perpendicular to the grain. The shearing strength along the grain is much lower than the shearing strength across the grain. Timber before being used must be treated with a suitable preservative, such as creosote or zinc chloride, to withstand decay. Softwood is used for general construction work, hardwood is used for interior finish, furniture, cabinet work etc.

Concrete

Cementing materials have, when mixed with water to a paste, the property to harden into a solid. Cementing materials are gypsum, lime and cement. The most important of these is cement. Cement mixed with sand or gravel and water gives concrete. It is a brittle material and is much stronger in compression than in tension. Plain concrete i. e. concrete alone without any other construction element is used for massive work carrying only compressive loads e. g. for foundations, walls, dams, arches, etc. When tensile stresses occur in the concrete constructions, steel bars are embedded in the concrete which supply the resistance to these stresses. Such material is called „reinforced concrete". Beams, columns, footings, bridges, chimneys, tanks, etc. are examples of constructions using reinforced concrete.

Plastics

Plastic is an artifical material which can be moulded to most forms. Most of the plastics are organic compounds in which coherence and strength have been brought about by a process of polymerization in which a new compound is made whose molecular mass is larger than the molecular mass of the original substance. Plastics are divided into two general classes: the thermoplastics which after heating must be chilled to harden and which resoften after reheating and the thermosetting plastics which are hardened by a heat process and do not soften when reheated. In order to become strong and stiff some constituents must be added to the plastics. One constituent is the binder which cements the various constituents together. Another ingredient is the filler, usually shredded textile material which gives the plastics resistance to shock. A third ingredient is the plasticizer which is used to improve the flow quality of the material. Especially in thermosetting plastics a catalyst is added to accelerate the polymerization reaction.

Plastics are formed by casting, by compression moulding using thermosetting plastics, or by injection moulding in which the plastic is injected by pressure into the moulds. Plastics have a large field of use in engineering construction. Thus housings, electrical insulating parts, surfaces of metal parts, gear wheels, handles etc. may be made of plastics.

7. Technical Drawing

Technical drawing is the language of engineers, by means of which they record their ideas. In the technical drawing, completed with specifications, by the design engineer the whole machine or structure is fixed and the manufacturer should be able to carry out the work by means of the design without any additional instructions from the designer. The draugthsman should have a complete knowledge of the details of the construction and should be thoroughly familiar with the materials employed for the machine or the structure.

In order to attain a uniformity in engineering production standard specifications are issued by each nation. Thus the British Standards Institution has issued standards specifications which have been adopted by most of the nations of the British Commonwealth. The German Standards Institution has also issued standards specifications. There is also an International Standard Orgnisation (ISO) which recommends the replacement of national standard specifications by international standard specifications. In this book the German Standards Specifications (DIN) will be used. The DIN standards specifications also include the international ISA standards specifications. The DIN standards specifications are divided into three classes: fundamental standards specifications, standardized materials and dimensions of standardized machine parts. Factories use their own working standards specifications which are DIN standards specifications adapted to working conditions.

7.1. Drawing Equipment

Technical drawings are executed with the aid of drawing instruments on drawing paper. The most important drawing instruments and the sizes of drawing paper are now described.

Drawing board

The drawing board is used for holding and supporting a sheet of paper flat whilst a drawing is made upon it. The board is made of wood of spruce or pine or of other soft wood. To prevent warping it is preferable that the board is not made of one piece but of strips glued together. The more expensive boards have on the lefthand edge an ebony strip inserted for the stock of the T-square to slide upon. The drawing board when used should be tilted of about 15° with the aid of wooden blocks.

T-square

The T-square is used for drawing long lines perpendicular to the left edge of the drawing board. It is composed of a long strip called the „blade" and a short strip called the „stock". The upper edge of the blade and the inner edge of the stock are the working edges. These edges are at right angles to each other. T-squares are usually made of hard wood. They should be varnished to protect them from moisture, and so they are prevented from warping. The blade has celluloid edges, so that the draughtsman can see the lines drawn on the paper. Sometimes T-squares of bakelite are used.
Parallel-rule attachments can be substituted for T-squares and are much in use. In industrial draughting rooms draughting machines are used.

Set squares

Set squares are right-angle triangles made of wood or celluloid. A pair of set squares is used in draughting, one having angles of 45°–45°–90° and the other one angles of 30°–60°–90°. They are used in conjunction with the T-square or other set squares for drawing short lines parallel, perpendicular or at an angle of the set square to a given straight line.

Scale

Scales are made of boxwood with white celluloid edges and are either flat or triangular. The triangular scale has the advantage of combining several scales on one instrument.

Compasses

Compasses consist of two legs, the yoke and the handle. The ends of the legs can be adjusted and are held rigidly by friction produced by screws on each side of the yoke. The other ends of the legs are provided with socket joints which permit the insertion of alternate parts. Thus one instrument may be used as pencil compasses, ink compasses or as a divider. A lengthening bar is useful for increasing the radius of the compasses.
Compasses may be purchased separately or in sets in velvet-lined cases. A set contains the following instruments: 1 pair of compasses with a pencil attachement, a pen attachement and a lengthening bar, 1 divider, 1 spring bow for pencil, 1 spring bow for pen, 1 spring bow dividers, 2 ruling pens and 1 case of leads for the compasses.

Pencils

Drawing pencils are made chiefly of graphite mixed with varying quantities of clay to produce different degrees of hardness. There are 18 degress of hardness: 7 B (very soft), 6 B, 5 B, 4 B, 3 B, 2 B, B, H B, F, H, 2 H, 3 H, 4 H, 5 H, 6 H, 7 H, 8 H, 9 H (very hard). For drawing 3 H and 2 H are usually employed and for lettering H and H B.

The following materials are also used in technical drawing:

The protractor for measuring or setting off angles other than those obtained with the set squares.

French curves and adjustable curves for drawing curves.

Engineer's triangle for drawing standard bolts, heads and nuts.

Drawing ink, pencil sharpener, rubber, erasing knife, paper fasteners thumbtacks, or draughting tape).

Drawing paper

Normally transparent paper is used in technical drawing. The initial size of the drawing paper is a rectangle the area of which is 1 m². The ratio of two sides of the rectangle is chosen in such a way that it remains constant when the rectangle is halved (Fig. 126).

The ratio of the two sides may be calculated from:

$$\frac{b}{a} = \frac{a}{2\,b} \quad \text{or} \quad 2\,b^2 = a^2$$

whence $\dfrac{b}{a} = \dfrac{1}{2}\sqrt{2}\,.$

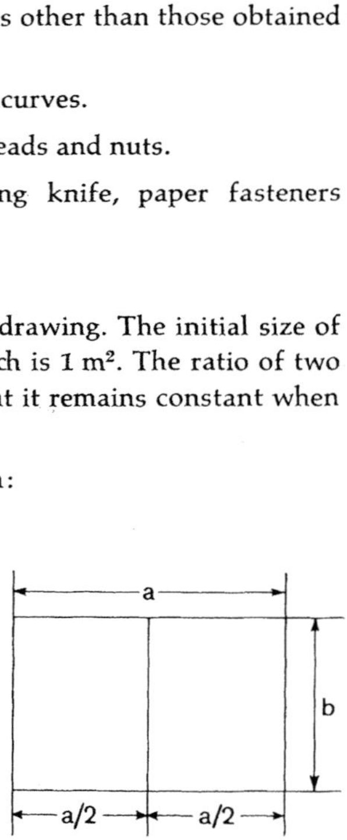

Fig. 126

The initial size of drawing paper may be calculated thus:

from $a\,b = \dfrac{a^2}{2}\sqrt{2} = 1\ \text{m}^2$

$$a = \sqrt[4]{2} = 1.183\ \text{m} = 1183\ \text{mm}$$

$$b = \frac{a}{2}\sqrt{2} = 841\ \text{mm}\,.$$

The other sizes may be obtained by halving the paper. 5 sizes of drawing paper are standardized.

number of size	size of paper	number of size	size of paper
A 0	841 × 1189	A 3	297 × 420
A 1	594 × 841	A 4	210 × 297
A 2	420 × 594	A 5	148 × 210

The dimensions are referred to the cut working paper.

7.2. Projections

In Fig. 127 different kinds of projection are represented. If the projectors converge to a point, a perspective projection is made. If the projectors are parallel, but inclined to the plane of projection, an oblique projection is made. If the projectors are parallel and are perpendicular to the plane of projection, an orthographic projection is made. In machine drawing orthographic projection is used exclusively.

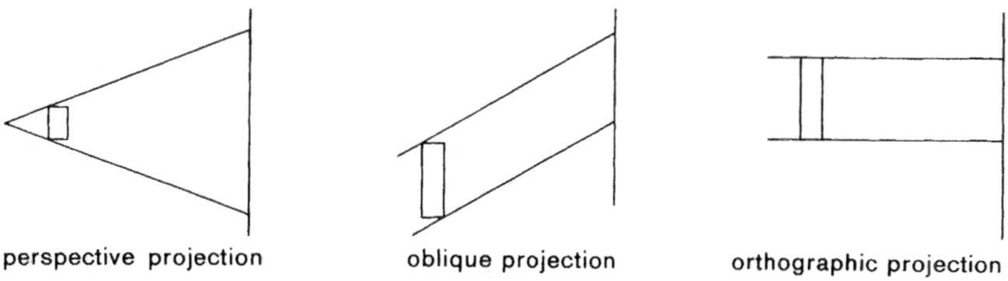

Fig. 127

There are two methods in current use of drawing orthographic projections: First Angle and Third Angle projection (Fig. 128). In First Angle projection the object is assumed to be placed in the first angle and projected on the two principal planes I and II. When the planes I and II are opened out, the front view and, below it, the top view are the result. In Third Angle projection the object is assumed to be placed in the third angle. The planes of projections III and IV are between the observer and the object and are

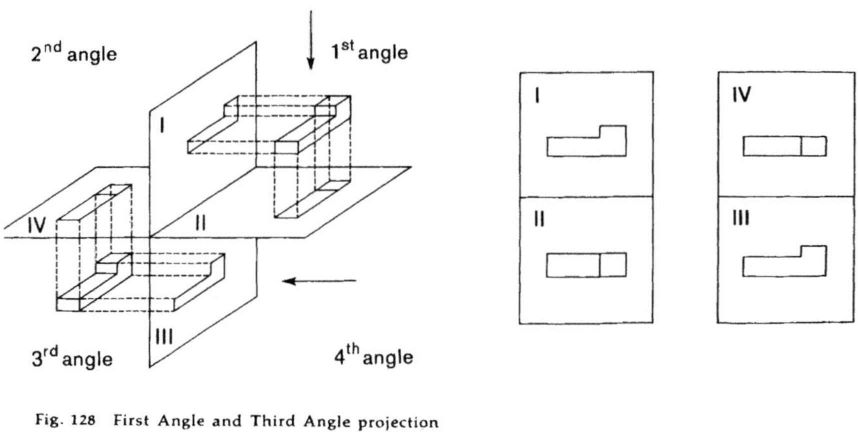

Fig. 128 First Angle and Third Angle projection

supposed to be transparent. The projections are viewed through the planes. When the planes are opened out, the top view and, below it, the front view are the result. First Angle projection is current in Europe, and Third Angle projection is used in North America. In Great Britain both methods are used. Here First Angle projection will be employed.

Theoretically an object is described fully in two views, because the extensions of the object in x, y and z direction are given, but usually, especially for more complicated objects more views are necessary to give complete information about the part. There are 6 views possible shown in Fig. 129.

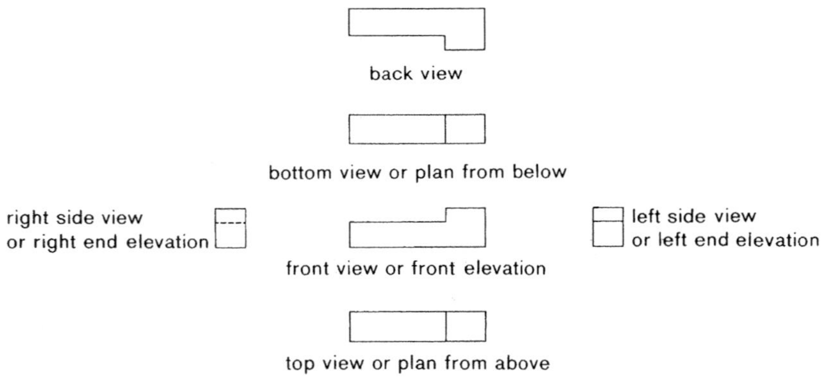

Fig. 129 The six views of a body

7.3. Pictorial Drawing

If we show an object in one view as it appears to the observer, a pictorial drawing is made. A disadvantage of the pictorial drawing is that it is not possible to measure the lines of the object directly as in the multiview

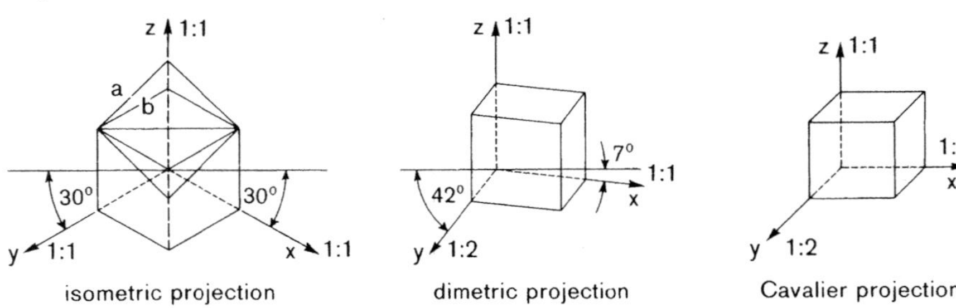

Fig. 130

drawing i. e. in a drawing represented in two or more views. Various types of pictorial drawing are used in catalogues and also in technical work to visualize objects for persons having no technical training.

There are two methods in use. The axometric projection and the oblique projection. In Fig. 130 two kinds of axometric projection, the isometric and the dimetric projection and one kind of oblique projection, the Cavalier projection are shown.

In the isometric projection the edges of a cube are all equally foreshortened. The relation between the foreshortened length b and the true length a of the edge of the cube in Fig. 130 is:

$$\frac{b}{a} = \frac{b\,c}{c\,a} = \frac{2}{\sqrt{3} \times \sqrt{2}} = \frac{\sqrt{2}}{\sqrt{3}} = 0.82\,.$$

In the dimetric and in the Cavalier projection only the edge in the y-direction is foreshortened. The foreshortened length is one half of the true length.

7.4. Types of Lines, Lettering and Scales

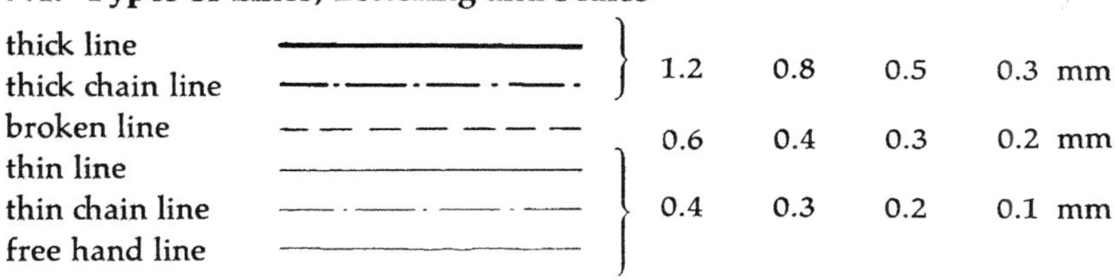

Fig. 131 Types of lines

In Fig. 131 the types of lines used in technical drawings and the thicknesses of lines are shown. The thickness of lines depends on the size of the drawing. For big drawings thick lines and for small drawings thin lines are used.

The thick line is used as outline and for visible edges. The thick chain line is used for indicating cutting planes, (Fig. 136). The broken line is taken for invisible edges, as root lines of bolts and for the root circle of gear wheels. Invisible edges are only drawn, if it is necessary for information. The thin line is employed as dimension and extension line and for hatching cut surfaces. The thin chain line is used as centre line, for the pitch circle of gear wheels and for indicating end positions of levels and handwheels. The free hand line is used as a break line to indicate the imaginary break in a partial view.

Inclined letters and numerals at an angle of 75° with the horizontal are used in lettering. The thickness of the letters and numerals should be about 1/7 of the height of the capitals. The space between the letters should be from 1/7 to 2/7 of the height of the capitals. The space between the words should be 4/7 of the height of the capitals.

The scale of the drawing must be indicated in the title strip. There are the following scales:

Full size:	1 : 1
Reductions:	1 : 2.5, 1 : 5, 1 : 10, 1 : 20, 1 : 50, 1 : 100
Enlargements:	2 : 1, 5 : 1, 10 : 1.

7.5. Framing of Drawings, Title Strip and Parts List

A working drawing is a drawing which completely describes the object represented so that no additional information will be needed by those who manufacture the object. Working drawings are classified as assembly and detail drawings. An assembly drawing is one which describes the complete structure and which shows the working relationship of the several parts. A detail drawing is one which describes a single part of a structure. Each part of the structure must be drawn as a detail drawing on one sheet. It is not admissible to draw several parts on one sheet.

All sheets are used broadside exept the size A 4 which is used upright. Each sheet must be framed with thick lines 5 mm inside the cut edges exept the left side where the frame is 20 mm inside. In the lower right hand corner of the sheet the title strip is placed. The following information should be given in the title strip: name of the object represented, name of the manufacturer, number of the drawing, material of the part (only for detail drawing), scale, allowable tolerances (only for detail drawing), name of the draughtsman, and the date, name of the checker and the date, and name of the checker of standards specifications and the date. On the left hand side of the title strip the alteration strip is attached in which modifications to drawings are inserted.

For each assembly drawing a parts list must be made. On the bottom of the parts list there is a title strip containing the same information as the assembly drawing belonging to it. In the parts list the following data for the single parts must be entered: name of the part, part number, number of parts required, material of the part and additional remarks.

7.6. Drawing of Views

For front view, that view is selected which gives the most information concerning the characteristic form of the object. Then the other views are selected. The object is not drawn in more views than is necessary to describe it fully. A more complicated object requires, of course, more views and often also special views such as partial views, auxiliary views or sections. Views with the least number of invisible lines should be preferred. The front view of the object is assumed in its normal upright position.

After the choice of the number and the arrangement of the views, the centre lines of the views are drawn. The outlines of the object and the visible edges in the various views follow. Then the invisible lines and the section lines are drawn. The views should be constructed simultaneously, invisible lines are only drawn, if they are necessary for information. If the object is completely represented in all views, the extension and dimension lines are drawn and the numerals are inserted. Then the machining symbols are inserted. A view does not need to be a complete projection. Sometimes a partial view or for symmetrical objects a half view may be drawn, if it describes the object sufficiently.

Often simplified views may be used in place of a complete view. Thus for flanges, it is sufficient to draw in place of a complete top view only the pitch circle and the holes. If a hole is not drawn in the right place, it must be indicated in the drawing.

If the true shape or size of an object is not shown in the ordinary projections, a projection is made in a plane inclined to the six ordinary planes. The view in this projection plane is called an auxiliary view. It is not always necessary to draw a complete auxiliary view, since a partial auxiliary view may be drawn as shown in Fig. 132, if it gives sufficient information about the object.

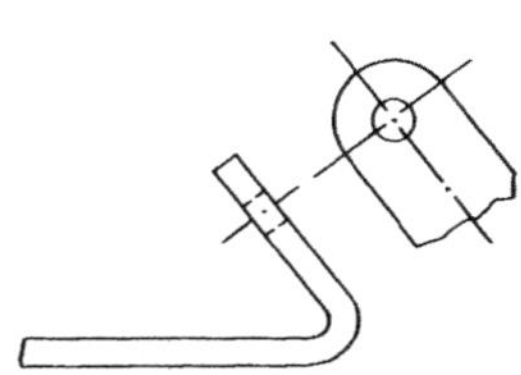

Fig. 132 Partial auxiliary view

Long bars may be drawn broken, as Fig. 133 shows.

Objects with a complicated internal structure are not represented in an external view, because the large number of invisible lines necessary to show the hidden edges are too confusing for the observer. Such objects are assumed

Fig. 133 Long bars drawn broken

cut apart by a plane and the portion in front of the plane removed so that the internal part can be seen. The resulting view is called a sectional view or a section. If the cutting plane is assumed lengthwise of the object, the resulting section is a longitudinal section, if it is crosswise, a cross section.

The surface actually cut is drawn hatched i. e. section lines at an angle of 45° with the border line of the sheet are drawn. The spacing of the section lines varies with the size of the area sectioned. Two adjacent parts should be sectioned in opposite directions.

The areas of a single part must be sectioned in the same direction, the areas of different parts in different directions, as shown in Fig. 134.

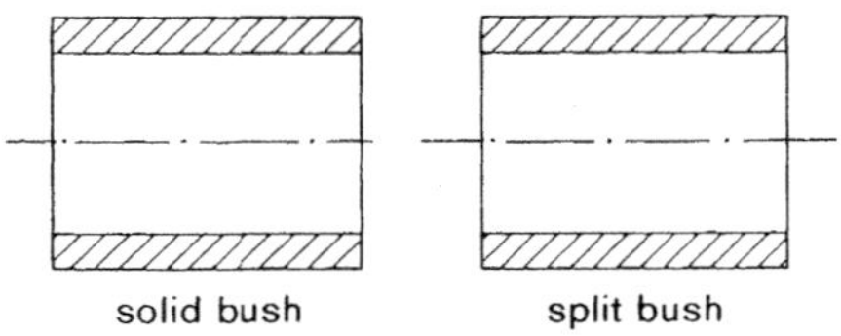

Fig. 134

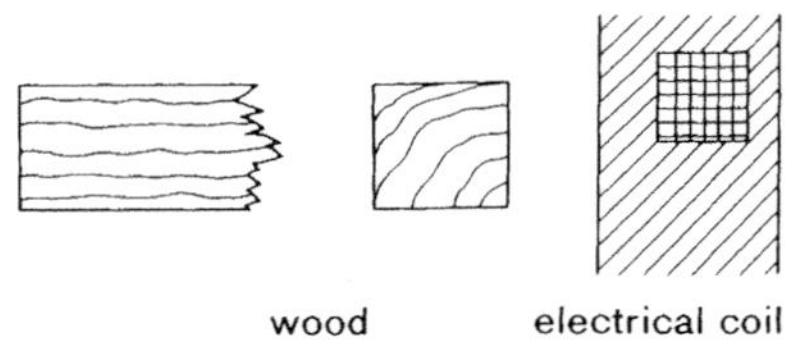

Fig. 135

Section lines should be interrupted for lettering. Small cutting planes are blackened. Longitudinal and cross sections of wood and cross sections of electrical coils are sectioned in a special way, as shown in Fig. 135.

Invisible lines should only be drawn, if they are absolutely necessary for clearness, otherwise they may be omitted.

If the section extends through the whole object, the section is called a full section, otherwise it may be a half section or a broken section.

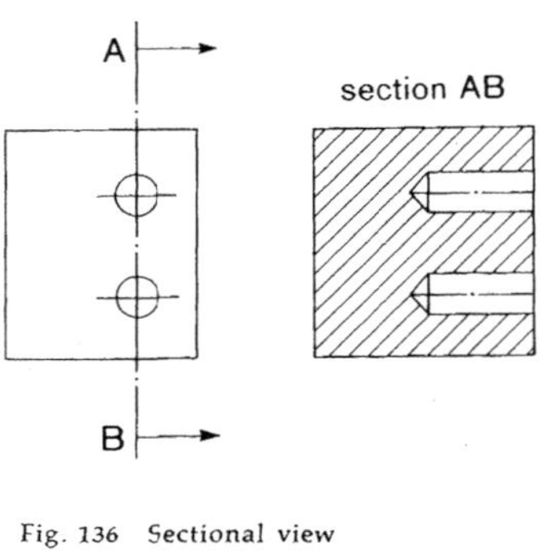

Fig. 136 Sectional view

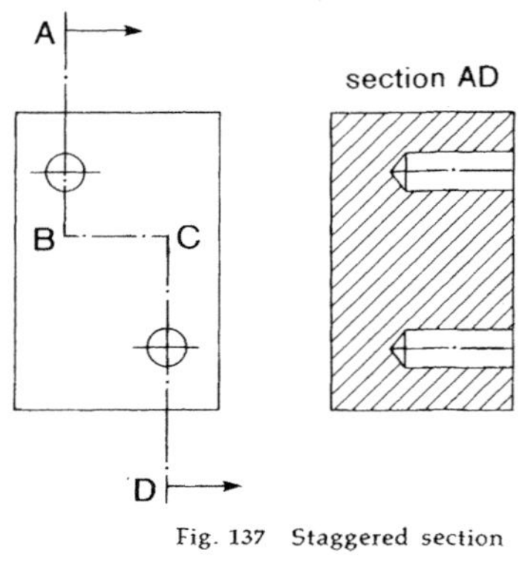

Fig. 137 Staggered section

The location of the sectioning plane should be indicated on the ends of the section by letters and by thick chain lines which have arrows pointing in the direction in which the object is viewed (Fig. 136). If the cutting plane coincides with the centre line, the representation of the cutting line is omitted.

If it is desired to show in one sectional view parts which are not in the same plane, a staggered section is employed. The path of the section must be exactly indicated (Fig. 137).

Objects which are symmetrical about a centre line may be drawn having one half in outside view and the other half in section. By drawing objects in this way, the construction of an additional view or the excessive use of hidden lines is avoided. An imaginary edge, as it is produced at a half section, may not be drawn and only the centre line separates the sectional from the unsectioned half.

Part or broken sections may be taken at convenient places to show interior parts which would otherwise be hidden. A break line may be used to limit the section. The break line may not coincide with an edge of the object.

The shape of the cross section of a bar, arm or other elongated object may be shown in the longitudinal view by means of a revolved section. Such sections are made by assuming a plane perpendicular to the centre line of the elongated part and revolving the plane through 90°. The outlines of such sections are thin lines (Fig. 138).

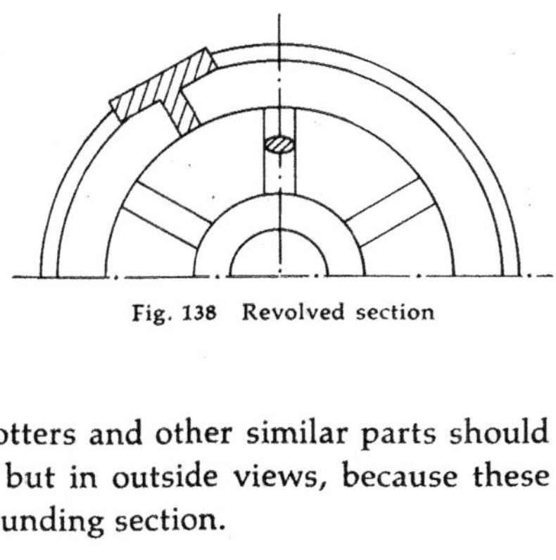

Fig. 138 Revolved section

Ribs, shafts, balls, rods, rivets, keys, cotters and other similar parts should not be drawn in longitudinal sections, but in outside views, because these views give a better contrast to the surrounding section.

7.7. Dimensioning

Introduction

Each drawing must carry all the dimensions needed in the workshop in a manner which will be convenient and practicable for the workman. The dimensions may be placed either inside the drawing by using a dimension line which bears the numeral denoting the distance between two edges of an object or outside by using a dimension line and two extension lines. If the dimension line has sufficient length, it is broken near its middle portion for inserting the numeral. For short dimension lines the numerals are written above the line. If the space between the extension lines is too small for drawing a dimension line, the arrows are made outside the extension lines and the numeral is inserted between the outside arrows. Extension lines extend about 2 mm beyond the points of the arrow heads. The length of the arrows is about 6 times the thickness of the outlines. The minimum distance from the dimension line to the edge of the object should be 8 mm and to any adjacent dimension line 5 mm. The height of the numerals should be equal to the length of the arrows. Invisible edges should not be dimensioned. The numerals are inserted in mm, fractions in decimal fractions, if mm are employed, and in common fractions if inches are employed. The dimension figures are at an angle of 75° to the dimension lines and must be read from the base or right hand side of the drawing. In a series of succesive dimensions the dimension figures should be staggered. Extension lines may not cross dimension lines, but dimension lines may cross dimension lines and extension lines may cross extenions lines. A dimension line may not continue an edge of an object.

Symbols for diameter, radius and square

In dimensioning a circle the symbol for diameter (⌀) should be given after the numeral, if the circle line does not appear in the drawing or if the dimension line has one arrow or if a leader is used (Fig. 139).

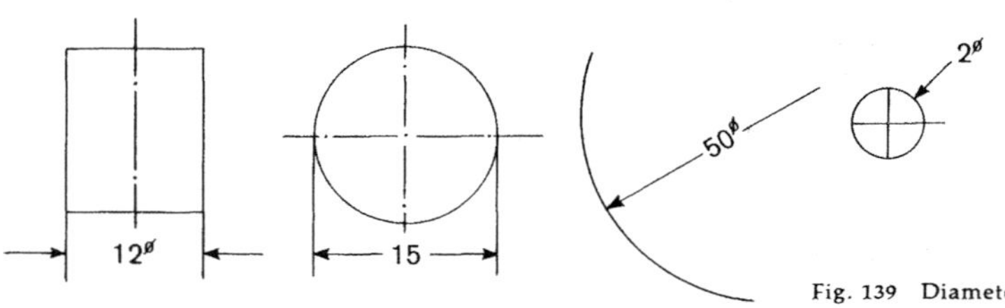

Fig. 139 Diameter symbol

The symbol for radius (*r*) is necessary in dimensioning an arc, if a leader is taken or if the dimension line is not drawn up to the centre of the arc (Fig. 140).

For dimensioning a bar of square cross section in a view in which the square does not appear the square symbol should follow the numeral. If a round bar has a plane surface on it, two diagonal lines are drawn on this plane surface (Fig. 141).

For dimensioning spherical surface the diameter symbol should follow the word „sphere".

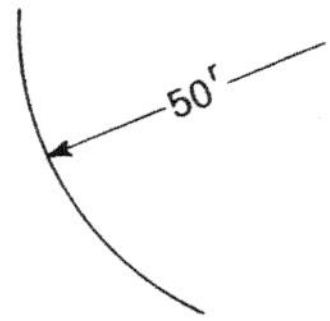

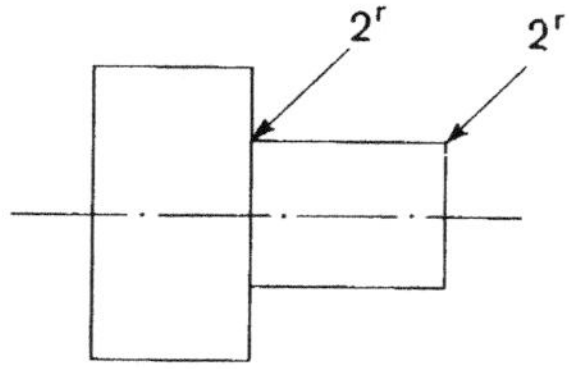

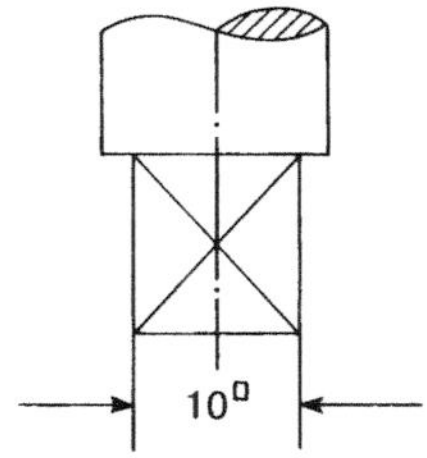

Fig. 140 Radius symbol

Fig. 141 Square symbol

All three symbols are placed slightly above and to the right of the numeral.

Taper and incline

On standardized cones or pyramids the taper must be shown and on objects with standardized slopes the incline must be indicated. The taper and the incline are defined as the ratio of alteration in diameter or width to the length of the part (Fig. 142).

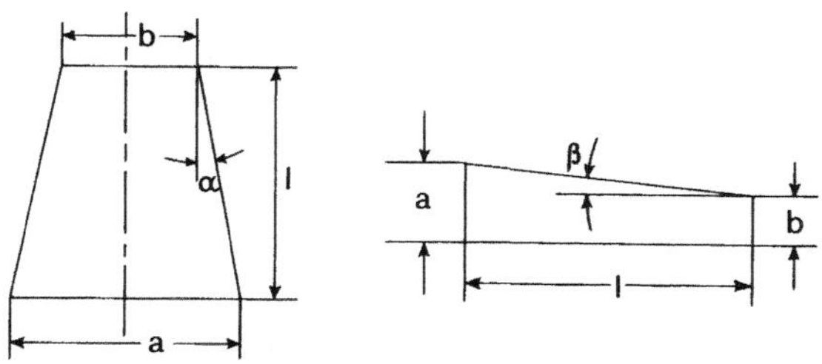

Fig. 142 Taper and incline

Expressed mathematically

$$\frac{a-b}{l} \qquad \tan\alpha = \frac{a-b}{2\,l} \qquad \tan\beta = \frac{a-b}{l}\,.$$

The taper is indicated on the centre line and the incline on the sloped edge (Fig. 143).

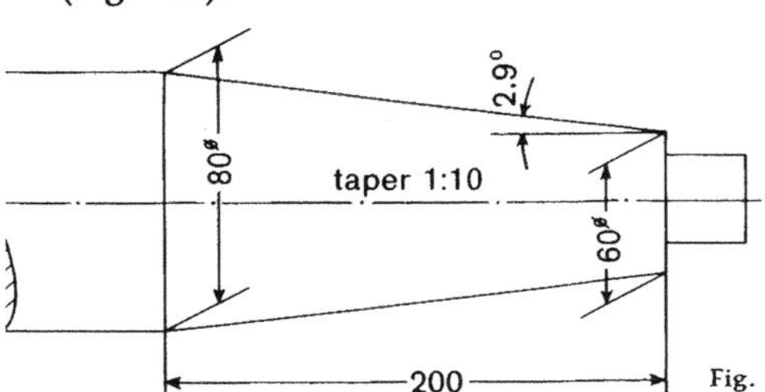

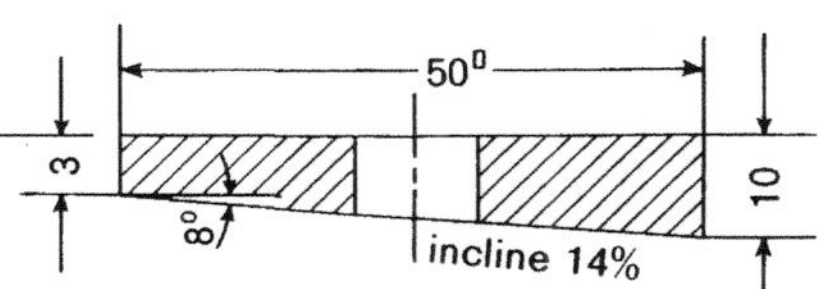

Fig. 143 Dimensioning of a cone and an incline

Principles of dimensioning

A general rule for the dimensioning of structural parts cannot be given. Every drawing has its own dimensioning problem. The most important feature is that the draughtsman makes the dimensions so clear that they can easily be understood by the craftsman. Each machine part may be regarded to be composed of parts taking the form of simple geometric solids. The dimensioning of engineering structures consists principally of two operations: first to give dimensions which locate the relative position of the simple solids and second to give dimensions which specify the sizes of these solids. The first dimensions are called "position dimensions" (p) and the second "detail dimensions" (s), (Fig. 144). The parts of a structural element which derive its form from machine operations may have a plane surface produced by planing, shaping, milling, etc. or a cylindrical form produced by turning, drilling, reaming, boring, etc. The position dimensions for machined surfaces should be referred to an axis of symmetry or to a finished plane surface and not to an edge or surface which is unmachined.

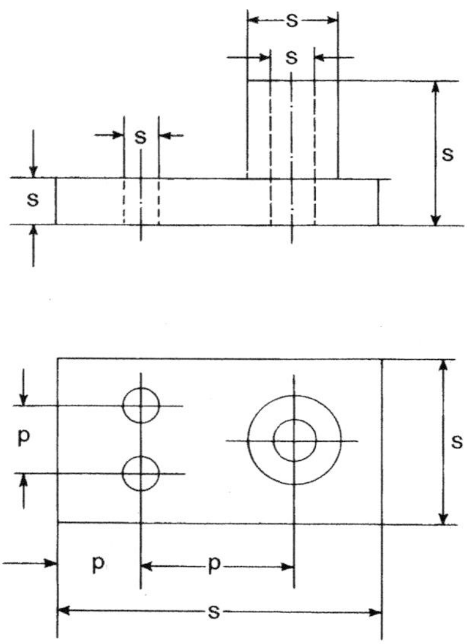

Fig. 144 Position and detail dimensions

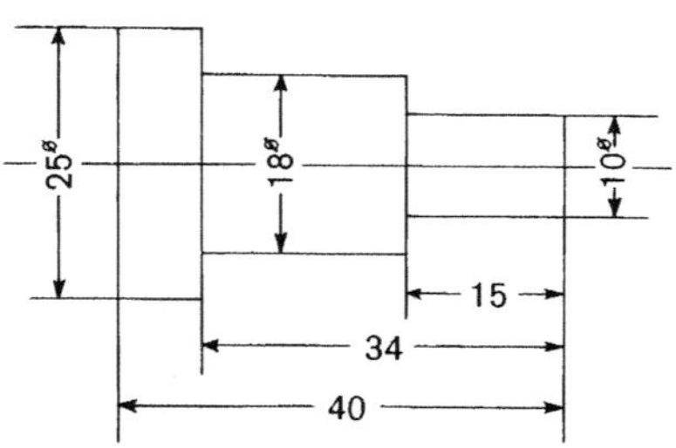

Fig. 145
Dimensioning of a stepped cylindrical part

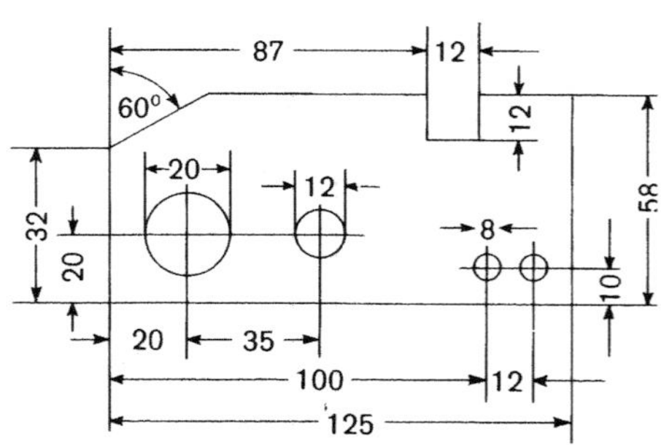

Fig. 146
Dimensioning of a sheet with holes

In Fig. 145 the dimensioning of a cylindrical part and in Fig. 146 the dimensioning of a sheet with holes is demonstrated. For cylindrical parts whose diameters are stepped the dimensions in the direction of the axis are referred to the smaller face, that is that face from which machining begins. For sheets with holes the position dimensions are referred to the base and to the left-hand edge.

7.8. Machining Symbols

The evenness of the surface of an object is indicated by a surface roughness number. The surface roughness number is defined as the distance from the highest to the lowest point of a roughness curve which is registered by a measuring instrument (Fig. 147).

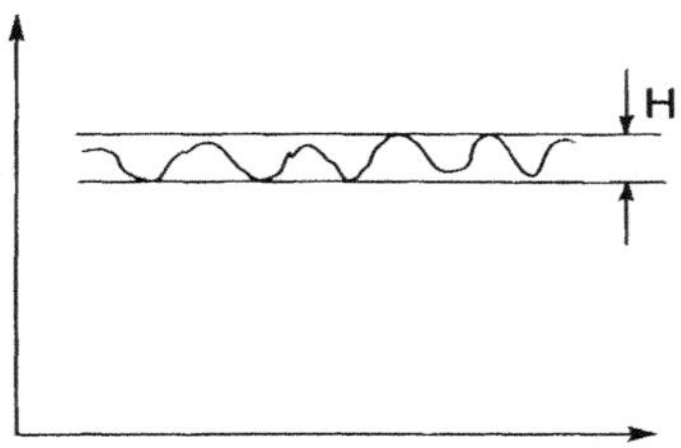

Fig. 147 Surface roughness

Surfaces which are produced by a non-machining process such as casting, rolling, forging, pressing, etc. are not marked. If the surface of a non-machined part should be a high degree of surface quality a special mark (~) is used.

Surfaces which are produced by a machining process such as turning, milling planing, shaping, drilling, grinding, etc. are marked by triangles. These are standardized in four classes of quality. The first and second classes are used for rough engineering, the third class for general engineering and the fourth class for precision engineering. Each class is divided into four roughness degrees.

	1	2	3	4		
rough machined	160	100	60	25	▽	for static parts
fine machined	40	25	16	10	▽▽	
finished	16	4	4	2.5	▽▽▽	for sliding and sealing parts
superfinished	–	1	1	0.4	▽▽▽▽	

The numbers in the table indicate the maximum allowable roughness in microns. In special cases the roughness number may be indicated instead of the machining symbol. On surfaces which are superfinished the machining mark as well as the superfinishing process should be indicated e. g. polished, scraped, lapped, honed, etc. On surfaces which are treated by hardening, annealing, normalizing, etc. or which are painted or galvanized it must be indicated as well as the machining mark also the process of treating, painting or galvanizing.

7.9. Tolerances and Fits

Tolerances

In order to enable interchangeability, engineering products must have a high accuracy in their dimensions. Parts which fit to other parts must be able to be replaced by another one without any additional matching operation.

A dimension given in the drawing cannot be observed exactly in manufacture. The true dimension (i) is either greater or smaller than the given dimension. Its exact value may not be determined. In order to indicate the accuracy of the dimension two limit dimensions, the maximum dimension (g) and the minimum dimension (k) are given (Fig. 148). The true dimension must be within these limits. The difference between the two limit dimensions is called "tolerance" (t). The smaller the tolerance, the more accurate is the dimension. The tolerance should be given the greatest possible value. If the tolerances are smaller than necessary for correct functioning of the parts, the subsequent cost will bee too high. The draughtsman must have a good familiarity with the principles of operation of the machine parts which should be produced as well as with the methods of manufacture to be able to provide the dimensions of the parts with the correct limits.

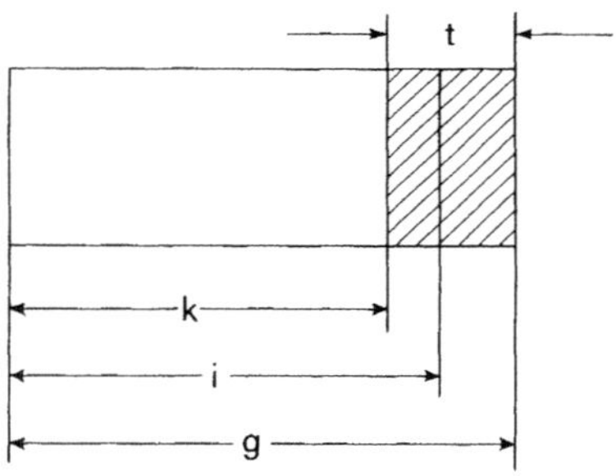

Fig. 148

It is customary to give the part in the drawing a basic dimension (n) and the difference dimensions between the maximum- and the basic dimension (a_o) and between the minimum and the basic dimension (a_u) (Fig. 149).

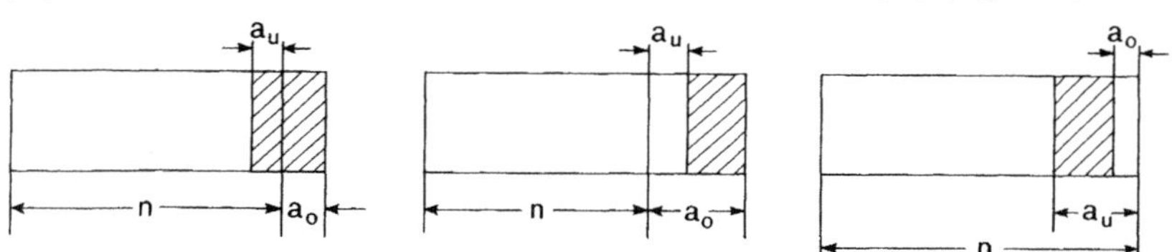

Fig. 149

The basic dimension may be inside or outside the tolerance area or may coincide with the maximum or the minimum dimension.

Gauges

Parts may be checked in their correct dimensions by means of adjusting gauges e. g. a slide gauge or micrometer or preferably by means of solid gauges. There are two types of solid gauges, shown in Fig. 150, the caliper gauge for checking shafts and the plug gauge for checking holes.

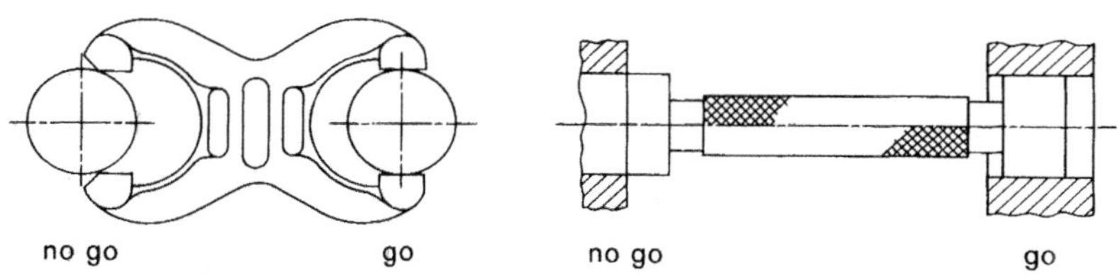

Fig. 150

The gauges have a "go" and a "no-go" end. The ends of the two gauges have the limit dimensions, thus the work piece will have the true size, if the go end of the gauge goes and the no-go end does not go on the shaft or on the hole, respecticaly. The go end of the caliper gauge must go, by its own weight, on the shaft.

Fits

In engineering products there are often two parts which are required to fit into one another. For these parts the dimensions of the matching parts must be in accord. Either the dimension *a* is determined by the dimension *b* or conversely the dimension *b* is determined by the dimension *a* (Fig. 151). There are plane fits for parts with plane matching surfaces and round fits for parts with round matching surfaces. If the parts must be fixed, then the dimension a must be greater than the dimension *b*, if the parts must be able to slide, then the dimension *a* must be smaller than the dimension *b*.

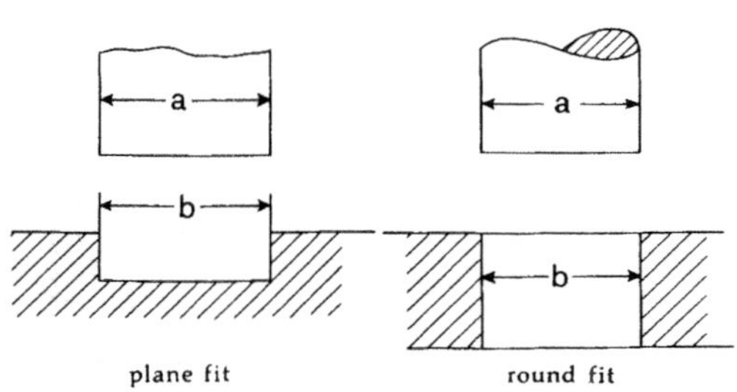

Fig. 151

There are three kinds of fits: clearance fit, interference fit and transition fit (Fig. 152).

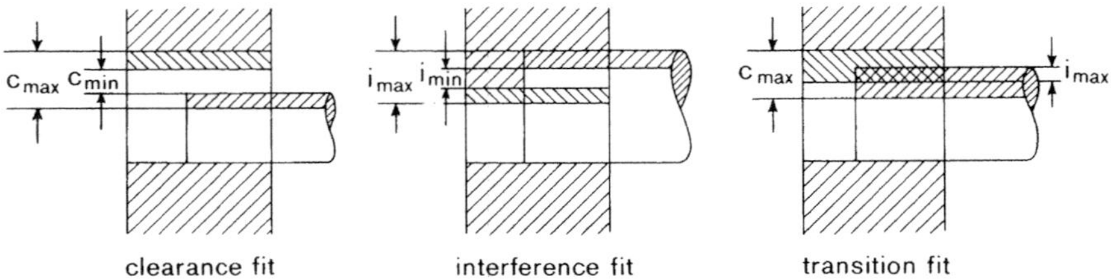

Fig. 152

The difference between the hole diameter and the shaft diameter is called "clearance" for the clearance fit and "interference" for the interference fit. For the transition fit either clearance or interference may occur. For the clearance or interference fit, respectively, the true clearance or interference, respectively, is between a maximum and a minimum clearance or interference, respectively. For the transition fit the clearance or interference, respectively is between a maximum clearance and a maximum interference. The clearance fit is used for sliding bearings, the interference fit for bushes and the transition fit for couplings, clutches and gear wheels.

There are two systems of fits: the basic hole system (hole basis system) and the basic shaft system (shaft basis system). For the basic hole system the hole is regarded as being constant in diameter and the various fits are obtained by varying the shaft diameters. For the basic shaft system the shaft is regarded as being constant in diameter and the various fits are obtained by changing the hole diameter.

Series of fits ranging from clearance to interference fit are shown diagrammatically in Fig. 153 on a hole basis and on a shaft basis.

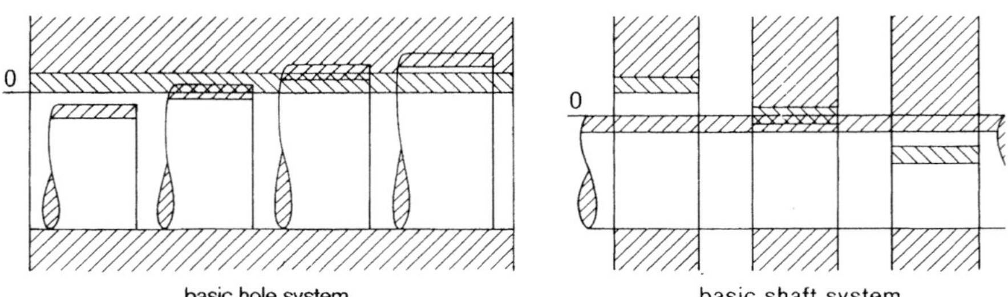

Fig. 153

ISA system

The various fits are fixed by the ISA system (international federation of national standardizing associations). For each dimension a unit tolerance is fixed. It increases with increasing diameter and may be calculated from the following formula:

$$i = 0.45\ \sqrt[3]{d} + 0.001\ d$$

where i = unit tolerance in microns
and d = shaft or hole diameter, respectively, in mm.

There are 18 standardized tolerance values indicated by numbers.

IT	5	6	7	8	9	10	11	12	13	14
	7	10	16	25	40	64	100	160	250	400

IT	15	16	17	18	
	640	1000	1600	2500	(number of units)

The tolerances IT 1 to IT 7 are for gauges.

The tolerances IT 5 to IT 11 are for work pieces produced by cutting processes.

The tolerances IT 12 to IT 18 are for work pieces produced by non-cutting processes.

The unit tolerances are standardized for a range of diameters for shafts and holes from 1 mm to 500 mm in 25 steps.

The position of the tolerance areas with respect to the reference line is indicated by letters which express the minimum distance of the tolerance areas from the reference line. The following designating letters are used: a, b, c, d, e, f, g, h, j, k, m, n, p, r, s, t, u, v, x, y, z. For holes capital letters and for shafts small letters are used. The letters a to h are for clearance fits and the letters j to z for transition and interference fits. In the basic hole system the reference line is the lower limit of the hole tolerance and in the basic shaft system the reference line is the upper limit of the shaft tolerance. In Fig. 154 various fits in the basic hole and in the basic shaft system are represented.

The j-tolerance is from IT 8, the J-tolerance from IT 9 symmetrical with respect to the reference line. The tolerances of holes with the letters A to H are symmetrical with the tolerances of shaft with the letters a to h. There is no symmetry for the other tolerances. For the tolerances of holes from the letter J to Z the letters do not express the minimum distance of the tolerance areas from the reference line.

To make manufacturing economical, often-used fits are selected and composed to a special ISA system which represents a combined system of the basic hole and the basic shaft system. For the shafts 15 tolerances and for the holes 9 tolerances are selected and combined into a system of 24 fits. In drawings not all dimensions are toleranced. In the title strip of a drawing the tolerances for non-toleranced dimensions (free tolerances) must be indicated. There are standardized free tolerances which may be used or the free tolerances may be chosen freely.

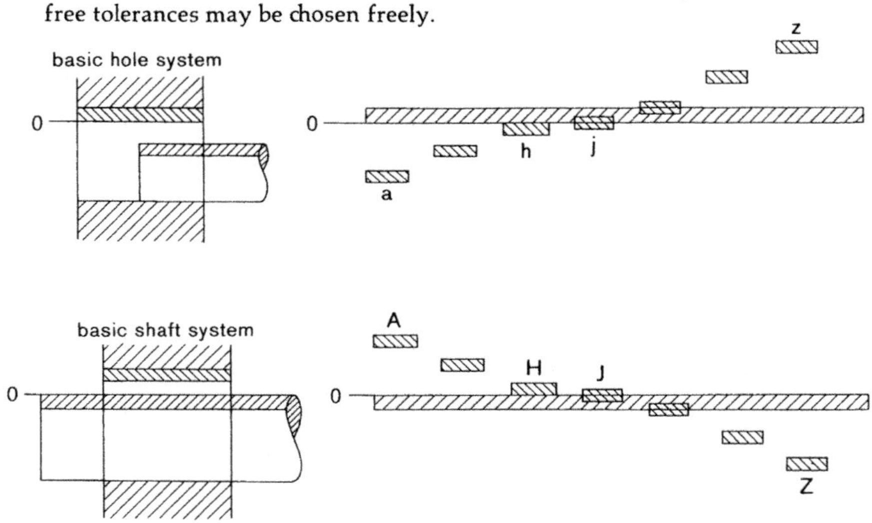

Fig. 154 Fits in the basic hole and the basic shaft system

8. Machine Elements

A machine is a combination of rigid members which have definite motions and which are capable of performing useful work. There are two types of machines: power engines which convert natural or electrical energy into mechanical energy and working machines which are driven by power engines and produce useful work. Natural energies are the energy of water dammed up in rivers or stored in natural or artificial lakes, the chemical energy of fuels and atomic energy. Electrical energy is a transformed energy which is used for the transport of energy. There are the following power engines: reciprocating steam engine, internal combustion engine, water-, steam-, and gas turbine and the electric motor. Examples of working machines are: pumps and compressors, machine tools, concrete mixers, elevators etc. and the electric generator.

A characteristic value of all machines is the "efficiency" which is defined as the ratio of the power output to the power input. There is an efficiency of a power engine and of a working machine which may be determined, as seen in Fig. 155. For the power engine

$$\eta = \frac{P_{\text{output}}}{P_{\text{input}}} = \frac{P_o - P_v}{P_o}$$

and for the working machine

$$\eta = \frac{P_{\text{output}}}{P_{\text{input}}} = \frac{P_o}{P_o + P_v}$$

where P_o = power inside the machine,
P_e = power outside the machine
and P_v = loss power.

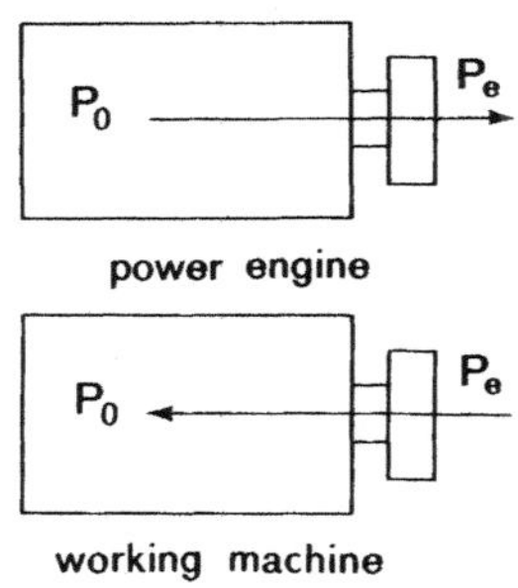

Fig. 155

Machine elements are the principal members or components of a machine. They are classified as joining elements, the elements of the drive, the elements of the slider crank mechanism, the pipes, valves and springs.

8.1. Joining Elements

To make a joint a physico-chemical adhesion process such as soldering, brazing or welding is used. Alternatively a mechanical method by using joining elements such as rivets, pins, bolts, keys, etc. is employed. Which kind of joint to be used depends on various conditions; if the joint has to be permanent or detachable, if forces are acting on the joint, or if the joint must be tight, etc.

8.1.1. Rivets

Steel structures, such as building frames, roof or bridge trusses, cranes, boilers and water tanks may be riveted. The plates or pieces of metal are connected by a number of rivets and make a permanent joint.

Methods of riveting

Matching holes are either punched, for very thin sheets, or drilled in the parts to be joined and rivets are inserted into the holes and the ends opposite to the heads are hammered until further heads are formed (Fig. 156). Riveting is carried out either by a cold or a hot process. Cold riveting is usually employed for steel rivet sizes up to 10 mm diameter and for rivets of non-ferrous metals.

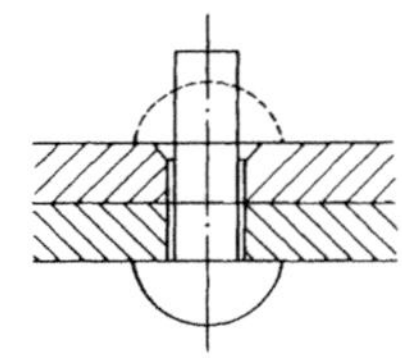

Fig. 156
Manufacture of a rivet joint

It is usually carried out by handhammering. Hot riveting is employed for steel rivet sizes more than 10 mm diameter and is executed by hydraulic or pneumatic hand tools or machines. The steel rivets, when heated, soften and during cooling they will shrink and produce tension in the shank and so the joined pieces grip tightly together and a considerable amount of frictional resistance is developed between the parts.

Rivet materials

Rivets are chiefly made of soft steel, but for some purposes especially for small work, they may be made of copper, brass or aluminium. The last named material, especially in alloyed form, is used for aircraft work.

Types of rivets (steel rivets)

There are three kinds of rivets: rivets for structures, rivets for boilers and tanks and small rivets less than 10 mm diameter. All the three types are standardized. The proportion of the rivet heads differ in minor ways from each other, as in Fig. 157.

Forms of joints (steel rivets)

There are two types of riveted joints: the structural joint and the boiler joint.

Structural joints

In Fig. 158 two examples of structural joints, the beam-to-girder connection and the beam-to-column connection, are shown.

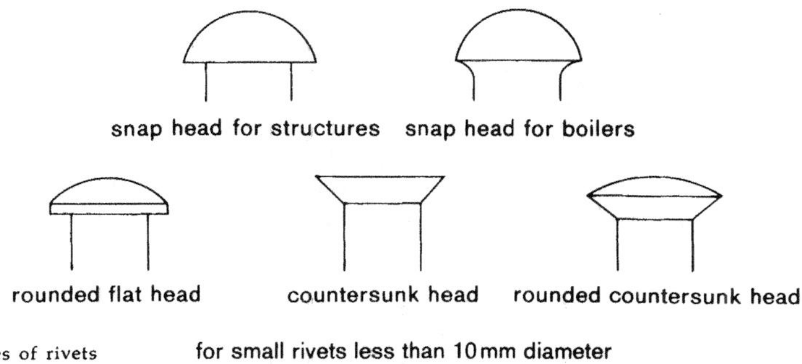

Fig. 157 Types of rivets

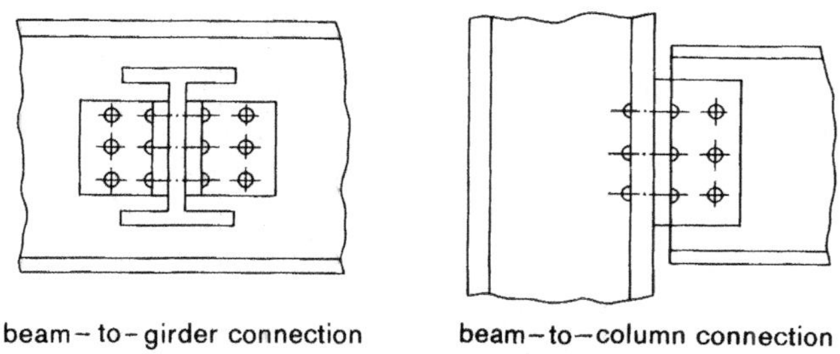

Fig. 158 Structural joints

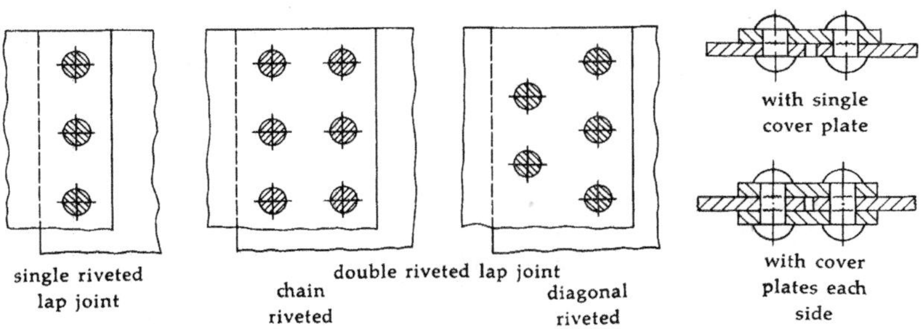

Fig. 159 Boiler joints

Boiler joints

If boilers or tanks are riveted, lap riveting or butt riveting is used. In lap riveting one plate overlaps the other, in butt riveting the plates are kept in alignment and a cover plate is placed over the joint and riveted to each plate. Sometimes two cover plates are used, placed one on each side of the plates. If the rivets are arranged in only one row, the joint is to be said single riveted, if there are two rows of rivets, the joint is called double riveted and if there are three rows of rivets, the joint is called triple riveted. Quadruple and quintruple riveted joints are also used, but only at butt joints. The rivets are usually placed zig zag, in what is called diagonal riveted. If the rivets are placed opposite to each other, it is called chain riveted. In Fig. 159 some kinds of boiler joints are shown.

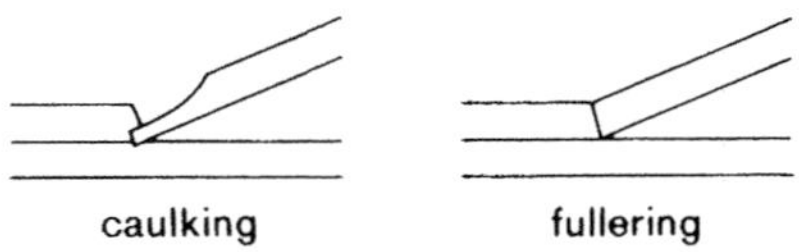

Fig. 160

To make boilers or tanks tight, the plate edges may be caulked or fullered, shown in Fig. 160. Caulking is the process in which the edges of the plates are driven in to form a leak proof joint. It is done by a bunt chisel-like tool called a "caulking tool" together with a hammer. The fullering process, similar to the caulking process is more effective in sealing the edges to the plates. The thickness of the fullering tool is equal to that of the plate. The lap edge is thickened and forced close to the lower plate.

Calculation of rivet joints

The force F acting on a rivet joint is taken partly by the friction between the plates and partly by the shearing resistance of the rivets, but the ratio of frictional to shearing force is not known.

For boiler joints the force F is assumed to be taken only by the friction of the plates.

$$F = \sigma_z \, \mu A \, m \, z = k_n \, A \, m \, z$$

where σ_z = tensile stress in the rivets
μ = friction coefficient
A = cross-section of one rivet
m = number of shearing sections in one rivet
z = number of rivets in one plate
$k_n = \sigma_z \, \mu$ = specific sliding restistance.

For structural joints and joints with aluminium rivets the force F is supposed to be taken only by the shearing resistance of the rivets.

$$F = \tau_s \, A \, m \, z$$

where τ_s = shearing stress in the rivets

When a rivet is closed, an indeterminate stress is set up in the shank owing to its tendency to shorten on cooling. For this reason rivets in a joint are usually arranged so that they are subjected to shearing forces only and not to axial forces (Fig. 161).

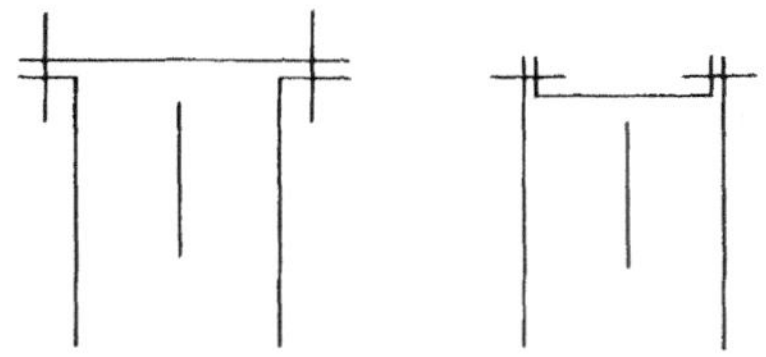

Fig. 161

Eccentrically-loaded rivet joints
When the line of action of the load F acting on the connection does not pass through the centroid of the rivets, the connection is said to be eccentrically loaded (Fig. 162). The force F is replaced by a force F of the same size as the original force acting parallel to the original one but passing through the centroid of the rivets together with a moment $M = F\,e$.

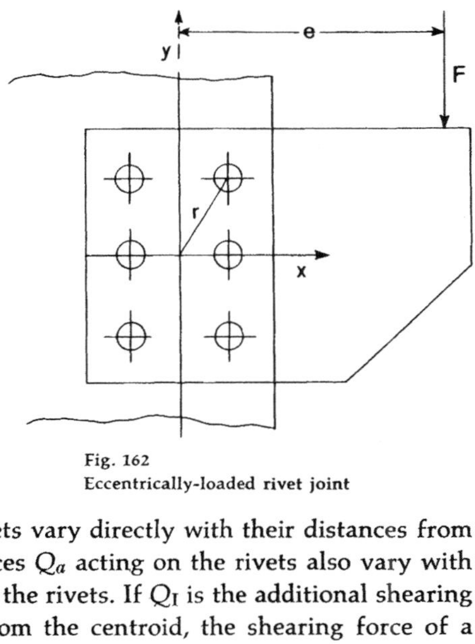

Fig. 162
Eccentrically-loaded rivet joint

The rivets are stressed by a shearing force $Q_s = F/n$ and an additional shearing force Q_a produced by the moment M. Since the deformations of the rivets vary directly with their distances from the centroid of the rivets, the forces Q_a acting on the rivets also vary with the distances from the centroid of the rivets. If Q_I is the additional shearing force acting at a unit distance from the centroid, the shearing force of a rivet at a distance r from the centroid is $Q_a = Q_I\, r$.

The moments of all shearing forces are in equilibrium with the moment M.

Therefore $F\,e = \sum Q_I\, r\, r = Q_I \sum r^2 = Q_I \sum (x^2 + y^2)$

$$Q_I = \frac{F\,e}{\sum (x^2 + y^2)} \tag{144}$$

where e = eccentricity of the force F
and x and y = coordinates of the rivets with respect to a coordinate system passing through the centroid of the rivets.

8.1.2. Welded Joints

To make a permanent joint without a machine element, soldering or welding may be used.

Soldering consists of bonding two pieces by means of a metal whose melting point is lower than that of the materials to be joined. Soldering involves only the melting of the materials which is employed to bind the two pieces together. The parent metal is not melted. Joining is therefore only a very slight diffusion of the solder in the parent metal. There are two principal methods: soft and hard soldering. The soft solder is composed of tin and lead in approximatively equal proportions. Soft soldering is used in the electrical industry for joining wires and in the sheet metal industry for uniting various articles of sheet metal. There are two principal hard solders: brazing alloy containing 10 to 50 % zinc with the remainder copper and silver solder containing between 16 and 52 % copper, 4 to 38 % zinc with the rest silver.

The most important process by which metals are joined is that of welding which is the process of making an intimate union of metal parts in the plastic or molten state. If the process is carried out with metals in the plastic state an application of pressure is necessary and therefore this process is called "pressure welding". If the welding operation is carried out with metals in the molten state without the application of pressure, the process is called "fusion welding".

Welding processes

In the fusion process the metals which are to be joined are heated to the fusion temperature and a filler metal is added. There are two principal methods of fusion welding: gas welding and electric arc welding.

In gas welding the heat for melting the metals is obtained by combustion of a gas. Normally acetylene gas (flame temperature 3100 °C) and oxygen mixed in a torch is employed. For thin sheets instead of acetylene gas hydrogen (flame temperature 2000 °C) is employed. A rod of filling metal is used which is melted and run into the joint. Gas welding may be used for dead mild steels and for mild steels with a carbon content up to 0.35 % without an additional material and for cast iron and most of the non-ferrous metals and their alloys with a welding powder.

In electric arc welding the heat is obtained from an electric arc set up between the base metal and an electrode. Normally direct current of about

400 amperes and 60 volts on no load and about 25 volts on load, is used. The temperature of the arc is 3800 °C for direct current and 3400 °C for alternating current. As the temperature of the positive electrode is greater than that of the negative electrode, more heat is liberated at the positive electrode. Therefore normally the work piece which requires more heat than the electrode is made positive and the electrode negative. Only in welding very thin materials is the work piece made negative and the electrode positive. The electrode may be without or with a coating. The coated or covered electrodes are wrapped with fluxing materials to prevent the penetration of destructive constituents of the atmosphere, nitrogen and oxygen, in the work piece. Nitrogen produces hardness and brittleness, oxygen produces inclusions and blow holes. Electric arc welding is used for dead mild steel and mild steel with a carbon content up to 0.34 % and for cast iron.

There are special welding processes for welding all metals, especially for high grade carbon and alloyed steel and for non-ferrous metals and their alloys. In these processes a flux or an inert gas is used to prevent the penetration of nitrogen and oxygen from the air into the work piece. There are two processes: the union melt process and the inert-gas shielded-arc welding process.

The union melt process is an automatic electric arc process requiring application of a special welding powder. The tip of the welding wire and the work piece are both covered by a layer of welding powder. Wire feed and driving motion of the wire are automatically switched and controlled electrically. The electric arc is invisible which eliminates the necessity of a protection against light or ultra-violet rays. In this process alternating current is used.

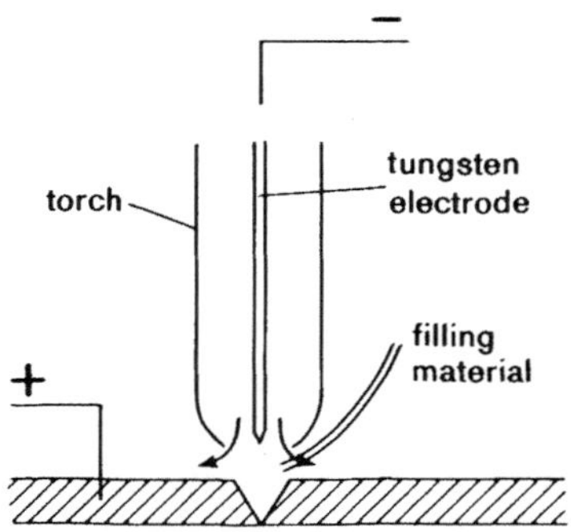

Fig. 163
Argon arc welding process

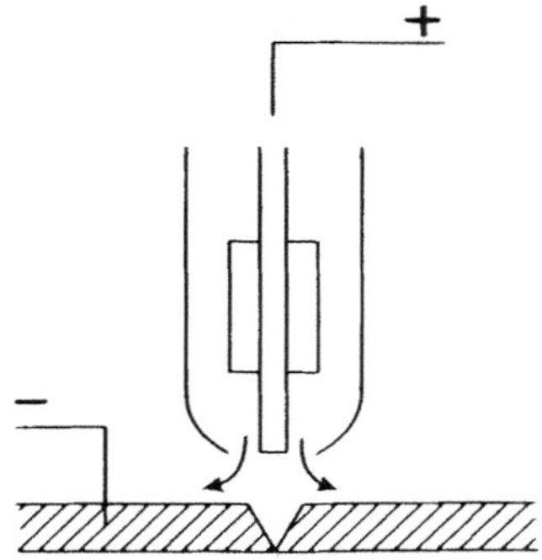

Fig. 164
SIGMA welding process

Inert-gas shielded-arc welding is an electric arc process with the application of an inert gas to protect the work piece from the penetration of oxygen and nitrogen. There are two processes: the argon arc process and the SIGMA process.

In the argon arc process normally direct current is used. A tungsten electrode which does not melt is used. The electrode is made negative and the work piece positive as in normal electric arc welding. For a filling material a welding rod is dipped into the arc. For the inert gas argon is used (Fig. 163). In the SIGMA process a welding rod is used as the electrode which is moved automatically. The work piece is made negative and the electrode positive. For the inert gas argon is used (Fig. 164).

Pressure welding, also called "electric resistance welding" is the process in which two parts are joined by heating them electrically to the plastic state and forcing them together under pressure. An alternating current is used up to 80000 amperes at up to 10 volts. There are three kinds of resistance welding: butt welding, spot welding and seam welding.

In butt welding the two parts to be welded are held by two heavy water-cooled electrode grips and forced together. An electric current flows through the jaws or grips and the resistance to flow produces the heat. Butt welding may be used for all metals exept cast iron and aluminium.

Spot welds can be made in plates up to 12 mm thick. The electrodes of spot welding machines, called tips, are made of copper alloy.

Seam welding is employed to unit two metals along a seam. The electrodes have the form of a roller. As the two pieces are in continual contact, the current must be regulated in a cyclic manner. The current is on for a fraction of a second and off for another fraction of a second. Seam welds can be made for sheets up to 5 mm thick. Spot and seam welding may be used for all kinds of steel and aluminium.

For cutting metals flame cutting may be used. A spot area along a line of cut is heated to ignition temperature (800 °C) by a preheat flame and a thin jet of high purity oxygen is directed at the spot. The jet quickly penetrates through the material and the torch is moved progressively foreward over the surface cutting a narrow slot along the desired line of severance. Normal flame cutting may be used for steels with a carbon content up to 0.34% and for cast iron. The thicknesses of the plates to be cut may be up to 300 mm and more.

For metals which cannot be cut by normal flame cutting, that is for steels having a high carbon content and for non-ferrous metals and their alloys, „argon-arc flame cutting" may be used. In this process the required heat is produced by an electric arc set up between a tungsten electrode and the work piece. It uses direct current. The work piece is made positive and the electrode negative. A mixture of 35 to 40 % hydrogen and the remainder argon or pure nitrogen is directed with high speed at the spot. The current is about 500 amperes at 30 volts. Argon-arc flame cutting may be used for plates up to 70 mm thick (Fig. 165).

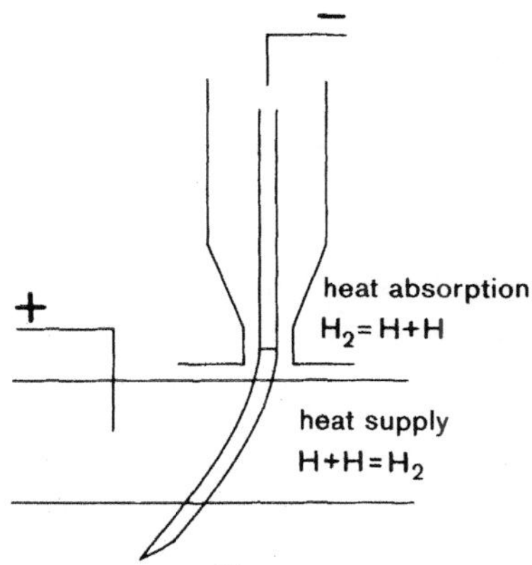

Fig. 165
Argon arc flame culting

Types of welds and welded joints

There are three basic types of welds: groove welds, fillet welds and bead welds (Fig. 166).

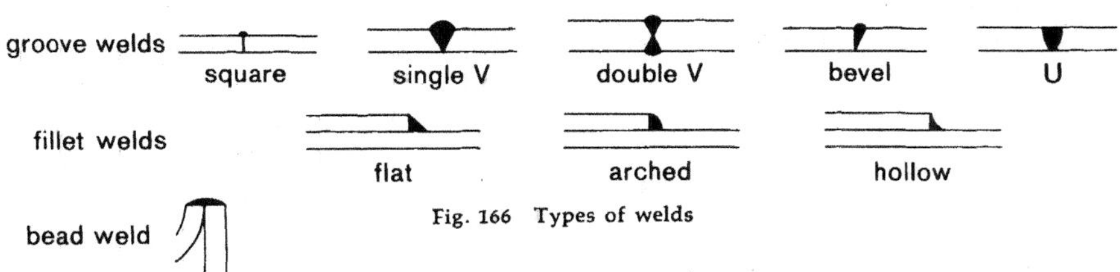

Fig. 166 Types of welds

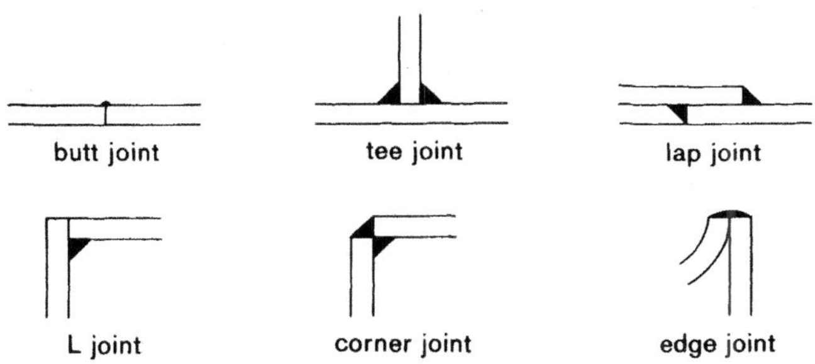

Fig. 167 Types of welded joints

Strength of welds

For constant loading, failure normally occurs in the basic material. For alternating loads, failure generally occurs close to the weld. The value of the strength of a weld cannot be exactly indicated. Several factors influence the quality of the joint. The stress concentration in the weld due to the shape of the work piece and the decrease of strength of the weld due to the weld's absorbing oxygen and nitrogen from the air are factors of influence which cannot be accurately determined.

For the calculation of a welded joint all these factors are included in the factor of strength α. The allowable stress in a welded joint is

$$\sigma_{\text{weld, all}} = \frac{\alpha\, \sigma_{\text{rel}}}{\nu}$$

where σ_{rel} = related stress which is for constant loading equal to the yield stress and for alternating loads equal to the endurance limit.
ν = safety factor.

8.1.3. Grooved Pins and Drive Studs

A grooved pin is a cylindrical pin with three longitudinal grooves. The three grooves are swaged into the cylindrical body to expand the pin within controlled standard limits.

When the pin is forced into a drilled hole of the same diameter, which the pin had before grooved, the hole wall will cause the expanded material to be compressed and thus a locking fit is produced, as Fig. 168 shows. Due to the locking force the grooved pin may resist severe shocks and vibrations. The grooved pins require only a straight drilled hole without the need for reaming, milling or tapping necessary for other fasteners. The same grooved pin may be reused many times. Grooved pins replace bolts, keys locatings pins, taper pins, cotter pins, lockwashers and rivets.

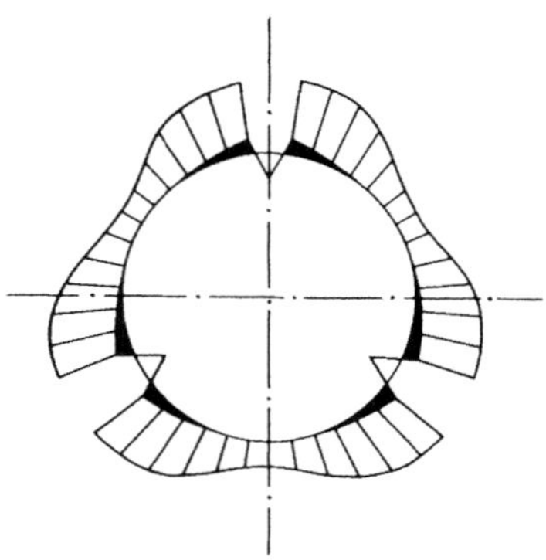

Fig. 168
Compressive stresses on a grooved pin

Drive studs embody the same principle applied in grooved pins. They have a head and are used wherever a headed fastening element is required and replace, screws, rivets etc.

Types of grooved pins and drive studs

In Fig. 169 some types of grooved pins and in Fig. 170 some types of drive studs are represented.

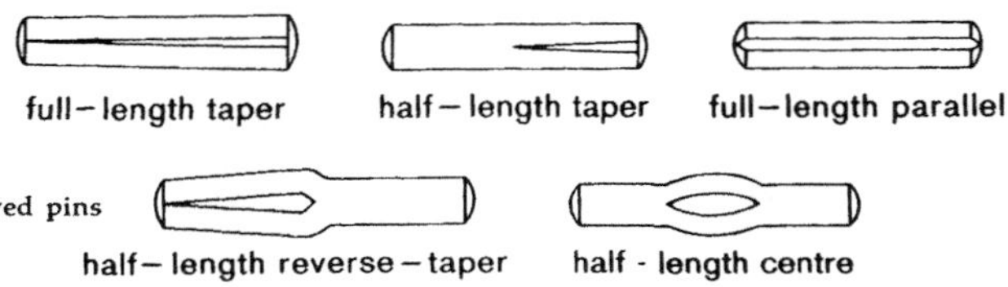

Fig. 169 Types of grooved pins

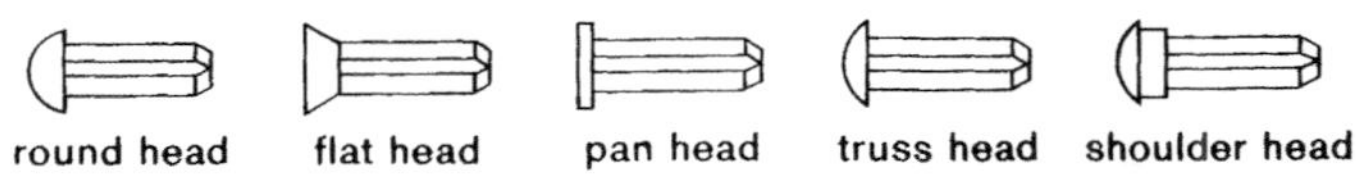

Fig. 170 Types of drive studs

Examples of applications

In Fig. 171 same examples of applications of grooved pins and in Fig. 172 one example of the application of drive studs are represented.

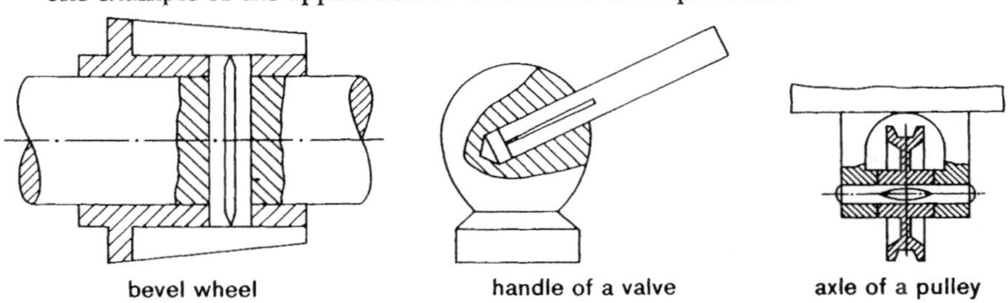

Fig. 171 Applications of grooved pins

Drive studs are used for fastening cover plates, name plates, decorative escutcheons, pipe clamps, etc.

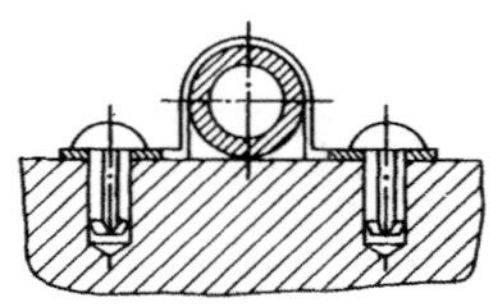

Fig. 172
Pipe clamp fastened by drive studs

8.1.4. Clamp Connections

Clamp connections are used for fastening divided or slotted hubs on shafts. By tightening the bolts which are inserted in the hub a pressure between the hub and the shaft is generated which has to resist the forces acting on the hub. In Fig. 173 a lever fastened to a shaft by means of a clamp connection is represented. The force S acting in the bolts may be calculated thus:

$$S\left(a+\frac{D}{2}\right)=Q\frac{D}{2}$$

whence $S=\dfrac{Q\,D}{2\,(a+D/2)}$.

From $2\,Q\,\mu\,\dfrac{D}{2}=F\,b$,

$$Q=\frac{F\,b}{\mu\,D}.$$

Substituting this value in the above equation gives:

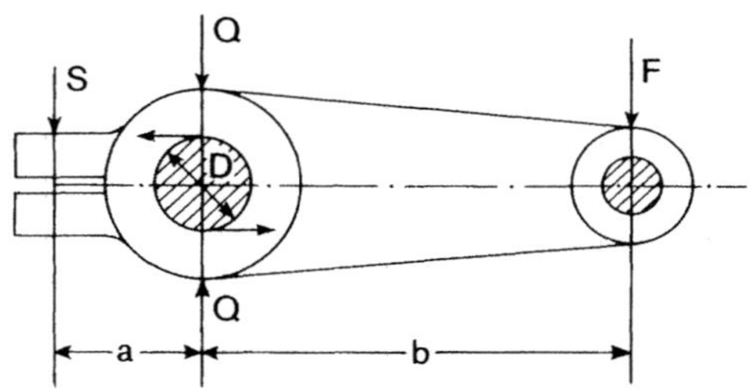

Fig. 173

$$S=\frac{F\,b}{2\,\mu\,(a+D/2)}. \qquad (145)$$

8.1.5. Connections with Shrunk Rings

Shrunk rings are used for fastening large and heavy machine parts on which forces with severe shocks and vibration are acting e. g. for connecting flywheels on shafts. The connecting parts must have an interference fit i. e. the shaft diameter must be greater than the hole diameter. For joining, the outside part must be heated up to 100 °C on a warming plate then up to 350 °C in oil. Warming must be such that the outside part may be moved easily over the inside part. Alternatively, the inside part may be supercooled, down to − 79 °C, in carbon dioxide snow and down to − 190 °C in liquid air.

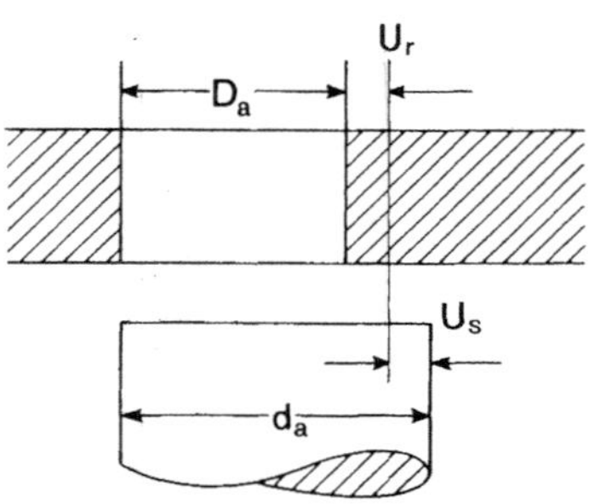

Fig. 174
Oversize at a shrunk ring connection

In order to move the shrunk ring over the shaft the shrunk ring must be heated so that its internal diameter is greater than the diameter of a solid shaft. After cooling the internal diameter of the shrunk ring is increased by U_r and the shaft diameter reduced by U_s, as Fig. 174 shows. The shrunk ring is stressed by tension and the shaft by compression. The difference between the shaft diameter and the internal diameter of the shrunk ring is called the "oversize",

$$U=d_a-D_i=U_r+U_s.$$

The maximum oversize U_{max} is obtained from the condition that the allowable tensile stress in the shrunk ring may not be exceeded, the minimum oversice U_{min} from the condition that the frictional force occuring between the shrunk ring and the shaft is equal to the force which should be transmitted by the shaft. The fit must be chosen so that the maximum and minimum interferences, i_{max} and i_{min}, are between the maximum and minimum oversize values, U_{max} and U_{min}, obtained by calculation.

8.1.6. Keys and Cotters

8.1.6.1. Keys

Keys are used to fasten pulleys, gear wheels, couplings, clutches and similar parts on a shaft. Keys normally have a rectangular cross section and have a taper on the surface projecting into the hub. The taper matches a corresponding taper in the hub keyway. The taper is generally 1/100. The wedging action forces the hub and shaft together on the side opposite the key and keeps the hub in its radial and axial position.

The load transmission between shaft and hub is produced by the friction between key and hub. The key is stressed by pressure and shearing, as shown in Fig. 175. Since the force for driving the key into the keyway is not known, the compressive stress and the shear stress in the key cannot be calculated.

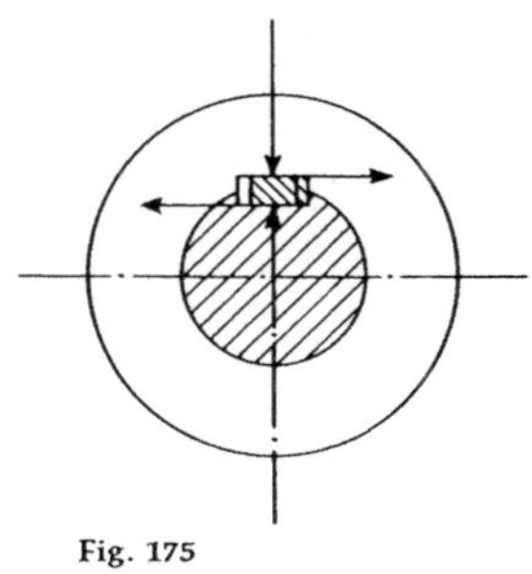

Fig. 175
Forces acting on a key

Feather keys have no taper. The load transmission between shaft and hub is produced by the shear force between the key and the hub. Feather keys are used for fastening light work and as a sliding arrangement in order to move a wheel along a shaft. There is a clearance in the keyway of the hub in the sides and at the top. The feather key has a lateral interference fit in the shaft keyway and is secured by screws if it is to be used as a sliding arrangement. The lateral fit for keys is normally D 10, for feather keys P 9 and for shafts h 9. Keys are made from special key steel.

Types of keys

In Fig. 176 the standardized types of keys and in Fig. 177 the standardized types of feather keys are represented.

Sunk keys

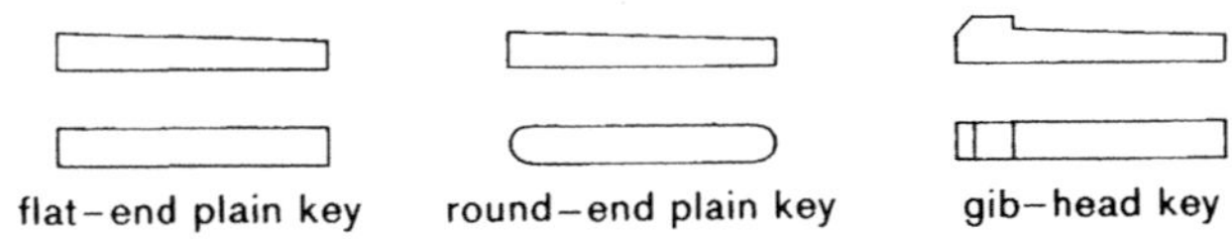

Flat and saddle keys

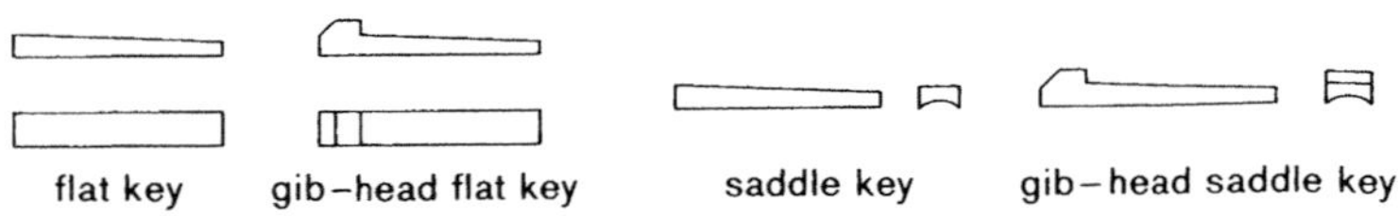

Fig. 176 Types of keys

Feather keys

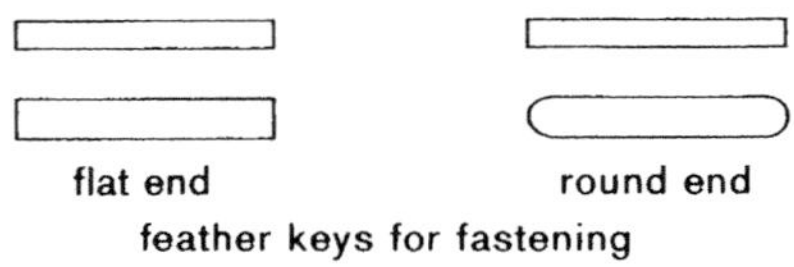

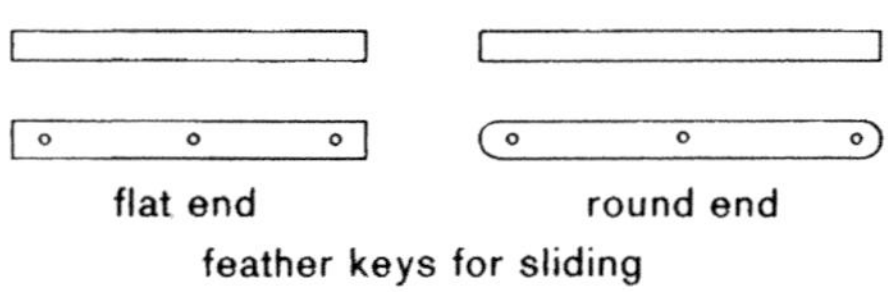

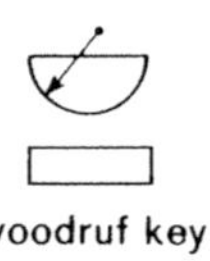

Fig. 177 Types of feather keys

Applications of keys

Sunk keys are employed, if the full twisting moment must be transmitted by the key.

The flat-end plain key is the most common. It is driven into a keyway cut half in the shaft and half in the hub (Fig. 178).

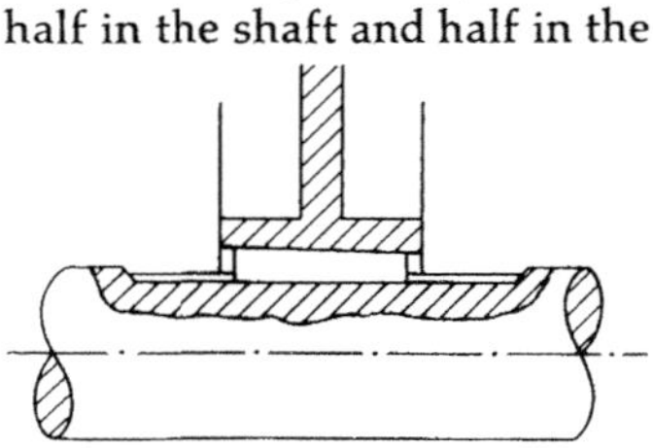

Fig. 178 Connection with flat end plain key

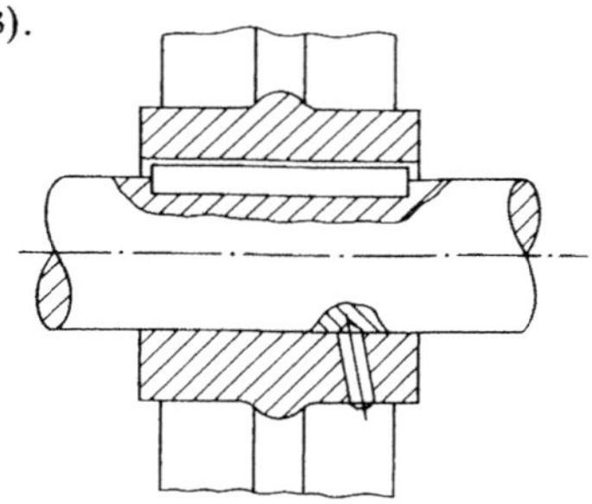

Fig. 179 Connection with round end plain key

The round-end plain key is inserted into the keyway of the shaft and the hub is driven over the key. This key is used, if a short keyway is desirable e. g. near a bearing or near another part mounted on the shaft (Fig. 179).

The gib-head key is used, if the key must be driven out from the same side that it was driven in, since on the other side of the hub another part may disturb the key from being driving out (Fig. 180).

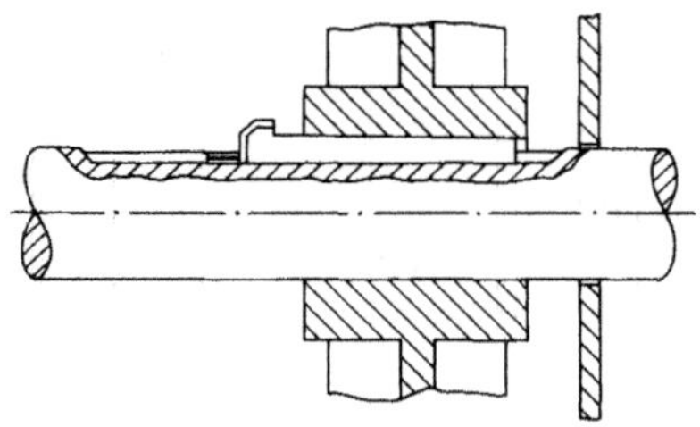

Fig. 180 Connection with gib head key

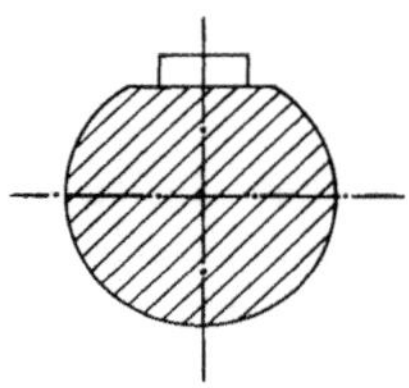

Fig. 181

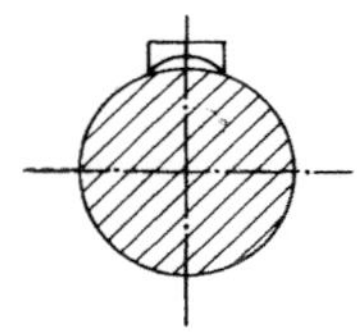

Fig. 182

Flat and saddle keys are only suitable for light work.

The flat key does not fit on a shaft keyway. It rests on a flat surface on the shaft, as in Fig. 181.

The saddle key has a curved surface which rests on the non-machined shaft. It holds only by friction (Fig. 182).

The depth of keyway in the hub is the same for the flat and saddle key as for the sunk key.

The Woodruff key serves for fastening light work. It has a half moon shape. Its curved base fits into a similar recess in the shaft. It is mostly used on small shafts or spindles with a taper, but it can also be used on parallel shafts (Fig. 183).

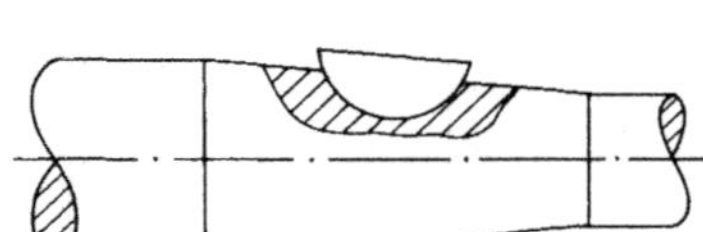

Fig. 183

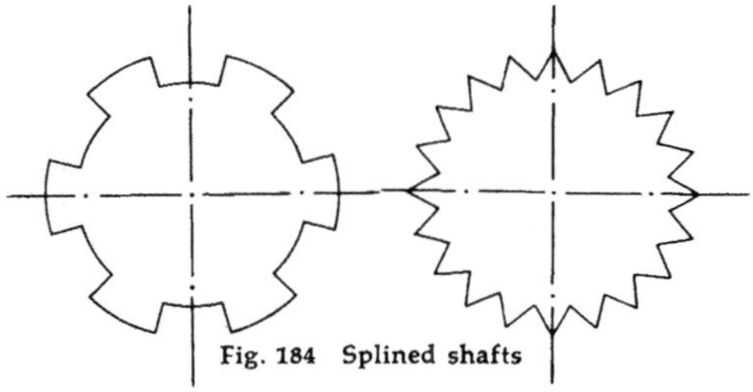

Fig. 184 Splined shafts

Instead of shafts with feather keys splined shafts may be used. The splines are of rectangular or of V formation in cross section, as shown in Fig. 184. Splined shafts are able to transmit larger twisting moments than shafts with feather keys. They are used in the manufacture of internal combustion engines, motor cars and machine tools.

For high loads or alternating loads or, if heavy shocks or vibrations are acting, tangential keys are used. Two pairs of keys having lateral taper are arranged at an angle of 120°, as shown in Fig. 185. The taper is from 1/60 up to 1/100.

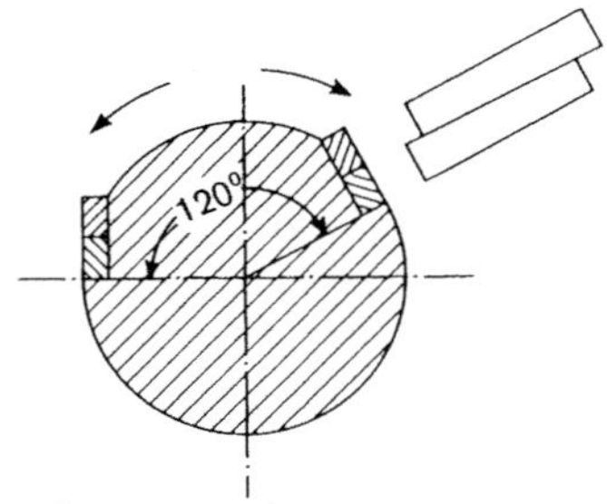

Fig. 185 Tangential keys

The dimensions of all keys and the corresponding keyways are standardized. They are equal for each shaft diameter. Only the length of keys may be varied. They are standardized in steps. If the full twisting moment of the shaft must be transmitted to the hub, the length of the key should be about 1.5 times the shaft diameter.

8.1.6.2. Cotters

Cottered joints are used to connect rigidly, rods which are subjected to axial forces only. The end of one rod and the end of the other rod which takes the form of a socket are slotted to take a cotter, a flat wedge shaped bar. The slots are made a little longer than the greatest width of the cotter, and their relative positions are such that the driving in of the cotter tends to force the end of one rod into the socket of the other rod as shown in Fig. 186. The taper of a cotter is from 1/15 up to 1/25. The relation between the force of driving in F, the force of driving out F' and the pressure force Q is given by equations (22) and (23).

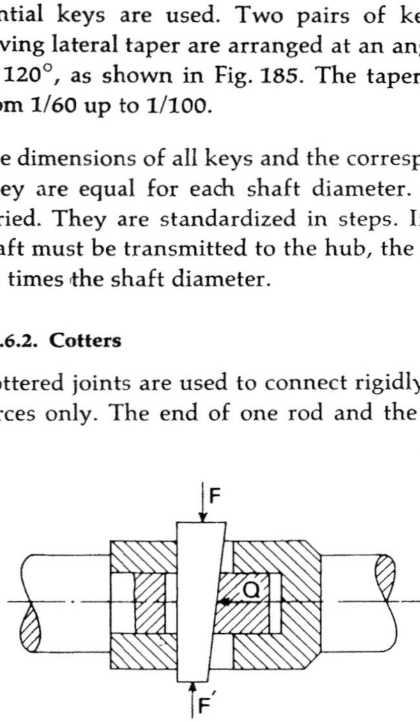

Fig. 186 Cotter connection

$$F = Q \tan(\alpha + \varrho)$$

$$F' = Q \tan(\alpha - \varrho)$$

where α = angle of cotter
and ϱ = angle of friction.

To prevent losing cotters, set screws or screwed cotters i. e. a cotter carrying a threaded section on its thinner end are used. The taper of a screwed cotter is 1/7.

8.1.7. Bolts and Screws

The most common joining or fastening elements are the bolt and the screw. The bolt is a pin having a head on one end and a thread on the other end and is used with a nut. The screw looks like a bolt, but it is used without a a nut. The term screw, besides, is used for threaded spindles and, as they transmit power, they are referred as "power screws". A pin threaded on both ends, which is sometimes used for fastening parts together, is called a "stud" Bolts, screws and studs are used for fastening together any two or more machine parts which may require dismantling quickly in an emergency, such as in repair work or for lubricating parts. Bolts and studs are often fitted with a plain round washer laid between the underside of the nut and the face of the work piece to form a cushion between the fastened parts and the nut.

Bolts, screws and studs are normally made of mild steel, but for special work they are made of high quality steel. There are also bolts, screws and studs made from brass, copper or other metals.

The shank of a bolt or a stud is screw-threaded for a length slightliy in excess of the height of the nut and the washer. In some cases in addition to an ordinary nut another nut is fitted to provide extra security. The latter is called a "lock nut" and is usually half the height of an ordinary nut. In this case the bolt or the stud must be longer and the shank must be threaded lower down to accomodate the nut, the lock nut and a washer, if one is to be used.

8.1.7.1. Screw Thread

If a point is moving around and along the surface of a cylinder with uniform velocity in both directions, a helix is generated. If the point is moving clockwise round the cylinder, then the helix is right hand, if the point is moving counter-clockwise, then the helix is left hand (Fig. 187). A screw thread is formed by moving a thread profile along the helix. The right-hand helix corresponds to a right-hand screw thread and the left-hand helix to

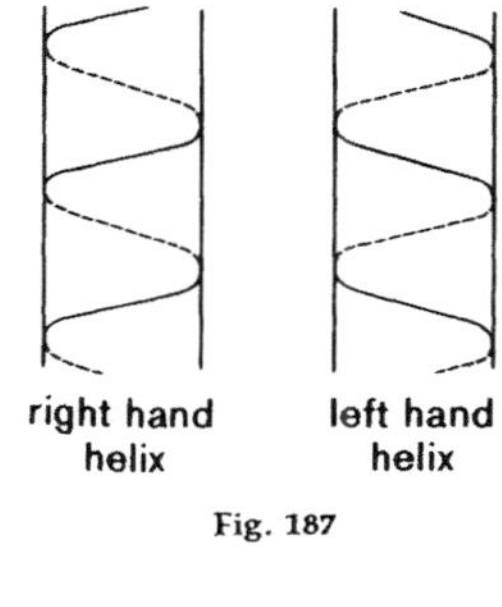

Fig. 187

a left-hand screw thead. For the right-hand thread the bolt advances into the nut when turned clockwise, for the left-hand thread the bolt advances into the nut when turned counter-clockwise. If a thread is not specified on a drawing, it is understood to be right hand.

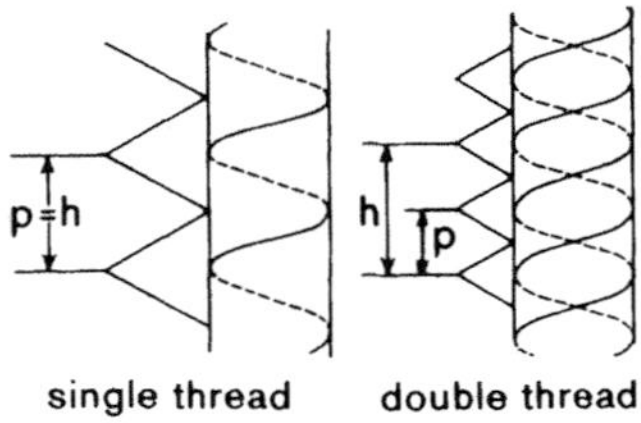

Fig. 188

If two, three or more threads of the same helix angle are arranged, a double, a triple or, expressed generally, a multiple thread is formed. For a bolt with a multiple thread the depth of thread becomes smaller than for a bolt with a single thread, if the helix angle is the same, as shown in Fig. 188. Multiple threads are used for power screws, if small depths of thread at large helix angles are desired.

Thread Notation

The following notation for threads is in common use (Fig. 189 and 190)

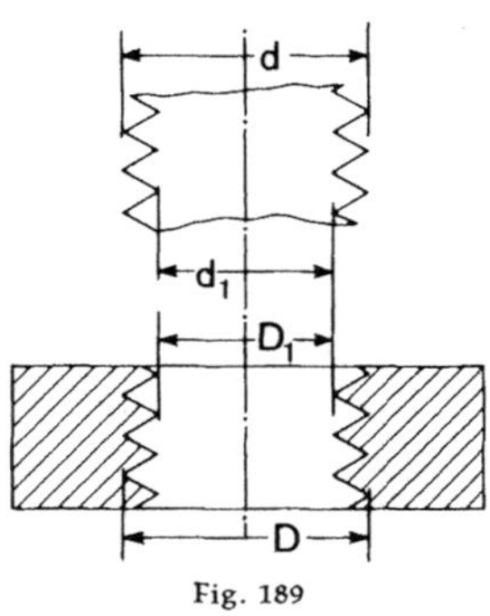

Fig. 189

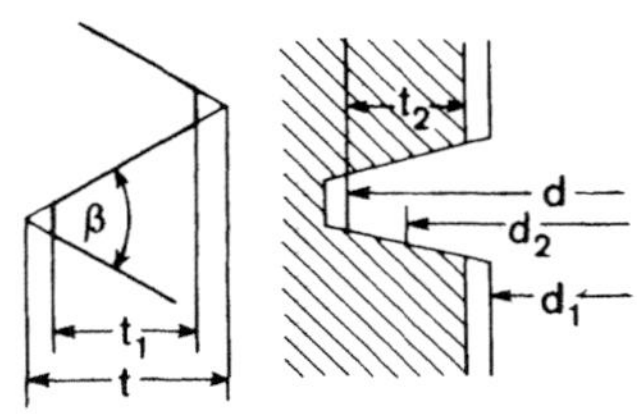

Fig. 190

d = major diameter of the bolt
d_1 = minor diameter of the bolt
D = major diameter of the nut
D_1 = minor diameter of the nut
p = pitch = distance from a point on the thread to a corresponding point on the next thread measured along a line parallel to the axis of the bolt.
h = lead = distance which a screw advances parallel to the axis in one turn.
$h = z\,p$, where z = number of threads

d_2 = pitch diameter
t = theoretical depth of thread
t_1 = true depth of thread
t_2 = depth of engagement
α = helix angle
β = thread angle

$$\tan \alpha = \frac{h}{d_2 \pi}.$$

There are threads for fastening bolts, studs and screws and threads for power screws.

Threads for fastening bolts, studs and screws

There are three kinds of threads for fastening bolts, studs and screws: coarse, fine and pipe.

Coarse threads

The metric screw thread, shown in Fig. 191, has a thread angle of 60°. The crest of the bolt and the nut are flat and the root of the bolt and the nut are rounded. This thread has been adopted for bolts, studs and screws from 0.3 mm to 68 mm diameter.

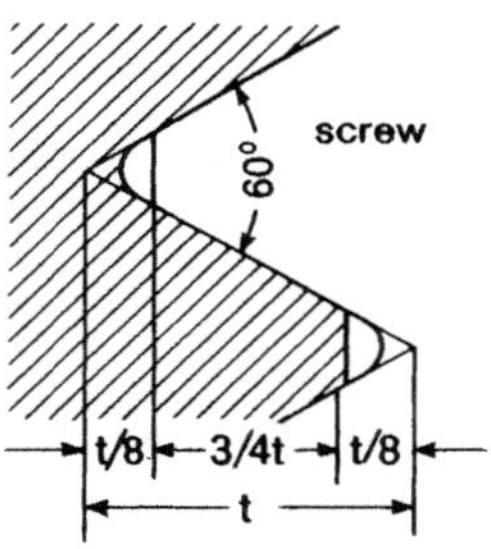

Fig. 191 Metric screw thread

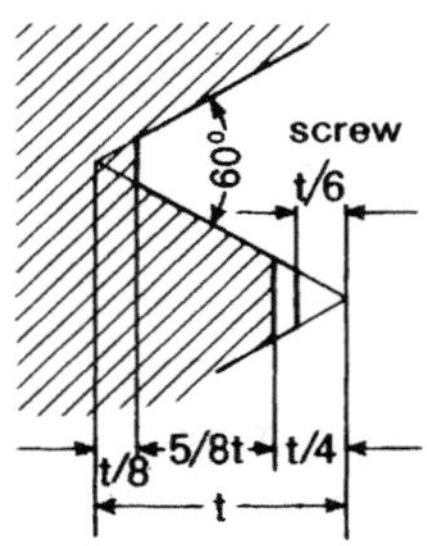

Fig. 192 American Standard screw thread

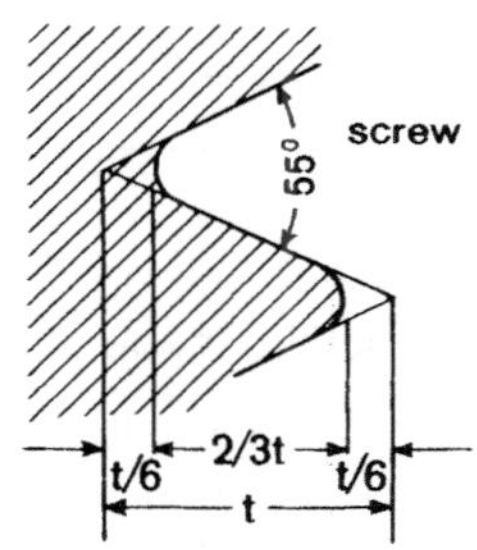

Fig. 193 Pipe thread

The American standard screw thread, shown in Fig. 192, has also a thread angle of 60°. The crest and the root of bolt and nut are flat.

Metric fine screw thread

The metric fine screw thread has the same profile as the metric coarse thread, only the helix angle and consequently the depth of thread are smaller.

Pipe thread

The British standard Whitworth screw thread, formerly used as a coarse and a fine thread, is nowadays employed as pipe thread. Its thread angle is 55°. The crest and the root of bolt and nut are rounded (Fig. 193).

Threads for power screws

The angle of thread is for all threads of power screws $\beta = 30^\circ$

ACME THREAD

The ACME thread is used in screw presses and in machine tools, where a disengaging nut is required (Fig. 194).

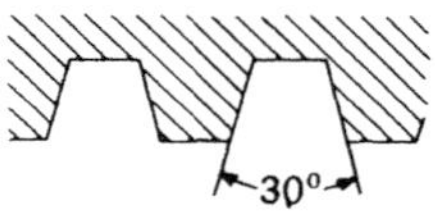

Fig. 194 ACME thread

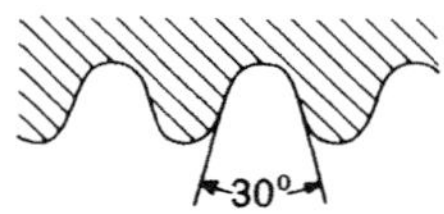

Fig. 195 Round or knuckle thread

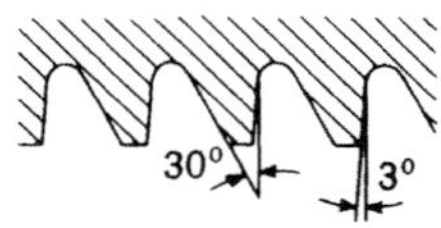

Fig. 196 Buttres thread

Round or Knuckle thread

The round or knuckle thread is an ACME thread with round edges. It is used instead of the ACME thread in dusty environments and if shocks are occuring (Fig. 195).

Buttress thread

The buttress thread is designed to transmit loads in one direction only (Fig. 196).

8.1.7.2. Types of Bolts and Screws

There are three kinds of screw connections: the connection with a bolt, the connection with a cap screw and the connection with a stud (Fig. 197). Normally a bolt with a nut is employed. If, however, one one side of the connection the presence of a bolt head or a nut is inadmissible, a cap screw or a stud may be used. A stud instead of a cap screw is preferred, if the connection is to be dismantled frequently, since the thread in the hole may break, especially, if the hole is drilled into cast iron.

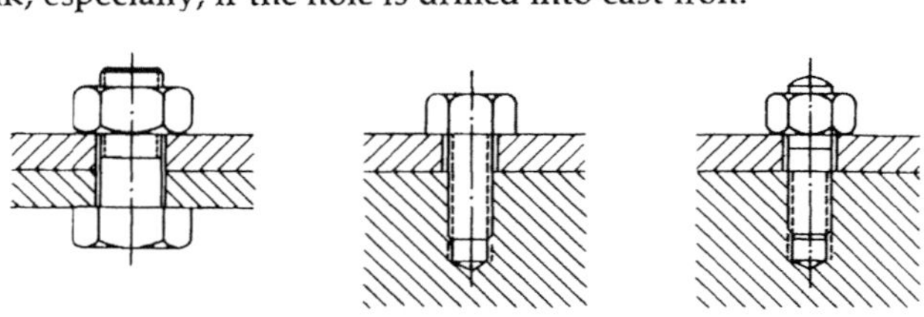

Fig. 197 Types of screw connections

Bolts

Bolts may be classified according to the form of their heads. There are bolts with hexagon, square, cup and countersunk heads (Fig. 198).

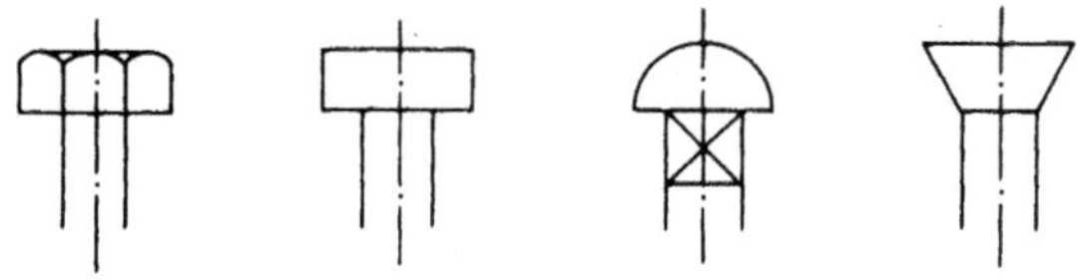

Fig. 198 Types of bolt heads

The hexagon head bolt is normally used. The square head bolt is used recessed into a square hole. It is thus prevented by the walls of the recess from turning during the process of having its nut screwed on. The cup head bolt is used where one of the parts to be joined together is timber. A square hole is made in the timber and the bolt neck is driven tightly into it. The countersunk head bolt is used, when it is desirable to sink the bolt head in the surface of the work piece.

A special bolt is the foundation bolt, shown in Fig. 199. This bolt is used in foundation work. Its tapered body of rectangular cross section has jagged cuts hot forged in it.

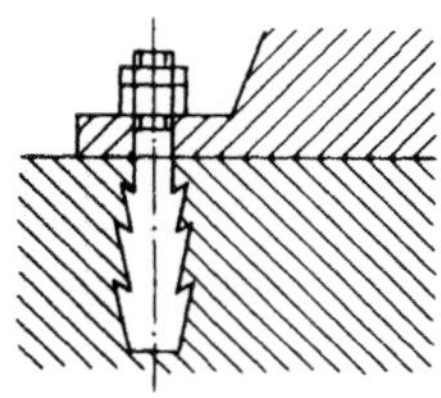

Fig. 199 Foundation bolt

Another special bolt is the elastic bolt, shown in Fig. 200. For this bolt a portion of the unscrewed part of the bolt is reduced in diameter. Thus this bolt has more elasticity than a normal bolt and can withstand better impulsive loads. Elastic bolts are not standardized, and for each connection a special design of the bolt is necessary.

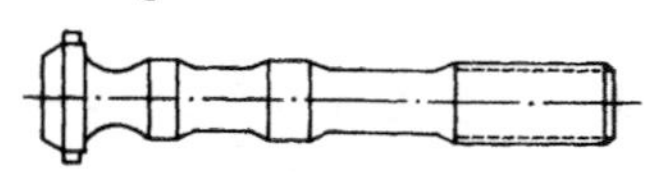

Fig. 200 Elastic bolt

Cap screws

There are cap screws with hexagon heads and socket heads and different kinds of cap screws with slotted heads as shown in Fig. 201.

Set screws

Set screws are used to prevent relative motion, usually rotary motion, between two machine parts, such as the motion of a hub of a pulley on a shaft. Set screws normally have slotted or socket heads. The ends of these screws have a point which may be flat, oval, conical, cupped or some other form, as shown in Fig. 202.

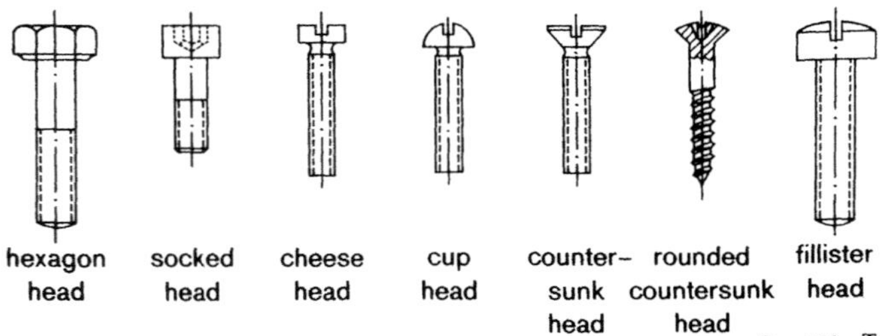

Fig. 201 Types of cap screws

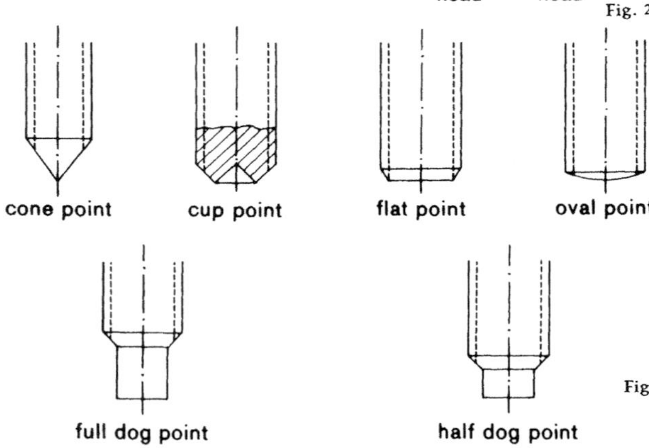

Fig. 202 Ends of set screws

Headless set screws, called "grub screws" are used one moving parts, where a projecting head would be dangerous. In Fig. 203 two kinds of grub screws are shown.

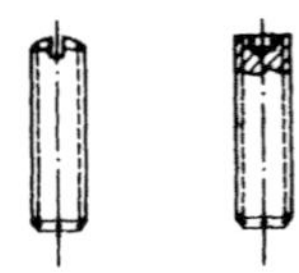

Fig. 203
Grub screws

8.1.7.3. Locking Devices for Nuts

There is a tendency for nuts to work loose, when the bolts to which they are attached carry varying axial loads or the bolts are subjected to vibrations. To prevent nuts from working loose various devices are used. There are two types of locking devices: positive devices and friction devices. In the first type, shown in Fig. 204, the working loose

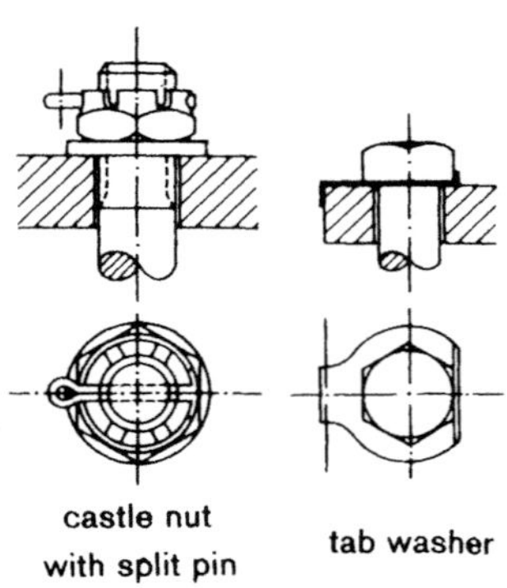

Fig. 204

of nuts is prevented by using split pins or tab washers. In the second type, shown in Fig. 205 lock nuts or elastic devices such as pring washers are used.

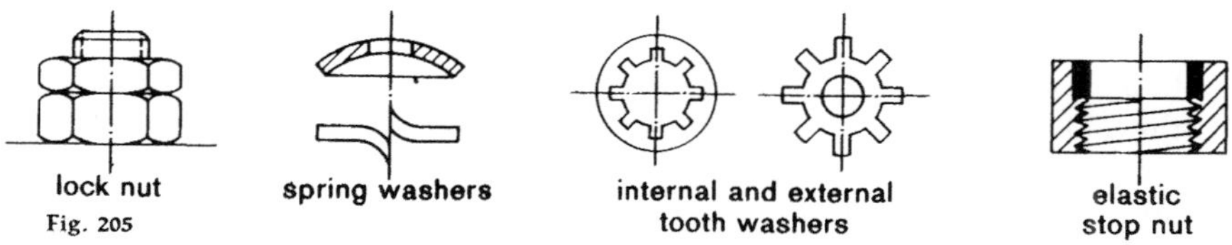

Fig. 205

Positive devices

Castle nuts are nuts with a cylindrical formation at the top which has recesses to accomodate a split pin. A tab washer is a thin washer which after the nut is screwed on, is bent up on one side and bent down on the other so that tabs are formed.

Friction devices

A lock nut which is usually half the height of an ordinary nut is used in conjunction with an ordinary nut. The lower nut, after having been tightened, is held by a spanner and the second nut is then screwed tightly on the lower nut. The pressure in the thread prevents the joint from working loose. There are many kinds of elastic washers. They have the effect of a spring which exerts, when compressed, pressure on the thread and prevents, like a lock nut, the joint from turning and working loose. The elastic stop nut has inside an elastic substance into which the thread of the bolt will be pressed when the nut is screwed on the bolt. The friction within the nut gives the locking effect.

8.1.7.4. Strength of Bolts

Screw diagram

If a bolt is tightened, an axial load called the "initial load" V will act on the bolt. The initial load causes a strain in the bolt λ_v and a compression in the plate δ_v, as shown in Fig. 206. If the prestressed bolt is loaded by an external force F, then an additional strain $\Delta\lambda$ is produced. At the same time the compression in the plate is reduced by the same amount. On the bolt there is a force $F_0 = F + V' = V + F_z$ and on the plate a force V' acting.

There are two characteristic values, one for the bolt and one for the plate called "bolt scale", $c_s = V/\lambda_v$ and "plate scale", $c_p = V/\delta_v$.

The following relation exists between the forces F_0, V and F:

$$F_0 = V + F \frac{c_s}{c_s + c_p}. \tag{146}$$

If the external load is changing between a maximum value F_{max} and zero, the additional force F_z in the bolt is changing between $F_{z\,max}$ and zero, as seen in Fig. 207. This force should be as small as possible. The additional force and consequently the additional stress in the bolt is smallest, if the value of c_s is very small, the value of c_p is very large and the pressure between the bolt and the plate is not cancelled out.

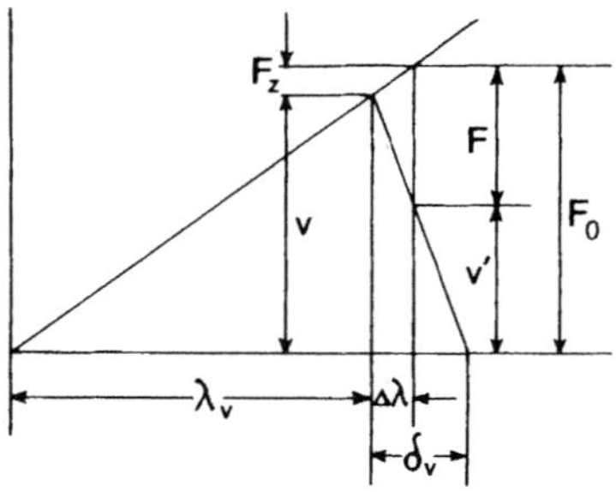

Fig. 206 Screw diagram for constant load

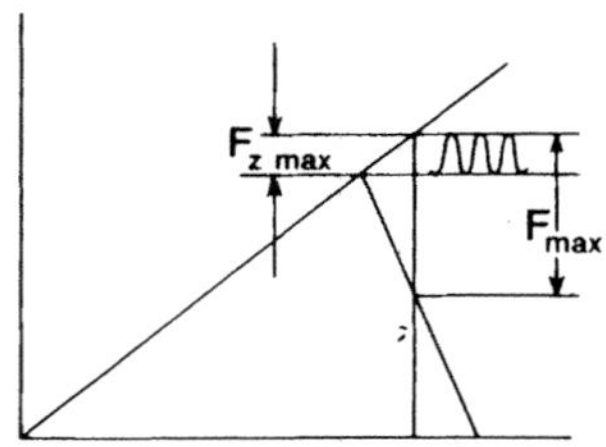

Fig. 207 Screw diagram for alternating load

The bolt scale may be calculated from following equation:

$$\frac{1}{c_s} = \frac{1}{E_s} \sum \frac{l_i}{A_i}. \tag{147}$$

The plate scale for a hollow circular cross-section is calculated from

$$c_p = \frac{E_p A_p}{l_p}. \tag{148}$$

where E_s and E_p = moduli of elasticity for the bolt and for the plate, respectively

l_i = lengthes of the parts of bolt having the same cross-section

l_p = thickness of the plate

A_i = cross-section of the parts of the bolt

A_p = cross-section of the plate.

Determination of the required thread of the bolt

The minimum cross-section of a bolt may be calculated from following formula

$$A_{min} = \frac{x\,F}{a\,\sigma_{rel}}$$

where $x = V/F = 2.75$ to 4.5
$a = 0.7$ for normal bolts and $a = 0.65$ for elastic bolts
$\sigma_{rel} = \sigma_S$ (yield stress) for a constant load
$\sigma_{rel} = \sigma_o$ (endurance limit) for an alternating load

$$A_{min} = A_{thr} = \frac{\pi}{4}\left(\frac{d_1 + d_2}{2}\right)^2 \text{ for normal bolts}$$

$A_{min} = A_{thr}$ or A_{shank} for elastic bolts.

After choosing the thread the maximum stress occuring in the bolt must be calculated. This stress is composed of the tensile stress, calculated from equation (67) and the torsional shearing stress, calculated from equation (82). The axial force depends on the kind of loading. It may be the initial load V, the external force F or the force F_0. The combined stress is found from equation (90 a). This stress must not exceed $0.8\,\sigma_{rel}$.

The friction moment in the thread which causes the torsional shearing stress in the bolt is found from the formula:

$$M_{thr} = V\,\frac{d_2}{2}\,\frac{p + d_2\,\pi\,\mu_{thr}}{d_2\,\pi - p\,\mu_A} \tag{149}$$

where V = *initial force*
d_2 = pitch diameter
p = pitch
μ_{thr} = coefficient of friction in the thread.

The moment for tightening the nut is composed of the friction moment in the thread M_{thr} and the moment M_A produced by the friction between the nut and the surface of the plate.

M_A is calculated from $M_A = \dfrac{d + s}{4}\,\mu_A$ (150)

where d = major diameter of the bolt
s = width across flats
μ_A = friction coefficient between nut and plate.

Pressure distribution in nuts

The pressure distribution in standard nuts is not uniform, as shown in Fig. 208. On the first thread i. e. the thread which is nearest to the load the pressure has its maximum. This thread must carry about 40 to 50 % of the load.

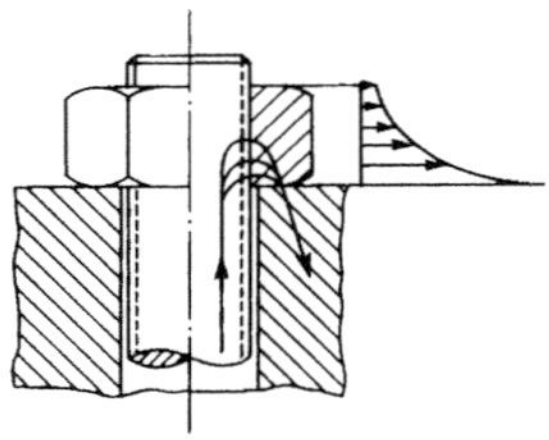

Fig. 208 Pressure distribution in a normal nut

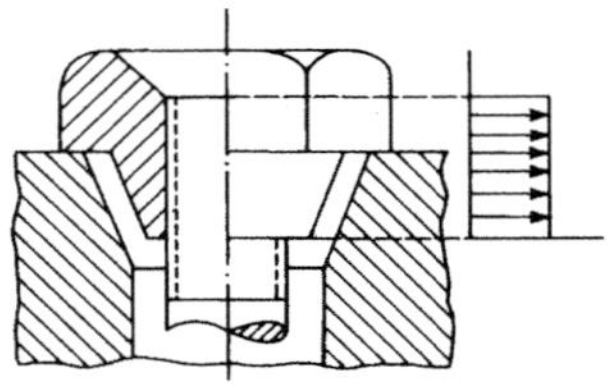

Fig. 209 Pressure distribution in a special nut

There are special nuts stressed in tension only, shown in Fig. 209, which may be loạded up to 30 % more than standard nuts. For these nuts the pressure distribution is nearly constant.

Efficiency of power screws

The efficieny of power screws occuring in working machines e. g. in screw presses may be calculated from the formula:

$$\eta = \frac{\tan \alpha}{\tan (\alpha + \varrho)} \tag{151}$$

where α = helix angle
and ϱ = angle of fricition.

8.2. Elements of a Drive

The elements of a drive are those elements which are used to convey or distribute power. Normally all these elements have rotary motion. For this reason these elements are also called "rotary elements". Such machine elements are the axles and shafts, the bearings, the couplings, the clutches, the belt and chain drive and the gear drive.

8.2.1. Axles and shafts

8.2.1.1. Axles

The axles carry rotating members. They could be fixed or be rotary, as seen in Fig. 210. Exemples of fixed axles are the axles of cars and funiculars, examples of rotary axles are the axles of railway carriages and trams. Axles may be of any cross-section, but the parts of axles which are supported in bearings must be of cylindrical shape.

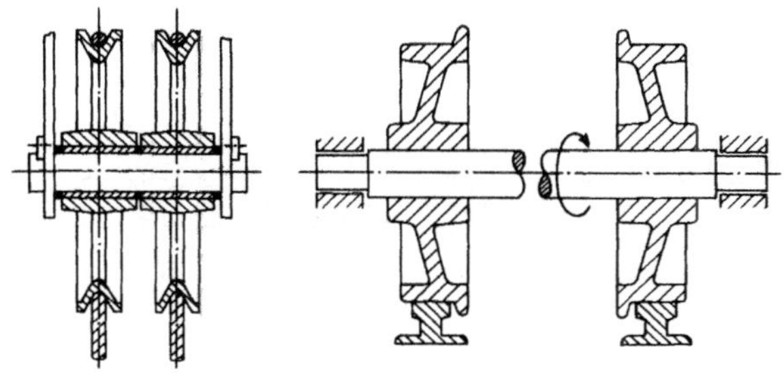

Fig. 210 fixed axle rotary axle

Calculations for axles

Axles are subjected to bending moments only. Axles may be made as axles of constant cross-section or as axles of uniform stress.

For axles of constant cross-section the required section modulus may be found from equation (74)

$$W_a = \frac{M_{b\,max}}{\sigma_{b\,all}}.$$

For axles of uniform stress the section modulus must be calculated for several places on the axle,

$$W_a = \frac{M_{bx}\,\beta_k}{\sigma_{b\,all}}.$$

The fatigue stress concentration factor β_k must be included, since the axles are normally made stepped.

8.2.1.2. Shafts

Shafts transmit power. On shafts are mounted pulleys, gear wheels, couplings, clutches, etc. Shafts are usually of solid cylindrical shape, but other shapes may be used. A shaft may also be of hollow construction. A thin shaft rotating with high speed is often referred to as a spindle. In addition to their being used for carrying gear wheels, pulleys, etc. shafts are used for other purposes such as for mounting the propeller on ship or aeroplane shaft, or for mounting the rollers on the shafts of grinding machines or mounting the cutters on the shafts of milling machines, etc.

To prevent the shaft moving longitudinally, loose or fixed collars or shoulders are employed. One bearing of the shaft must be constructed as a locating bearing, the other ones must be loose bearings, so that jamming may not occur (Fig. 211).

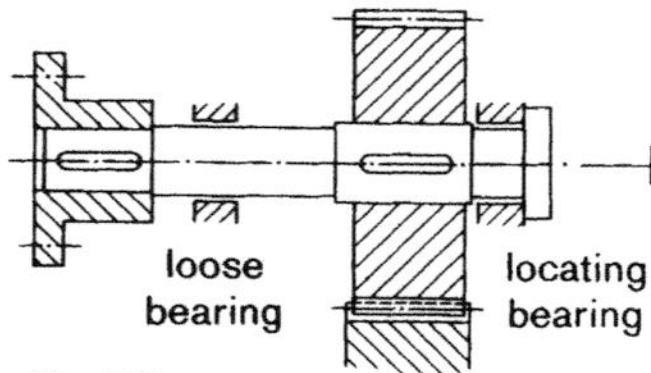

Fig. 211

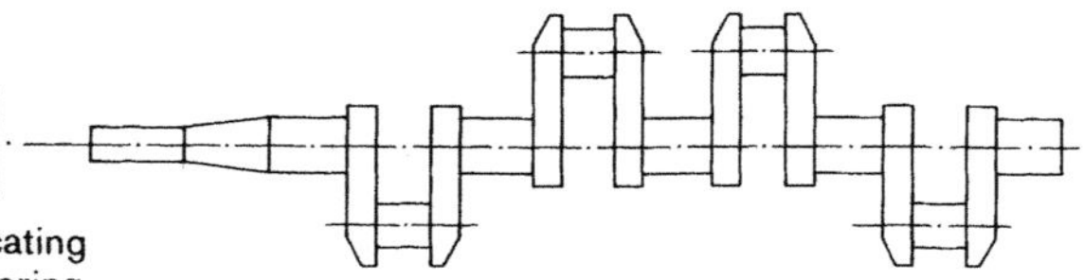

Fig. 212 Crank shaft

A crank shaft is a special form of shaft designed to convert reciprocating motion into rotary motion. It is used in reciprocating steam engines or internal combustion engines. Alternatively it may convert rotary motion into reciprocating motion, as used in piston pumps or compressors. The shafts have offsets called "cranks" positioned at certain distances from their centre line (Fig. 212). The distance from the shaft centre to the crank pin centre is called the "throw".

Another kind of special shaft is the camshaft. These shafts contain several cams which may be machined from solid steel or may be made by drop forging (Fig. 213). Camshafts are widely used for operating the valves of internal combustion engines.

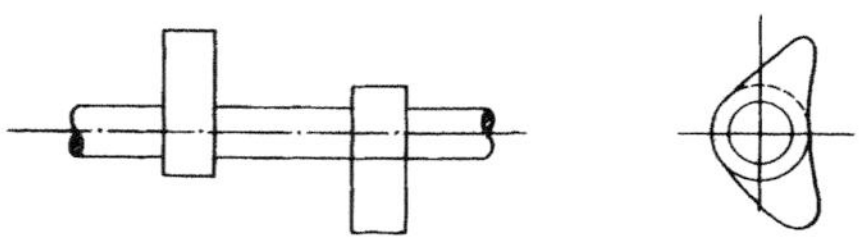

Fig. 213 Camshaft

Shafts are normally operated in a horizontal position, but for certain purposes shafts may be used in a vertical position.

Calculations on shafts

Shafts are stressed by bending and twisting moments.

The relation between power, speed and twisting moment is given by the equation

$$M_t = 71\,620 \frac{P}{n} \tag{152}$$

where M_t = twisting moment (cm kp)
P = power (PS) (1 KW = 1.36 PS)
n = speed (r. p. m.).

As for the construction of a shaft, normally only the transmitting power and the speed are known, so that the shaft cross-section at first may only be roughly estimated.

For solid circular shafts the following formula may be used:

$$d = c \frac{P}{n} \tag{153}$$

where d = shaft diameter (cm)
P = power (PS)
n = speed (r.p.m.).

c is a function of τ_{all} and is chosen according to the shaft diameter from a table.

After choosing the shaft diameter the combined stress must be calculated by means of the formula of (90 a) and it must be confirmed that the safety factor does not fall below the allowble value.

8.2.2. Bearings

Bearings are machine elements in which shafts or rotary axles are supported so that they may rotate freely. That part of the shaft or axle which is supported in the bearing is called the "journal". Bearings may be divided into two groups: the sliding bearing in which the furface of the sliding parts are separated by a small film of lubricant and the rolling contact bearing which has balls or rollers between the furfaces of contact. When a bearing takes a radial load, it is called a "journal bearing", and when it takes an axial load, it is called a "thrust bearing".

8.2.2.1. Sliding Bearings

Lubrication of sliding bearings

During rotation the loaded journal is supported by the oil pressure without any actual contact between the journal and the bearing. The oil pressure is built up in the clearance space, when the journal is rotating in the bearing. The factor most directly responsible for the load carrying ability of the oil film is the flow of the oil in a converging channel. The path may be either linear (at thrust bearings) or circular (at journal bearings). For thrust bearings the blocks in the bearing are sligthly inclined and the moving surface on the end of the shaft carries oil into the wedge shaped space. For journal bearings there is no oil film carried between the surfaces of the journal and the bearing, if the journal is at rest. If the journal is rotating, then oil is introduced in the converging channel between the journal and the bearing. The locus of the path of the centre of the journal is a semicircle. If the speed of the journal is infinitely great, then the centre of the journal coincides with the centre of the bearing (Fig 214). If a bearing must operate without sufficient supply of lubricant or too low speed to maintain the oil film thick enough to separate completely the journal and the bearing, then the bearing is starved and the operation is in the region of boundary lubrication. The maximum oil pressure is at that place where the thickness of the oil film is a minimum (Fig. 215). The smallest allowable thickness of oil film is dependent upon the surface finish of the journal and the bearing, the deflexion of the journal and the cleanliness of the lubricant.

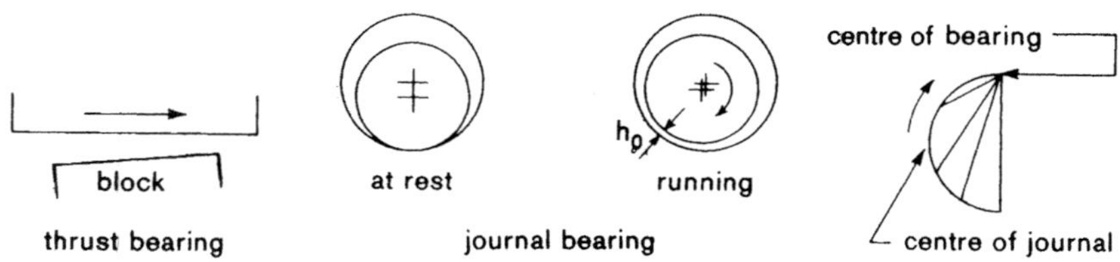

Fig. 214

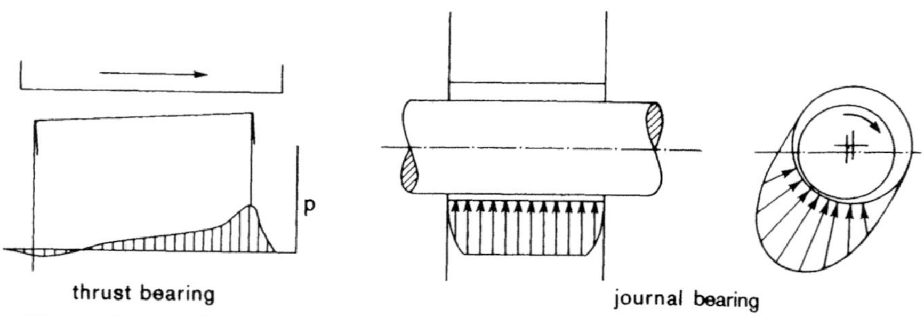

Fig. 215 Pressure distribution in sliding bearings

Properties of lubricating stuffes

For satisfactory operation of the bearings the lubricating stuffes, the oil or the grease must have certain properties. The most important property of a lubricating oil is its viscosity. The international unit of viscosity in 1 centipoise (cP) or 1 centistoke (cSt). The relation of the two units is $\eta = \varrho \nu$ where η is given in cP, ν in cSt and the density ϱ in kg/dm³. The average density of lubricating oil is 0.9 kg/dm³. It is measured by means of an areometer. Three other properties are important to describe the oil quality; the flash point which is that temperature at which the oil gases catch fir by foreign ignition, the point of self ignition, which is that temperature at which the oil gases begin to burn without foreign ignition and the setting point which is that point at which the oil begins to solidify. The latter one is only important for oils which are exposed to low temperatures.

Simple sliding bearings, ball and roller bearings and various parts of machinery may be lubricated by grease. Lubricating greases are made by dissolving natrium or lithium soaps in mineral oil. The most important property of a lubricating grease is its consistency which is represented by the value of penetration. For determining it, the point of a standardized cone is pressed in the grease to be checked and the depth of penetration in 1/10 mm is measured after 10 minutes. Another important property of lubricating grease is the dripping point, which is that temperature at which the grease begins to drip. For heavy machines and for rough work, if superfinish of the sliding planes is not possible, graphite grease is used. The graphite fills up the channels in the sliding surfaces of the bearing.

Lubricators for oiling and greasing

Lubricators for oiling

The methods of applying lubricant may be classified in four groups: gravity feed, ring lubrication, force feed and splash.

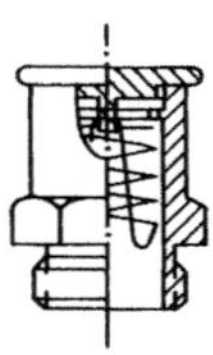

oil cup

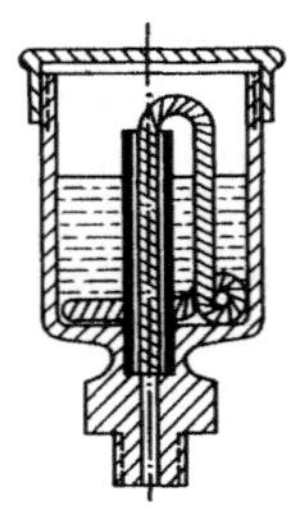

wick oiler

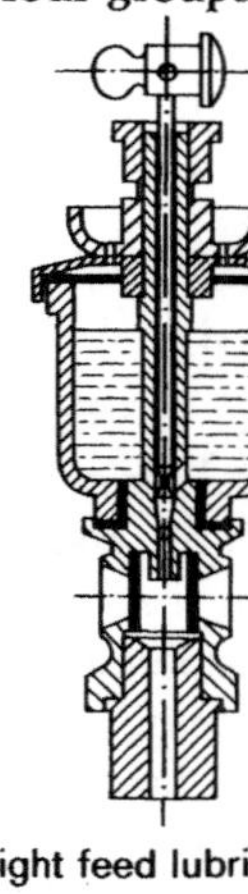

sight feed lubricator

Fig. 216 Lubricators for oiling

Gravity feed

In Fig. 216 three types of lubricators are represented, the oil cup, the wick oiler and the sight feed lubricator. The oil cup is the simplest lubricator. The oil cup is filled with oil by means of an oil can. For the second type, the wick in the small tube absorbs the oil which then drops through a hole into the bearing. This oiler cannot be turned off. The last one, the sight feed lubricator, consists of an oil container with a small valve, by which the oil quantity which is flowing to the bearing may be regulated and turned off. The oil flow may be observed by a sight glass.

Ring lubrication

There are two types of ring oil bearings, the bearing with one or two loose rings and the bearing with a ring fixed at the journal (Fig. 217). For the first type a loose ring put over the shaft dips into a reservoir of oil. If the shaft rotates, it causes the ring to turn also and the oil adhering to it is brought up and deposited in the bearing, from which it is led back into the reservoir. For the second type a ring fixed on the shaft dipping into the oil rotates with the shaft. The oil adhering to the ring is stripped off by an oil scraper and it flows into the bearing. In the ring oil bearing the oil must be maintained at a proper level and this is ensured by a spout in the bearing, so that no superfluous oil can be in the bearing, as the level is prescribed. Bearings with fixed oil rings are used for small shaft diameters and, if the speed of the shaft is low.

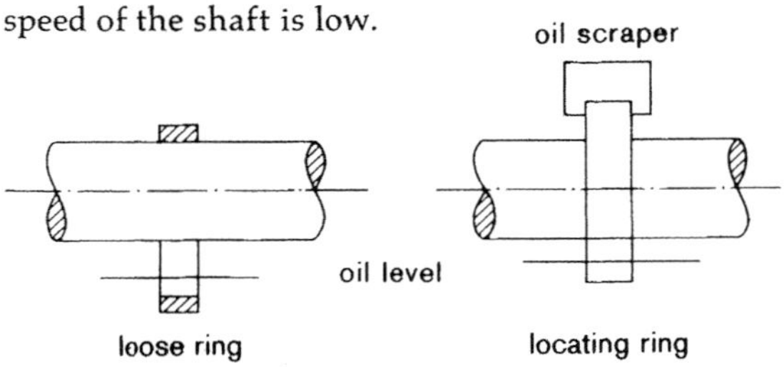

Fig. 217

Force feed

While for the gravity feed and the ring lubrication each bearing is lubricated for itself, at the force feed several bearings are lubricated at the same time. An oil pump driven by the machine whose bearings are to be lubricated delivers oil to the bearings. The oil is then drained from the bearings to a sump from which it is recirculated by a pump (Fig. 218).

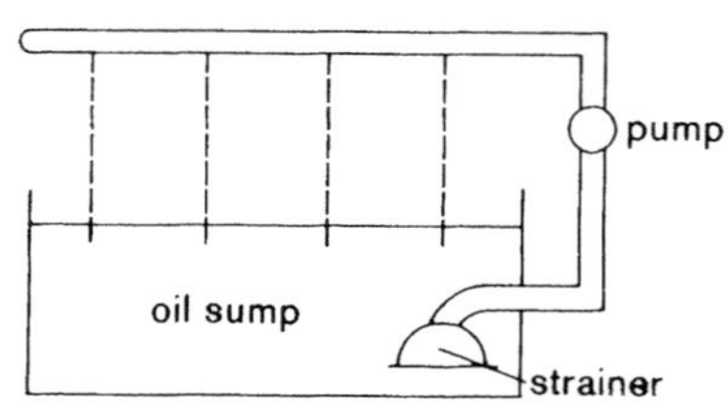

Fig. 218 Lubrication by circulation of the oil

Splash lubrication

In this method a wheel or other member attached to the shaft dips into the oil and a stream of lubricant is continually splashed round the parts requiring it. This method is employed for gears and for bearings inside the gear drives.

Lubricators for greasing

In Fig. 219 two types of grease lubricators are represented, the grease nipple and the Stauffer screw-down greaser. The first one, the grease nipple, is the simplest kind of grease lubricator. The grease is pressed into the nipple by means of a grease gun. The second one, the Stauffer screw-down greaser is filled with grease and the top is screwed down forcing the grease into the bearing. For larger bearings grease lubricators are used which are driven by the machine whose bearings are to be lubricated.

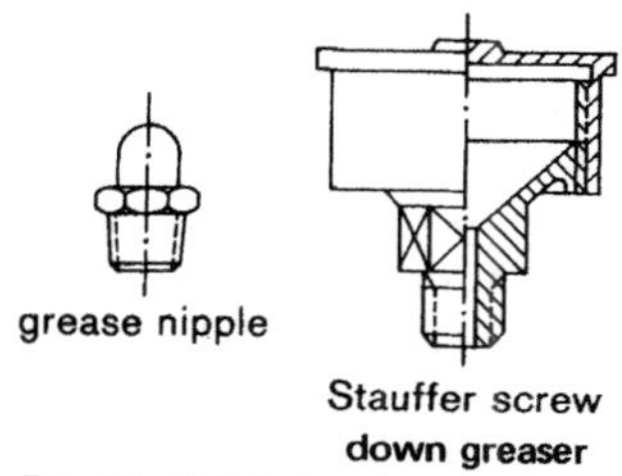

Fig. 219 Lubricators for greasing

Bearing materials

Bearing materials must have sufficient compressive strength to prevent extrusion or other permanent deformation of the bearing. Bushes and steps must have, moreover, a sufficient bending strength. The particles of dust, grit etc. which may not be prevented from entry into the bearing act, if the bearing material is hard, as an abrasive. Therefore the bearing material must have sufficient embeddability, i. e. is must be soft enough for the particles to become embedded without pressing against the journal. A high thermal conductivity is desirable to permit the rapid removal of the heat generated by friction. The coefficient of thermal expansion must be considered for bearing materials because of the effect of reducing the clearance in the bearing at high temperatures, if the operating temperature is relatively high or, if the bearing must operate over a wide range of temperatures. The lining material must adhere well on the steps and also the lubricant must adhere well on the lining material. Unfortunately no single material posses-ses all the attributes to a desirable degree. Many different alloys and methods of manufacturing have been used in the attempt to secure the best possible combination of properties for a given application.

For simple bearings of cast iron the cast iron itself is used as the bearing material. These bearings may be of one piece or of two pieces and are employed for low speeds and low pressures in the journals. Bearings consisting of one piece bushes may be pressed in. In bearings consisting of two parts steps are laid which consist either of the bearing material itself or of cast iron or steel lined with a bearing material.

The following materials are used for bushes and steps: Cast iron with perlite structure, red brass, brass, lead bronze, aluminium, graphited bronze, sintered metal and plastics. The sintered metal is made by pressing a mixture of powdered iron and copper (5 to 10 % copper, the remainder pure iron) or copper and tin (9 % tin, the remainder copper) and after that by sintering (annealing at high temperature and high pressure). Sintered metals are porous and may be impregnated with oil to the extend of 30 % of their volume. Sintered metals are used as bearing metals with lubrication with oil or for small bearings as oil-less bearings. Plastic materials are tough and resilient and have good friction and wear properties. The oil adheres well on the web. Therefore the surface film of these materials must be removed. Their coefficient of thermal expansion is very high, about ten times of steel, and thus the clearance must be enough to accomodate this expansion. The thermal conductivity of plastics is very bad, thus these bearings must be cooled well. The allowable bending stress for plastics is very small, so the plastic bushes or steps must lie completely along their surfaces. The bearings may be lubricated by oil or by grease. The temperature of plastic bushes or steps may not exceed 80° C.

For lining the bearings white metal (babbit metal) (10 or 5 % tin, and the remainder, antimony and copper) is normally used. The thickness of the lining varies from 0.1 to 3 mm for white metal according to the diameter of the step. Sometimes lead bronze is used for lining the bearings. The thickness of the lining varies from 0.2 to 3 mm for lead bronze according to the diameter of the bearing. Pouring in is for lead bronze more difficult than for white metal because of the higher melting temperature of this metal.

Mechanical design details

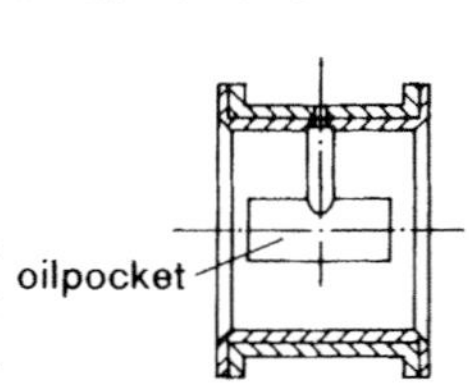

Fig. 220
Supply of lubricant of a sliding bearing

A major problem for bearings is that of ensuring an adequate supply of lubricant. If the load is acting always in one direction, the oil hole should be located on the unloaded side of the bearing. The loaded side of the bearing must have an even surface. The oil flows through grooves or holes to oil recesses called "oil pockets" which are in the middle of the bushing, from which it is sucked under the journal (Fig. 220).

Oscillating journals must run in bearings having bushings with several grooves from which the oil flows under the journal (Fig. 221). If the deflexion of the journal is more than can be tolerated by the bearing material, it is necessary to resort to a self-aligning bearing. These bearings are either made with spherical external surfaces, which are held in a spherical seat in the housing or as a bearing with tilting bushes or steps (Fig. 222).

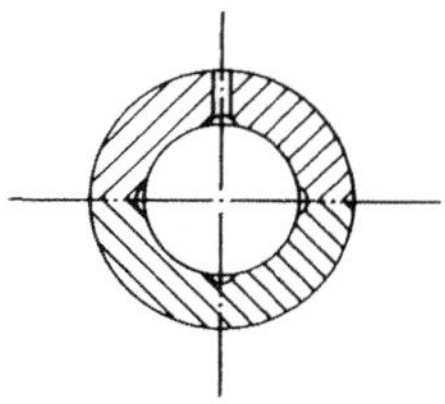

Fig. 221
Bushing for oscillating journal

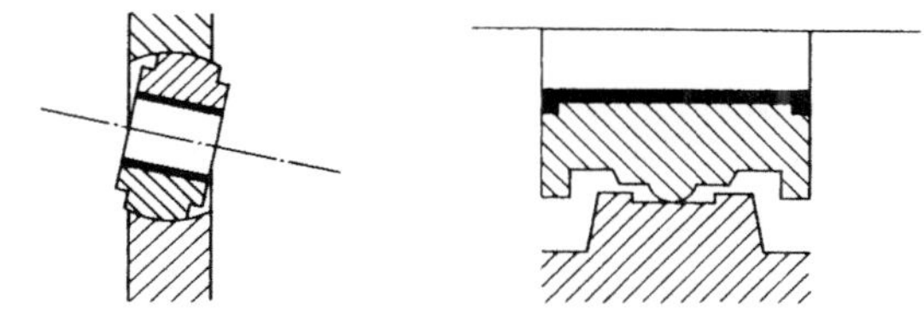

Fig. 222
Self-aligning sliding bearings

Types of bearings

Journal bearings

Simple bearings for shafts rotating at low speed are lubricated by grease. There are two types of grease-lubricated bearings, the solid bearing made in one piece shown in Fig. 223 and the bearing made of two parts shown in Fig. 224. All these bearings are made of cast iron. Solid bearings are made either without bushes and are called „plain bearings" or with bushes and are called „bushed bearings". In order to prevent the bush from rotating a set screw is screwed in or a pin is driven in tightly. There are two types of

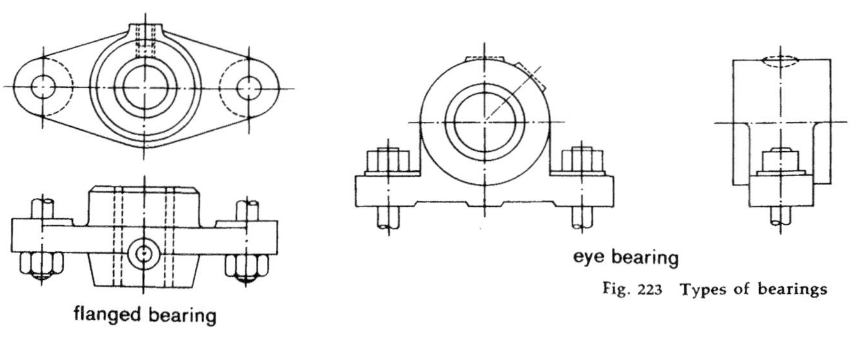

Fig. 223 Types of bearings

solid bearings, the "flange bearing" and the "pedestal bearing". A solid pedestal bearing is called an "eye bearing". A more satisfactory construction for a pedestal bearing is one which consists of two parts.

The main part of the housing is called the "pedestal body", and the upper part of the housing is called the "cap". The cap is held either by bolts or by studs. These bearings are made either without or with steps. The steps are formed in two halves. They are referred to as the "top step" and the "bottom step". They are fitted in the pedestal body.

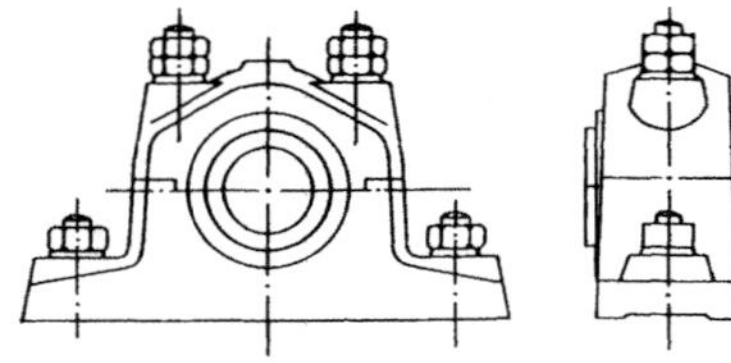

Fig. 224 Pedestal bearing

For higher speeds and higher pressures ring oiled bearings are made. There are ring oiled bearings with loose and with located rings as shown in Fig. 225. Both types of bearings may be constructed as loose and as located bearings (see Fig. 221). The loose and located rings are made of cast iron. All the parts of a ring oiled bearing are made of cast iron, the inner halves are lined inside with white metal.

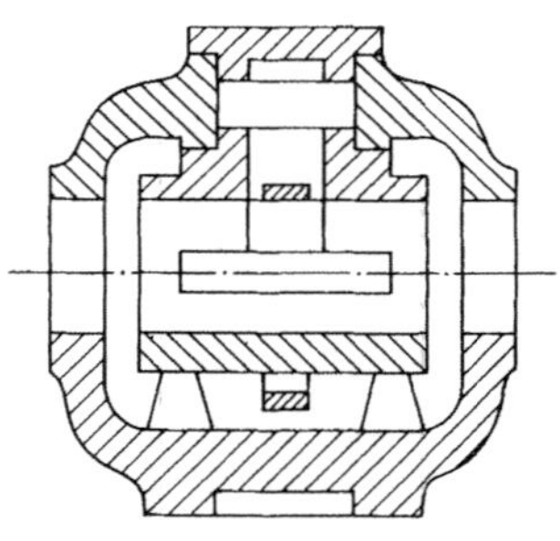

bearing with loose ring

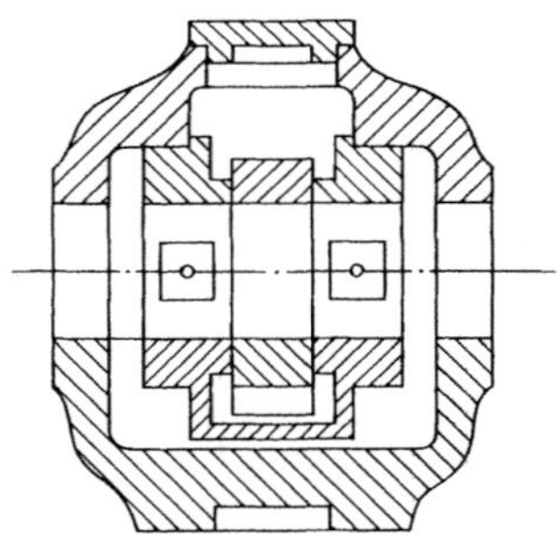

bearing with located ring

Fig. 225 Ring oiled bearings

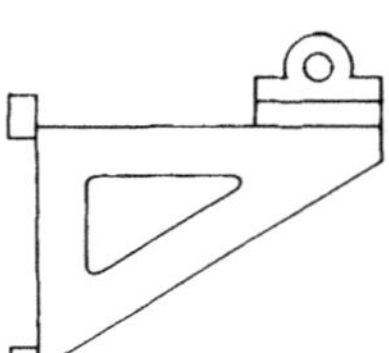

wall bracket

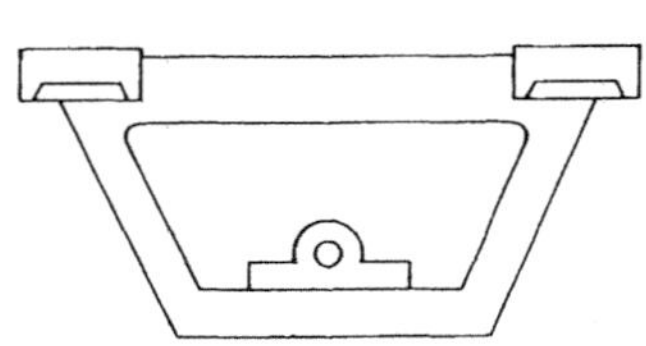

hanger bracket

Fig. 226 Brackets for bearings

A shaft may be supported from a wall by "wall brackets", or from the ceiling by "hanger brackets". The brackets are used for carrying pedestal bearings (Fig. 226). They are fixed with bolts on the walls or on the ceiling, respectively.

Thrust bearings

A simple type of thrust bearing is called a "footstep bearing" or a "pivot bearing", shown in Fig. 227. This bearing is only suitable for a slow-running slightly loaded shaft. The shaft is guided in a bush pressed into the pedestal body and is prevented from turning by a pin. The end of the shaft is supported by an adjustable "breast plate" made of steel or bronze. Lubrication is provided by oil which is carried under pressure into a hole inside the pivot and then between the surfaces of rubbing contact.

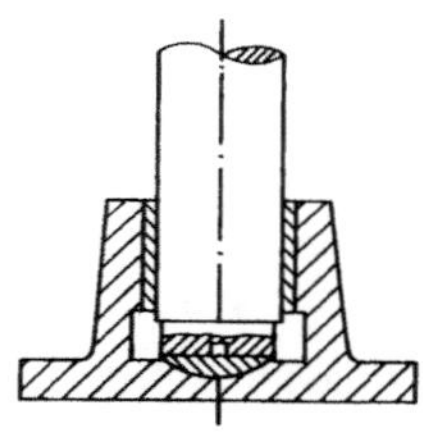

Fig. 227 Footstep bearings

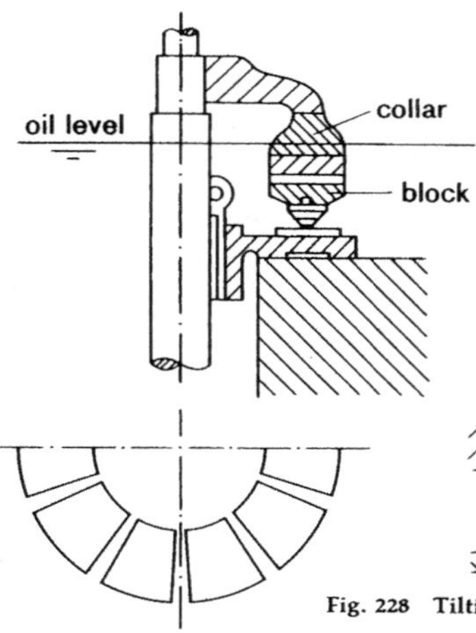

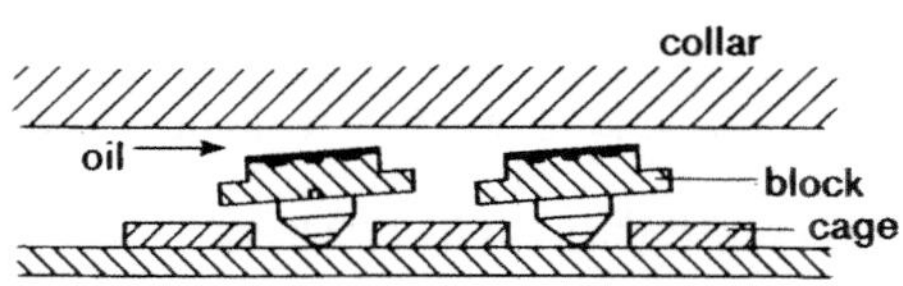

Fig. 228 Tilting pad thrust bearing

If reliable lubrication is necessary, the thrust bearings are built as tilting pad thrust bearings (Fig. 228). These bearings have all-round lubrication as described in "lubrication of sliding bearings" in section 8.2.2.1. Tilting segments produce converging oil channels under high pressure which carry a ring fixed on the shaft.

8.2.2.2. Rolling Contact Bearings

Ball and roller bearings substitute rolling friction for sliding friction. The coefficient of rolling friction is lower than that of sliding friction and is practically the same at starting as in motion. The sliding bearing has, when starting, a high coefficient of friction, because the bearing works in the region of boundary lubrication. The coefficient of friction is reduced, if the speed of the journal increases, because pure liquid friction takes place. Another important feature of the rolling contact bearing is that several types of ball and roller bearings combined can carry radial and axial loads.

Rolling contact bearings require less space than comparable sliding bearings. Another major advantage is that it is easier to provide lubrication for rolling contact bearings than for sliding bearings. This is particulary true for low speed bearings, where grease may be used. The grease is retained by simple and inexpensive seals and the initial supply of lubricant often lasts for the service life of the bearing. On the other hand the service life of a rolling contact bearing is shorter than that of a sliding bearing. Also, rolling contact bearings are noisy after long use and cannot be used for high loads.

The rolling contact bearings are classified according to the shape of the rolling members either as ball or roller bearings. Another classification for the rolling contact bearings is into those designed to carry radial loads and those to carry thrust loads.

A rolling contact bearing consists of a grooved inner and outer race and of a number of rolling elements (balls, rollers or needles). The races and the rolling elements are made of high carbon chrome steel. The radius of a ball is slightly less than the radius of the curvature of the grooves in the races. The rolling elements are retained in position by a cage which is either a one-piece pressed cage made of steel sheet or a one-piece solid cage made of steel, brass, aluminium or bakelite. Bakelite is used to reduce noise.

Types of rolling contact bearings

Radial bearings

1. Deep groove ball bearing

   The most common type of ball bearing is the deep groove ball bearing, shown in Fig. 229. It is made either with a single or with a double row of balls. The bore of both types is made cylindrical. During assembly the races are offset and the balls are placed between the races. After the races are centred and the balls symetrically located by use of a cage. This bearing can also carry axial loads.

2. Angular contact bearing

   This bearing is also made either with a single or with a double row of balls (Fig. 230). The bearing with a single row of balls can carry an axial load only in one direction. The bearing with a double row of balls can carry axial loads in both directions. The bore of both types is cylindrical.

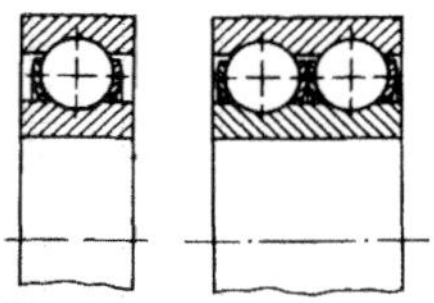
Fig. 229
Deep groove ball bearings

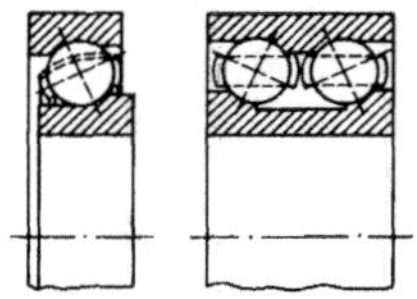
Fig. 230
Angular contract bearings

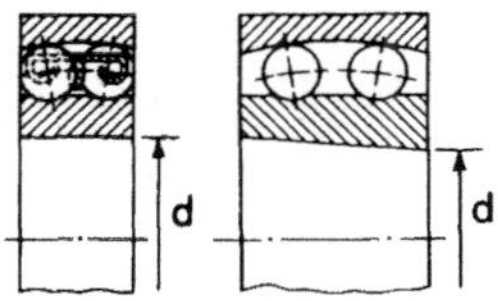

Fig. 231
Self-aligning ball bearings

3. Self-aligning ball bearing

The normal clearance in a ball bearing is too small to accomodate any appreciable angular misalignment of the shaft. For shafts having an angular misalignment, self-aligning ball bearings, shown in Fig. 231, are used. The inner surface of the outer race of these bearings is made spherical. Consequently the inner race may be displaced through a small angle. These bearings are always made as double row bearings with cylindrical or conical bores. They can also carry thrust loads in both directions.

4. Cylindrical roller bearing

When a greater load capacity is required in a given space, it becomes desirable to substitute the line contact in a roller bearing for the point contact in a ball bearing. The rollers may be cylindrical, spherical or conical.

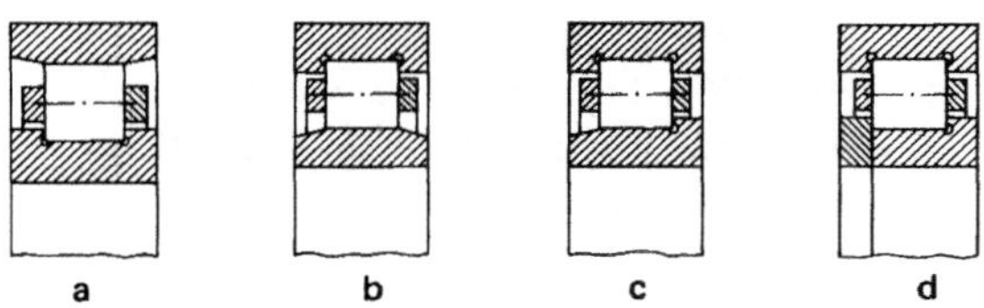

Fig. 232 Cylindrical roller bearings

There are four types of cylindrical roller bearings, shown in Fig. 232.
cylindrical roller bearing with flanges on the inner race (a)

cylindrical roller bearing with flanges on the outer race (b)

cylindrical roller bearing with flanges on the outer race and with one flange on the inner race (c) and

cylindrical roller bearing with flanges on the outer race and with one flange on the inner race together with an end washer (d).

The first two types cannot carry thrust loads. The third type can carry a thrust load only in one direction. The fourth type can carry thrust loads in both directions. Cylindrical roller bearings are all made with a single row of rollers and with a cylindrical bore.

5. Needle bearings

The rollers of needle bearings are relatively slender. Needle bearings are ideally suited for use when heavy loads with oscillatory motion occur e. g. for supporting piston pins in internal combustion engines. They are also often used for replacement of bushes. Firms which manufacture needle bearings produce three kinds: needle cages with needles but without races, needle bearings with the outer race only and neddle bearings with an outer and an inner race. Ordinary needle bearings are made with a single or a double row of needles. There are also self-aligning needle bearings. The bore of all needle bearings is cylindrical (Fig. 233). Needle bearings cannot carry thrust loads.

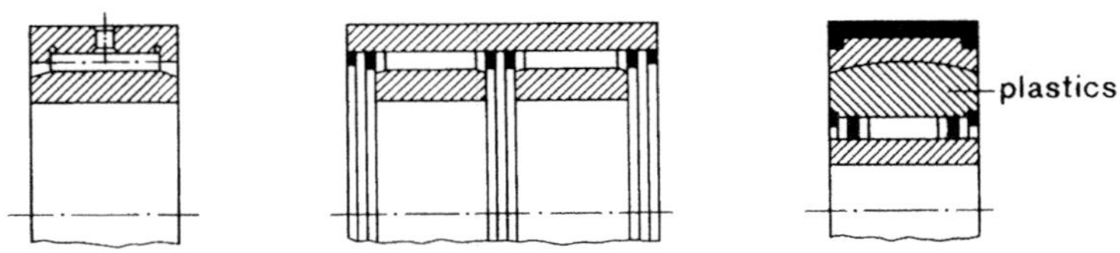

Fig. 233 Neddle bearings

6. Taper roller bearing

The rollers and the outer race of a taper roller bearing are frustums of cones. This bearing can carry both radial and thrust loads. They are always made with single row of rollers. Their bores are always made cylindrical (Fig. 234).

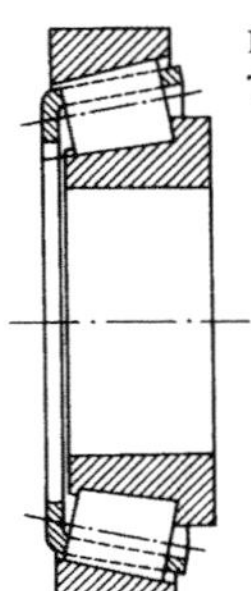

Fig. 234 Taper roller bearing

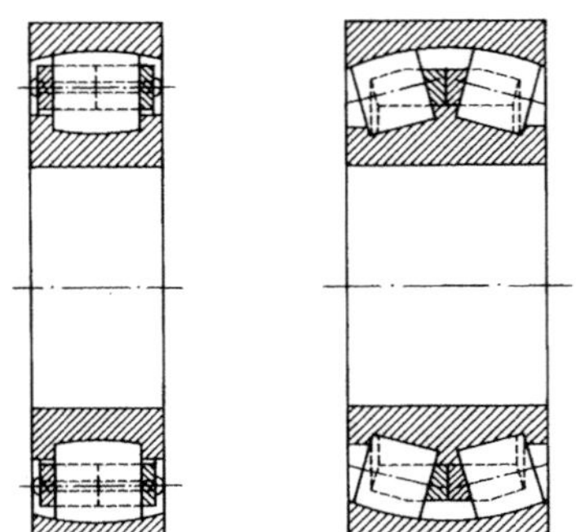

Fig. 235 Spherical roller bearings

7. Spherical roller bearing

The sherical roller bearing, shown in Fig. 235, is a self-aligning bearing like the self-aligning ball bearing. The inner surface of the outer race is spherical, so that the inner race may be displaced through a small angle. These bearings are always made with a double row of rollers and have cylindrical or conical bores. The bearings can carry thrust loads in either direction.

Thrust bearings

1. Ball thrust bearing

There are ball thrust bearings with a single row of balls which transmit a thrust load in one direction only (Fig. 236) and those with a double row of balls which transmit thrust loads in either direction. Ball thrust bearings cannot take radial loads. The centering of the shaft must be taken up by radial bearings. The thrust is taken by the shaft washer and transmitted by the balls on the housing washer and then on the housing. The housing washers have either plane or spherical surface. A spherical surface is used for compensating for inaccuracies in alignment of the housing.

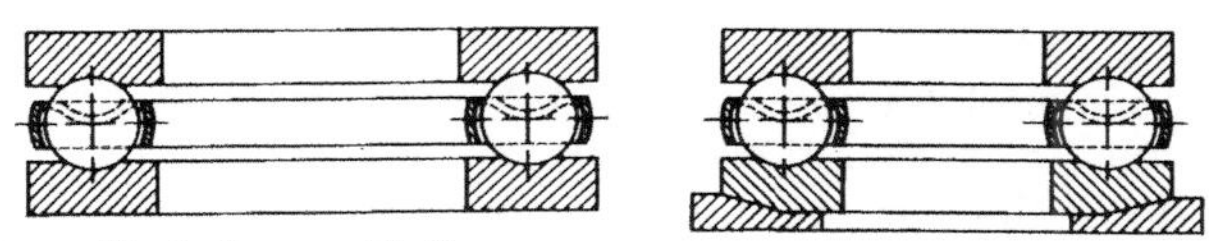

Fig. 236 Ball thrust bearings with single row of balls

2. Roller thrust bearing

The roller thrust bearing, shown in Fig. 237, is a self-aligning bearing for heavy loads. It can also carry radial loads.

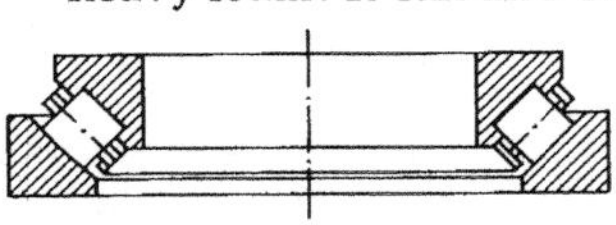

Fig. 237
Roller thrust bearing

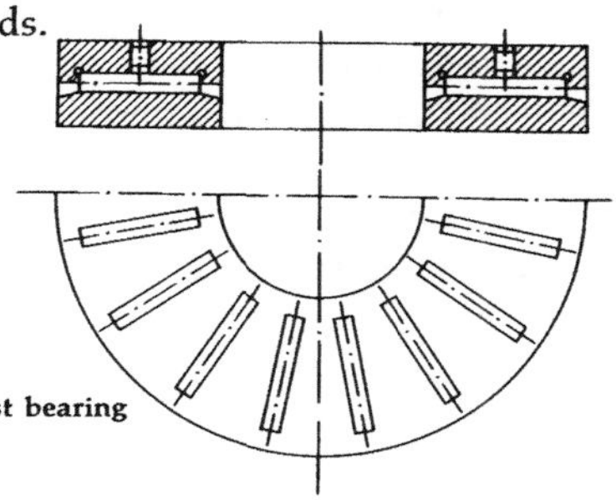

Fig. 238
Needle thrust bearing

3. Needle thrust bearing

The needles in a needle thrust bearing, shown in Fig. 238, have due their cylindrical form a rolling motion as well as a sliding motion, but as the needles are relatively short (about 1/6 of the cage diameter) the grinding of the needles has little effect.

Compound rolling contact bearing

When a bearing is constructed in such a way that a radial and a thrust bearing is combined, it is referred to as a combound bearing. Thus a needle radial bearing combined with a ball thrust bearing is produced (Fig. 239).

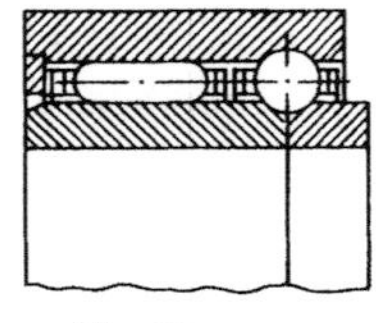

Fig. 239
Compound rolling contact bearing

Calculations for rolling contact bearings

The calculation of the load capacity of rolling contact bearings as for other direct contact mechanisms such as gear teeth and cams is based upon the Hertz equations. However, the rating must be also confirmed by experimental tests and the correlation of these test data with theoretical results gives formulae for practical use.

Bearings are often required to carry loads at rest and while rotating. The load at rest is spoken of as a static load, and the static load rating is limited by the requirement that the permanent deformation of the balls and the races must be less than that produced in rotating operation.

Ball and roller bearings have finite service life and the problem becomes one of determining the load corresponding to the desired service life i. e. to the number of revolutions which a bearing can perform. The major difficulty in rating the load capacity is that the failure point is different for single bearings, as seen in the curve, demonstrated in Fig. 240, which represents the results of many tests. According to this curve the average service life is attained by 50 % of the bearings. The longest service life rarely exceeds four times the average service life. For catalogue values of load that service life is taken which 90 % of the bearings attain. It is about 1/5 of the average service life. As relative service life is defined the ratio of the service life indicated in the catalogues i. e. the service life which 90 % of the bearings attain to the true service life of the bearings. For each relative service life the percentual number of bearings which attain this service life can be determined from the curve in Fig. 240, e. g. nine times the catalogue service life are attained by 30 % of the bearings.

In the catalogue, loads indicated correspond to a service life of 10^6 revolutions. For other loads the corresponding service life may be calculated by using the following empirical formula:

$$\frac{L}{L_0} = \left(\frac{C_{\mathrm{dyn}}}{F}\right)^p$$

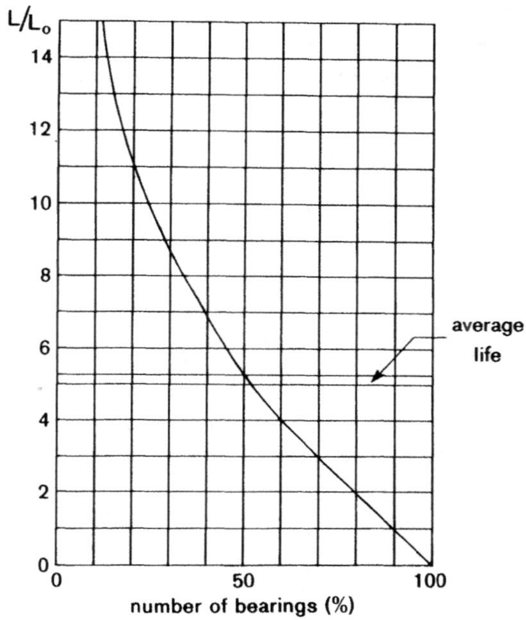

Fig. 240
Diagram of service life

where L = service life in revolutions corresponding to any load F
and $L_0 = 10^6$ revolutions corresponding to the catalogue load C_{dyn}.

If 10^6 is substituted for L_0 then is $L = 10^6 \left(\frac{C_{dyn}}{F}\right)^p$.

For calculations the service life is taken in hours.

Since $L = 60\, n\, L_h$,

$$L_h = \frac{L}{60\, n} \quad \text{or}$$

$$L_h = \frac{10^6}{60\, n} \left(\frac{C_{dyn}}{F}\right)^p \tag{154}$$

where L_h = service life in hours
n = speed (r.p.m.)
C_{dyn} = dynamic catalogue load
F = load acting on the bearing
p = 3 for ball bearings
p = 10/3 for roller bearings.

The load rating given in catalogues is based for radial bearings upon a pure radial load and for thrust bearings upon a pure thrust load. The load rating for radial bearings is based on the inner race rotating with the outer race stationary. If for radial bearings the outer race rotates and the inner race is stationary and if the load is not directly radial, then an equivalent radial load is calculated.

If the inner race of a radial bearing rotates, a circumferential load acts on the inner race and a point load on the outer race. On the other hand, if the outer race rotates, a point load acts on the inner race and a circumferential load on the outer race. A point load is more harmful than a circumferential load. As the specific pressure on the inner race is higher than that on the outer race, as demonstrated in Fig. 241, the service life gets smaller, if a point load is acting on the inner race i. e. if the outer race rotates.

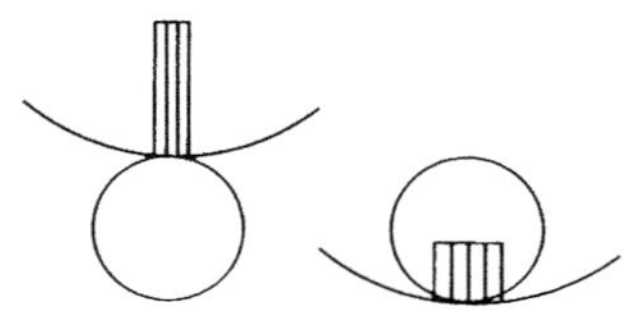

Fig. 241
Pressure distribution on races

The equivalent radial load for all radial bearings exept cylindrical roller bearings is calculated from:

$$F = X\,V\,F_r + Y\,F_a \tag{155}$$

where F_r = radial load
F_a = thrust load
X = radial factor
Y = axial factor
V = factor of rotation.

The values X, Y and V may be obtained from catalogues for rolling contact bearings.

Mounting arrangements for rolling contact bearings

Radial bearings

Radial bearings must be fixed on the shaft and in the housing so as to prevent radial, tangential and axial displacement. Fixing against radial and tangential displacement is produced by the fits of the races. That race on which a circumferential load is acting must have an interference or transition fit to prevent movement of the race. For that race on which a point load is acting, a transition or clearance fit is sufficient, because for this race movement does not occur. The fixing of the inner race on the shaft against axial displacement is produced at the end of the shaft by shoulders, nuts, snap rings, withdrawal sleeves or other arrangements and in the middle of the shaft by snap rings or adapter sleeves. Fixing of the outer races against axial displacement is produced by shoulders, snap rings etc (Fig. 242). For loose bearings either that race in which a point load is acting gets a clearance fit so as to be able to move or a cylindrical roller bearing with two flanges on the inner or outer race is used in which the inner or outer race, respectively, is able to move freely.

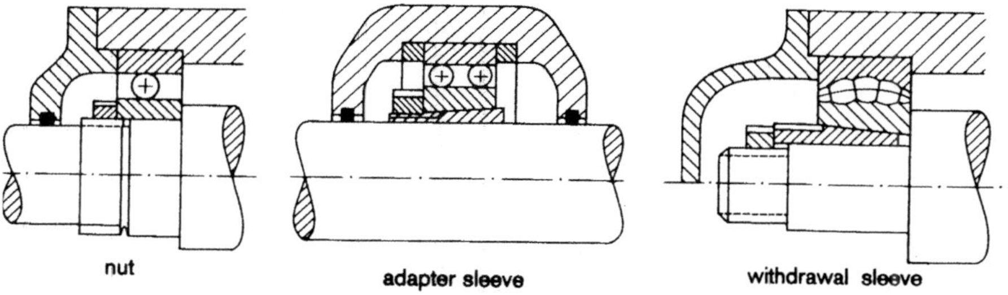

Fig. 242 Fixing of the races of radial bearings on the shaft

Thrust bearings

In Fig. 243 the assembly of a thrust bearing with a single row of balls for transmitting a load in one direction and the assembly of a thrust bearing with two rows of balls for transmitting loads in two directions are represented. Thrust bearings require loose radial bearings to accomodate axial displacements. The inside of the shaft washer and the outside of the housing washer must have an interference or a transition fit. The other sides of the washers must have a clearance fit to accomodate axial displacement.

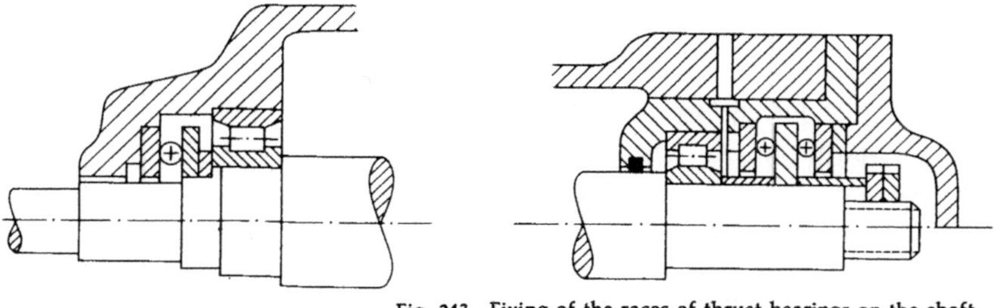

Fig. 243 Fixing of the races of thrust bearings on the shaft

Snap rings

To locate the races of radial bearings on the shaft or in the housing snap rings are often used. There are two types of snap rings; snap rings of uniform width which are only used for locating needle bearings that require snap rings with small width and snap rings of uniform stress which are wider than the other ones and which are used for locating the other kinds of radial bearings which have wider races than those of the needle bearings. The snap rings of uniform width have the same shape for shafts and for bores. The snap rings of uniform stress for shafts (external snap rings) are different from those for bores (internal snap rings) (Fig. 244).

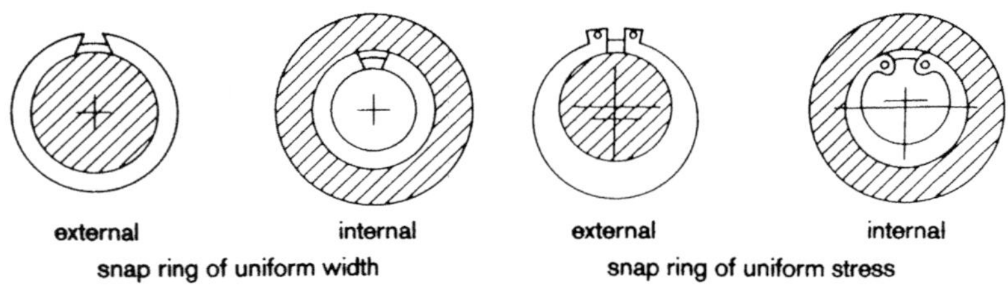

Fig. 244

Housings for rolling contact bearings

Rolling contact bearings are either mounted in machine housings or in separate housings which are fixed on machine frames or on foundation plates. There are solid flanged housings and solid and split pedestal housings, shown in Fig. 245.

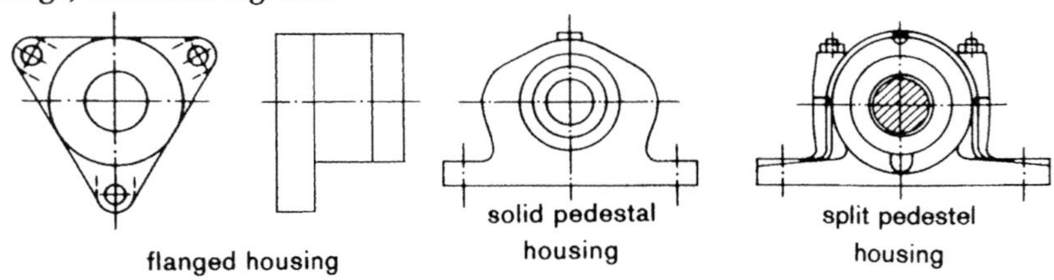

Fig. 245

Lubrication and sealing of rolling contact bearings

Theoretically for the rolling action of the balls or rollers the rolling contact bearings do not require lubrication. Lubrication is only used to reduce friction and wear between the cage and the balls or rollers, respectively, which slide in the cage. Also, lubrication is necessary to prevent corrosion in the bearing, to exclude penetration of dirt and to dissipate heat. The bearings may be lubricated with grease or with oil. The type of lubricant depends on the operating conditions, particularly on the working temperature. Grease lubrication is usually preferred as bearings using this form of lubrication require very little maintenance. Only special bearing greases should be used; cup grease is not suitable. The range of temperatures is from − 30 °C up to 50 °C for lime-based grease, up to 90 °C for sodium grease, up to 100 °C for sodium potassium grease and up to 120 °C for lithium-based grease. Grease has only a limited life in which it keeps its lubrication quality. In catalogues of lubricants there are curves representing the service life of grease as a function of the shaft diameter and of the shaft speed for the single bearing. If oil lubrication is used, special attention must be given to ensuring a steday supply of oil to the bearings and to sealing against leakage. The oil level should be at the centre of the lowest ball or roller. High speed bearings are often lubricated by a mist of oil and air. The oil used for rolling contact bearings should have a viscosity of 12 to 22 cSt at 50 °C during operation.

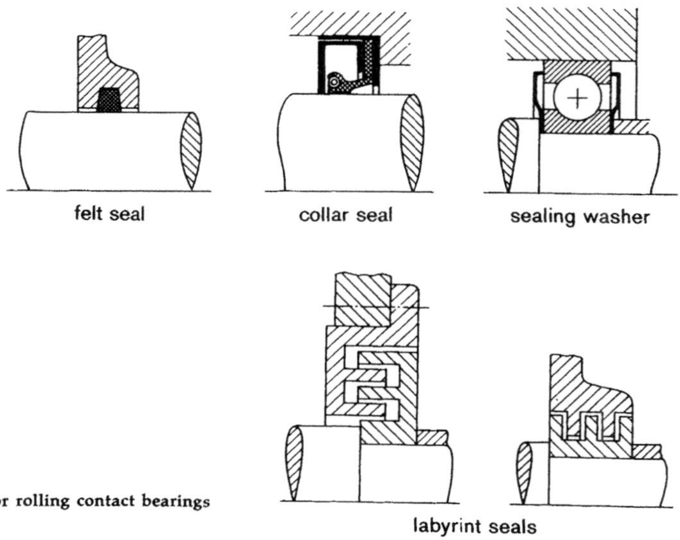

Fig. 246
Types of scals for rolling contact bearings seals

Rolling contact bearing must have seals to prevent entrance of contaminants and to prevent loss of lubricant. There are various types of seals for grease and oil lubrication, shown in Fig. 246. There are rubbing seals and non-rubbing seals. Rubbing seals are felt or collar seals, both suitable for grease or oil lubrication and sealing washers only suitable for grease lubrication. Non-rubbing seals are labyrinth seals suitable for grease or oil lubrication.

8.2.3. Couplings, Clutches and Brakes

8.2.3.1. Couplings

The coupling is used to connect shafts. It is used for shafts which have the same speed and the axes of which coincide. In order to connect shafts which have the same or different speeds and the axes of which do not coincide chain or gear drives are used. For shafts having different speeds and the axes of which coincide, special gear drives may be used (Fig. 247). The term "coupling" is taken for a device used to make a permanent connection. For devices which permit a rapid connection and disconnection the term "clutch" is used. There are two types of couplings, the rigid coupling which does not accomodate a misalignment of the shafts and the flexible coupling which accomodates misalignments of the shafts.

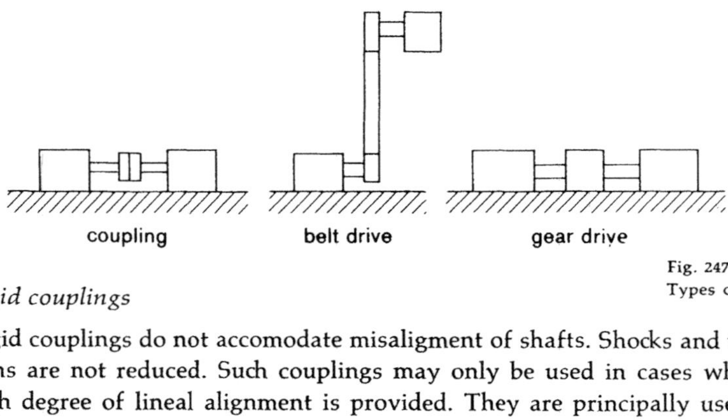

Fig. 247
Types of drive

Rigid couplings

Rigid couplings do not accomodate misaligment of shafts. Shocks and vibrations are not reduced. Such couplings may only be used in cases where a high degree of lineal alignment is provided. They are principally used for transmission shafts to connect the single parts of the shafts. Two types of rigid couplings are standardized: the split-muff coupling and the flanged coupling.

Split muff coupling

The split muff coupling, represented in Fig. 248, is made of cast iron and is designed for light and medium drives. The coupling is recessed for nuts and bolt heads. Around the nuts a clearance is provided for the spanner. On the recesses for the bolt heads, shoulders are left to prevent the bolts from turning, when the nuts are screwed on. The bolts on each side are inserted from opposite directions for convenience when tightening up. By tightening the bolts the two parts of the coupling are pressed against the shaft ends and a clamp connection is made. The transmission of torque occurs only by friction. For shafts over 50 mm diameter feather keys are used to guarantee a secure drive of the coupling. The coupling may be dismantled easily.

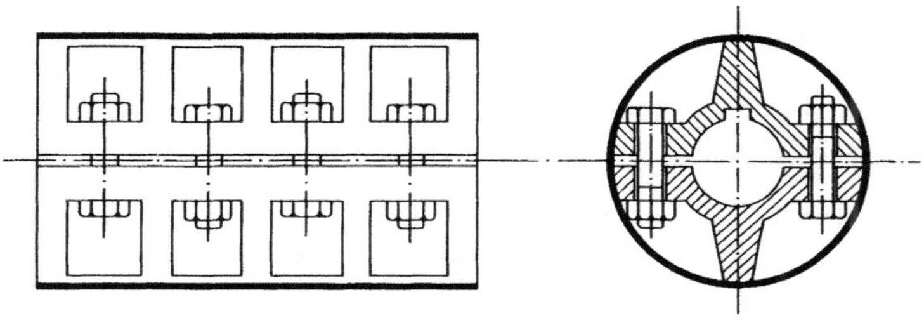

Fig. 248
Split muff coupling

Flanged coupling

The flanged coupling, represented in Fig. 249, is widely used for connecting shafts which transmit large torques. It is usually made of cast iron, but it is sometimes made of steel. The two halves of the coupling are either shrunk or forced by hydraulic pressure on the shaft ends and securely keyed. The recommended fit is U 7 for shrinkage and K 7 for cold pressing. The machined faces are held together by bolts. Centering is obtained by the provision of a turned taper projection on one flange which fits into a corresponding recess in the other flange. A disadvantage of this coupling is that the shaft must be displaced when being dismantled. For shafts which must be dismantled frequently a coupling having a split centering plate is designed. For this type a shaft may be dismanted after removing the split plate without displacing the shaft ends. The flanged coupling transmits the torque by friction in the contacting flange faces. Thus the bolts are only loaded in tension and not in shear.

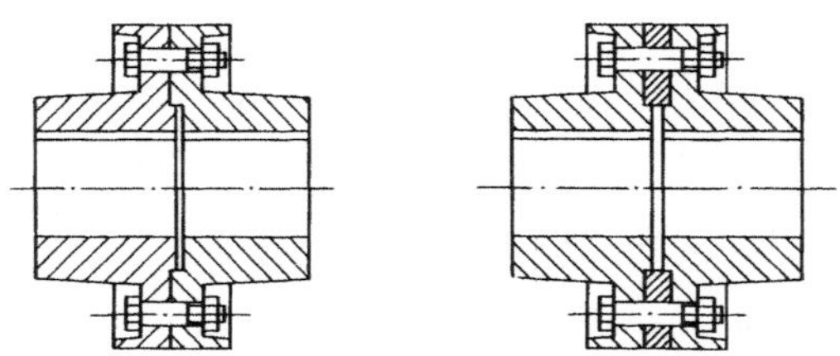

Fig. 249
Flanged couplings

Flexible couplings

Flexible couplings accomodate misalignments in the shafts. Axial, radial and angular misalignment may occur. Misalignment may be caused by errors in mounting the shafts, by the wear in the bearings, by thermal expansion in the shafts or by sinking of the foundation. Also, there may be a desirable angular misalignment, if shafts are operating at an angle to each other. There are two types of flexible couplings. The first type, the flexible non-elastic coupling, consists only of rigid parts. It may accomodate according to its construction either axial, radial or angular misalignment together or only angular misalignment. The second type, the elastic flexible coupling, has elastic members mounted between the coupling halves. It may accomodate small axial, radial or angular misalignment together.

Flexible non-elastic couplings

Floating centre coupling (Oldham coupling)

The floating centre coupling, also called the "Oldham coupling" is represented in Fig. 250. It consists of three pieces, two flanged half coupling units and a centre piece. The flanged parts are made of cast iron or steel, the centre piece is made of steel. The flanged parts are grooved in order to accomodate the lugs which protrude from the centre piece. The lugs are set at right angles to each other. The lugs slide in the grooves and must be lubricated. The coupling accomodates considerable axial misalignment, about 5% of the shaft diameter radial misalignment and up to 3° angular misalignment. The coupling is suitable for transmitting heavy loads for slow shaft speeds.

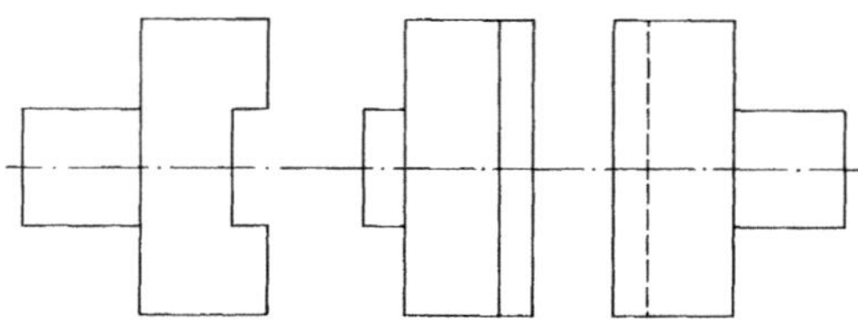

Fig. 250
Oldham coupling

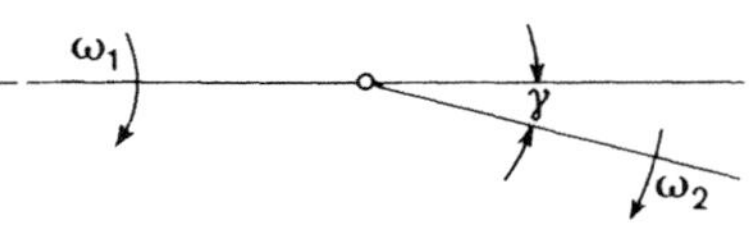

Fig. 251

Cardan joint

The Cardan joint is used to connect shafts which are set at an angle (γ) to each other. This angle may be varied during rotation of the shaft, but it should be less than 30° (Fig. 251).

The angular velocity of the first shaft (ω_1) is assumed to be constant. The angular velocity of the second shaft (ω_2) varies between a maximum and a minimum value.

$$\omega_{2\,\max} = \frac{\omega_1}{\cos\gamma} \quad \omega_{2\,\min} = \omega_1 \cos\gamma \,.$$

The coefficient of cyclic variation may be calculated from:

$$\delta = \frac{\omega_{max} - \omega_{min}}{\omega_1}.$$

After substituting the values for $\omega_{2\,max}$ and $\omega_{2\,min}$ this becomes

$$\delta = \frac{1 - \cos^2 \gamma}{\cos \gamma}. \tag{156}$$

The Cardan joint consists of two forks and an intermediate cross piece having journals which are embedded in needle bearings inserted in holes of the forks (Fig. 252).

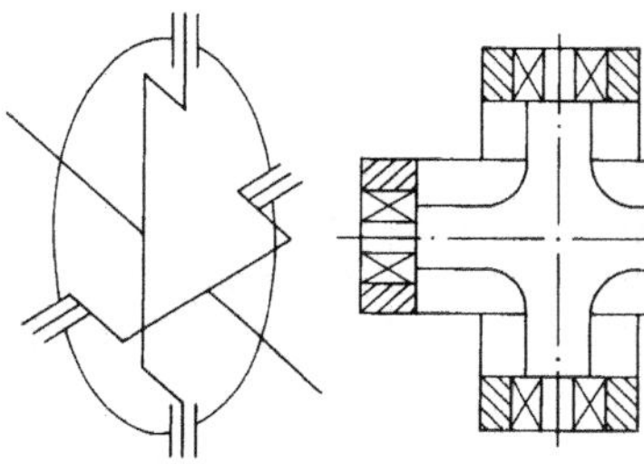

Fig. 252 Cardan joint with centre cross and needle bearings

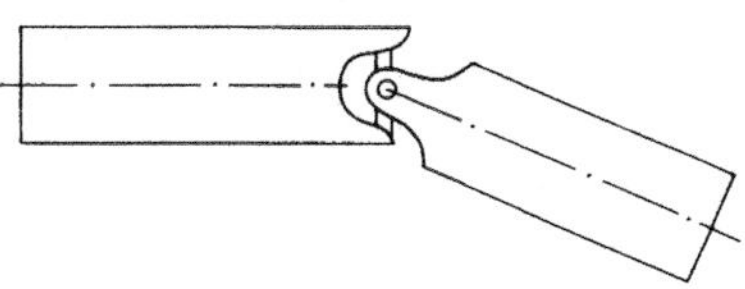

Fig. 253 Cardan joint with intermediate block

For small shaft diameters a simpler construction is used. Instead of the cross piece, a block in the form of a cube is inserted between the forks and held by pins put in the holes of the forks (Fig. 253).

The variation in the velocity ratio may be eliminated by using two Cardan joints and an intermediate connecting shaft. The two shafts must be equally inclined to the connecting shaft and the forks on the connecting shaft must lie in the same plane. This arrangement is known as a "double Cardan joint" (Fig. 254).

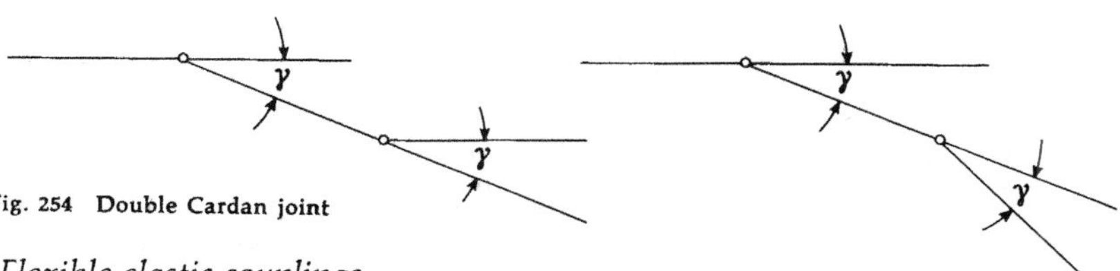

Fig. 254 Double Cardan joint

Flexible elastic couplings

These flexible couplings accomodate small axial, radial and angular misalignment, compensate for variations in the twisting moment and damp torsional oscillations in the shaft. Variations in the twisting moment are diminished partly by absorption and partly by the elasticity of the elastic members in the coupling. In Fig. 255 two diagrams, (a) the $M_t - \varphi$ and (b) the $M_t - t$ diagram are drawn. The included area in the $M_t - \varphi$ diagram represents the

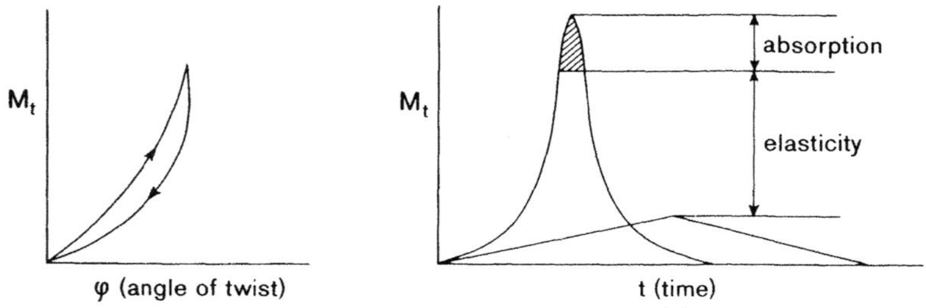

Fig. 255

absorbed energy which is transformed into heat. In the $M_t - t$ diagram the diminution of a shock by the absorption and elasticity of the elastic material is shown. According to a theorem of mechanics (section 2.2.2.4.) the angular impulse $\int M_t \, dt$ must be constant, if no external forces are acting. The critical speed of a shaft is dependent on the ratio M_t/φ. If this ratio is constant, then the critical speed has a finite value, but, if this ratio is a function of the angle of twist, as shown in Fig. 255 (a), then the critical speed has a certain range and the value of the maximum oscillation amplitude becomes smaller, as shown in Fig. 256.

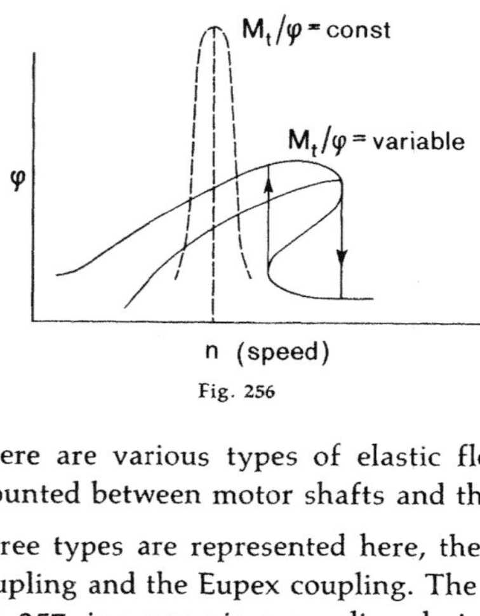

Fig. 256

There are various types of elastic flexible couplings. They are generally mounted between motor shafts and the shafts of driven working machines.

Three types are represented here, the neoprene flexible coupling, the Elco coupling and the Eupex coupling. The neoprene flexible coupling, shown in Fig. 257, is a one piece coupling designed only for small powers. The Elco coupling, shown in Fig. 258, has for elastic members grooved rubber bushings fitted in steel pins spaced in circular formation on one half of the coupling. It is made in different sizes and may transmit small and large twisting moments according to its size. For the Eupex coupling, shown in

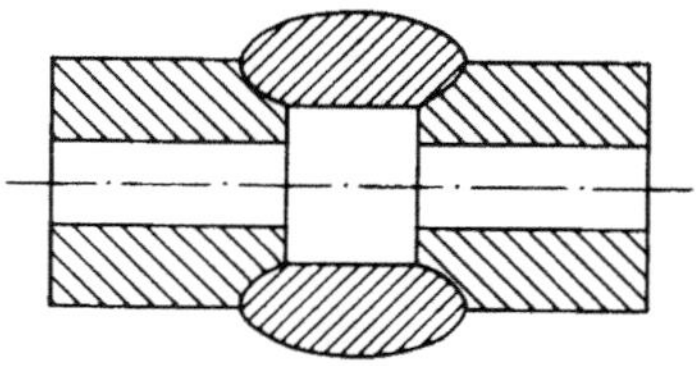

Fig. 257 Neoprene flexible coupling

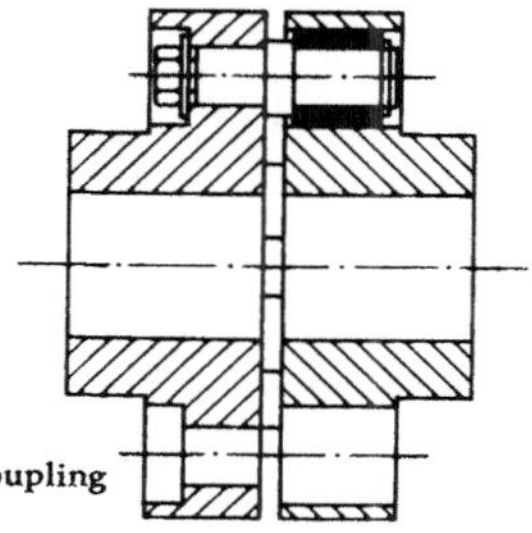

Fig. 258 Elco-coupling

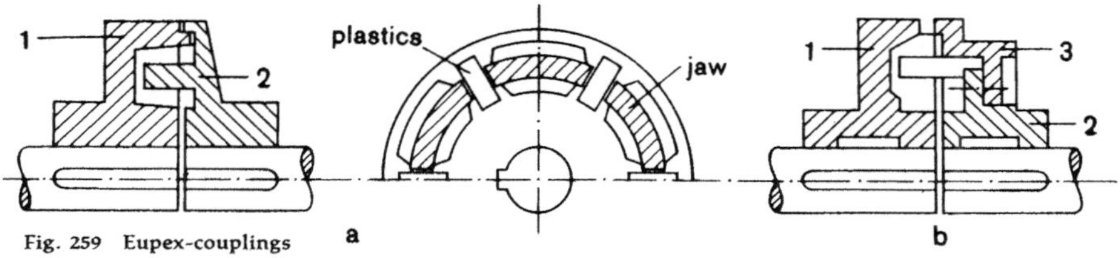

Fig. 259 Eupex-couplings

Fig. 259, one half has on the face protruding jaws which are placed between elastic members of rectangular formation inserted in recesses of the other half. As well as the normal construction with two parts (a), there is another construction with three parts (b). This type is constructed in such a way that a shaft for dismantling need not be displaced. The two types are made in different sizes for transmitting small and large twisting moments.

The rating of the size of an elastic flexible coupling is based on the maximum twisting moment which is calculated from

$$M_t = c\,71\,620\,\frac{P}{n} \quad \text{where} \quad c = \text{service factor.}$$

The service factor c is dependent upon the driving engine and upon the driven machine and may be obtained from tables which are in the catalogues of the firms manufacturing these elastic flexible couplings.

8.2.3.2. Clutches

Clutches are devices which permit connection or disconnection of two shafts. There are two types of clutches: the positive and the friction clutch. For the first type the transmission of power is effected by jaws or teeth. Such clutches may be engaged or disengaged only if the shafts are at rest. For the second type the transmission of power is effected by friction. These clutches may be engaged or disengaged, if the shafts are at rest or are rotating.

Positive clutches

Two types of positive clutches are described here, the jaw clutch and the tooth clutch.

Jaw clutch

The jaw clutch, shown in Fig. 260, consists of two halves, one is pressed and keyed on the driving shaft, the other one slides along a feather key on the driven shaft. Both halves have projecting jaws on the face which fit into spaces between the jaws of the other half. In order to engage or disengage the clutch, the sliding part of the clutch is moved by a forked lever which fits into two pins fixed on a ring embedded in the sliding part. For centering, the driving shaft enters the driven part of the clutch and remains in it, when the clutch is disengaded.

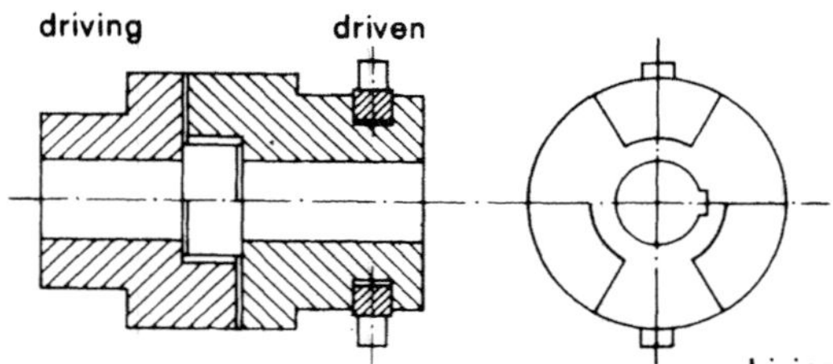

Fig. 260 Jaw clutch

Tooth clutch

The tooth clutch, shown in Fig. 261, consists of three parts. Two parts having an external gear are fixed on the shaft ends. The third part, a sleeve having an internal gear, accomodates the external gears. The clutch is engaged, if the sleeve is put over the two external gears of the two fixed parts.

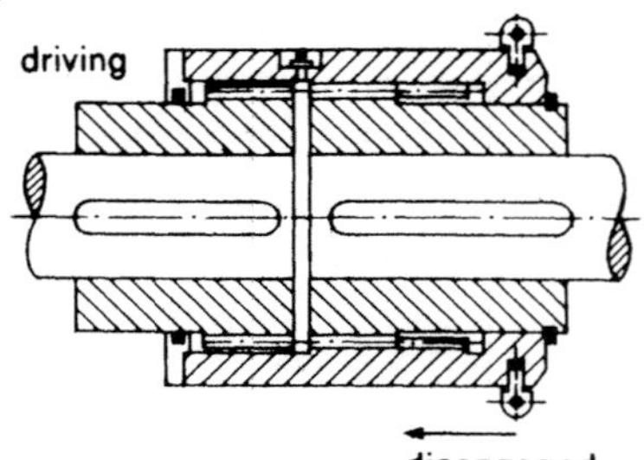

Fig. 261 Tooth clutch

After disengaging the gear of the sleeve remains in the gear of the driven part.

Friction clutches

Friction clutches have two or more rotating concentric surfaces which are forced together so that the tangential friction force transmits torque from the input to the output shaft. A principal advantage of the friction clutch is that the engagement may be smoothly made by a gradual increase of the frictional force. There are clutches which have dry rubbing surfaces and clutches the rubbing surfaces of which are lubricated. In the first type the rubbing surface of one member is coated with fabric, asbestos, etc. The rubbing surface of the other member is made of cast iron or steel. In the second type plain flat steel discs are used which require lubrication. The rubbing surfaces may be plane or conical. The taper of the conical surfaces should be not too small, since otherwise clamping or engaging by jerks occur. The pressure force is always applied from two sides to prevent a free axial force. All friction clutches must be designed so that the distance of the rubbing surfaces may be adjusted. Springs arranged between the rubbing plates enable the plates to separate when the clutch is disengaged.

The rating of the size of a friction clutch is obtained in the same way as for elastic flexible couplings. Also the same service factor c, valid for the elastic flexible couplings, is taken.

Types of friction clutches

There are many types of friction clutches. A few are discussed here.

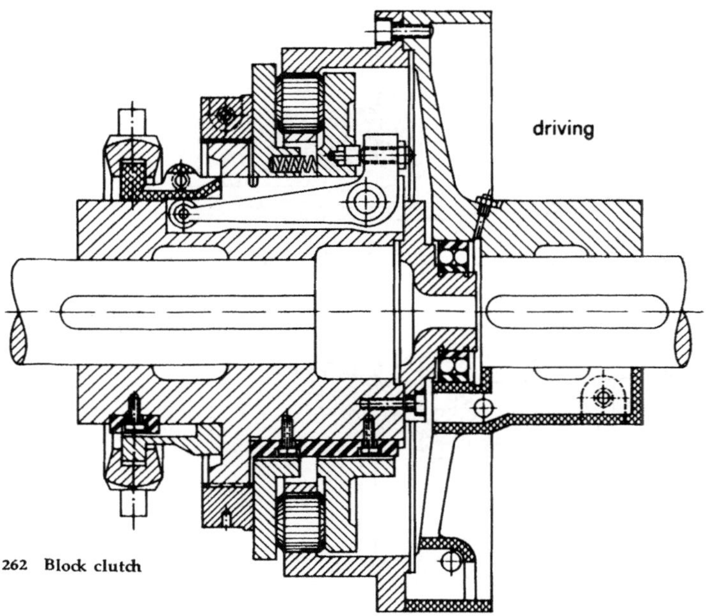

Fig. 262 Block clutch

Block clutch

The block clutch, shown in Fig. 262, consists of two cylindrical plates loosely attached on the driven part of the clutch and a cylindrical plate fixed on the driving shaft in which rubbing blocks are inserted. When the sleeve is moved against the clutch, the wedges in the sleeve force the three levers down and the plates are pressed against the blocks. The distance of the loose plates may be adjusted by turning a threaded adjusting ring.

Multiple disc clutch

A shaft may be driven by another shaft which lies on the same axis as the driven shaft by means of a clutch the two parts of which are attached on the ends of the shafts. But it is also possible to drive a shaft from a rotating wheel which is mounted loosely on the shaft. In this case one part of the clutch is keyed to the shaft and the other one fixed on the rotating wheel. Many different kinds of clutches may be constructed like this.

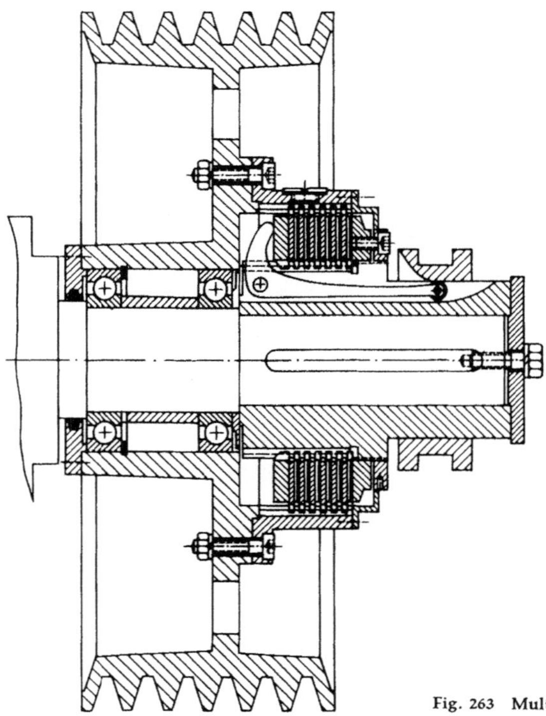

Fig. 263 Multiple disc clutch

The multiple disc clutch, shown in Fig. 263, has several steel discs, some of which have tongues fitting in recesses in the driving cup and the others have tongues fitting in recesses in the driven body of the clutch. When the sleeve is moved against the clutch, the steel discs are pressed together. The required clearance between the plates may be adjusted by rotating a threaded adjusting ring as in in the block clutch. Rotation of this ring on the clutch during operation is prevented by a pin which enters one of the numerous holes in the end plate. The discs are slightly bent to enable separating of the discs when the clutch is disengaged. The discs must be lubricated by oil to prevent gripping.

Toggle lever clutch

The toggle lever clutch, shown in Fig. 264, has two internal conically-shaped rubbing surfaces or one conical and one plane surface against which external rubbing surfaces are pressed. The pressing operation is made by three toggle levers. Working loose of the clutch is prevented by levers fixed on the sleeve.

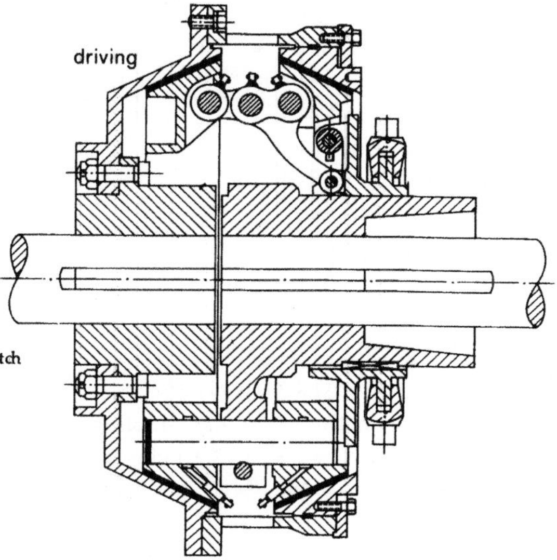

Fig. 264 Toggle lever clutch

Electromagnetic clutch

There are two types of electromagnetic clutch, represented in Fig. 265. The first one is constructed as a dry clutch having dry rubbing surfaces and the second one is constructed as a multiple disc clutch with lubricated discs. The force required for pressing the surfaces together is produced for the electromagnetic clutch by an electromagnet which works at 24 volts. The electromagnetic clutch has the following advantages when compared with other types of clutches: Engagement and disengagement occurs completely smoothly even, if the shaft is loaded. No levers are recessary. The clutch may be switched remotely. A disadvantage of this clutch is the permanent consumption of energy. For clutches operated mechanically by levers, the sleeve and the levers are always mounted on the driven part, whereas for the electromagnetic clutch the electromagnet and the collector rings are always mounted on the driving part.

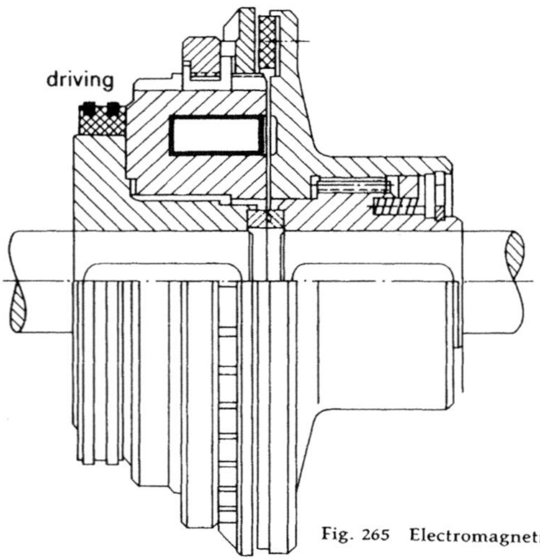

Fig. 265 Electromagnetic clutch

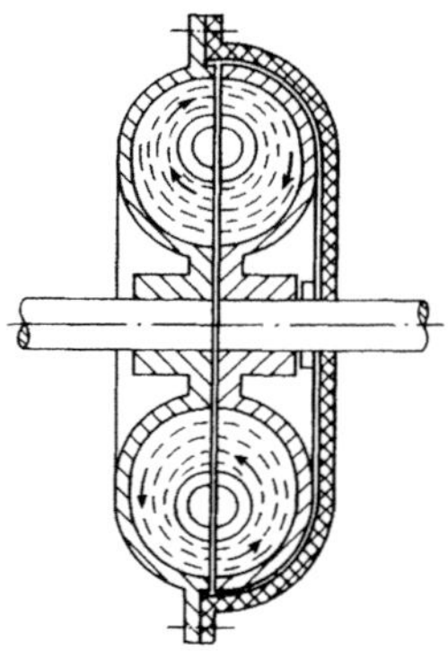
Fig. 266
Hydraulic coupling

Special clutches

Hydraulic or fluid coupling

Since the fluid in the hydraulic coupling acts like an elastic member, this coupling may be considered as an elastic flexible coupling. But since it may be engaged or disengaged at rest or during rotation it may also be considered as a clutch.

The fluid coupling is represented in Fig. 266. It consists in general principle of two main parts, the impeller which is a disc fitted with radial vanes and the runner of similar construction. Both parts are mounted in a casing sealed by a metallic seal. There is no mechanical connection between the impeller and the runner, the connection occurs only by the streaming fluid. The coupling is filled with oil. The impeller fixed on the driving shaft acts as a pump and pumps the oil in the runner which acts like a turbine. The runner is rotating at a lower speed than the impeller. The difference between the two parts is expressed as the slip $s = (n_1 - n_2)/n_1$. The slip is about 2 to 3%. The efficiency of the coupling is about 97 to 98%. Engaging and disengaging occurs by filling and emptying the coupling. The coupling reduces shocks and vibrations and damps oscillations to a high degree. Similar to the hydraulic coupling is the torque converter which has an additional stationary member to change the direction of the oil flow. The difference of the torques in the impeller and the runner is accomodated by this stationary part.

Slip clutch

The slip clutch, one type shown in Fig. 267, is used as a security clutch to prevent the driven shaft from being overloaded. The rubbing surfaces are loaded by springs. By compressing the springs any torque may be obtained. If the torque gets greater than the adjusted torque the clutch begins to slip.

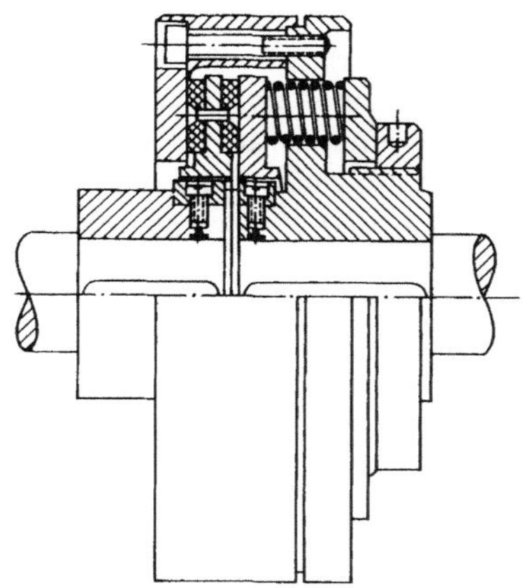

Fig. 267 Slip clutch

Centrifugal clutch

The centrifugal clutch, shown in Fig. 268, is designed to give automatic engament to the driven part at a certain speed of the driving part. The shoe segments in the inner

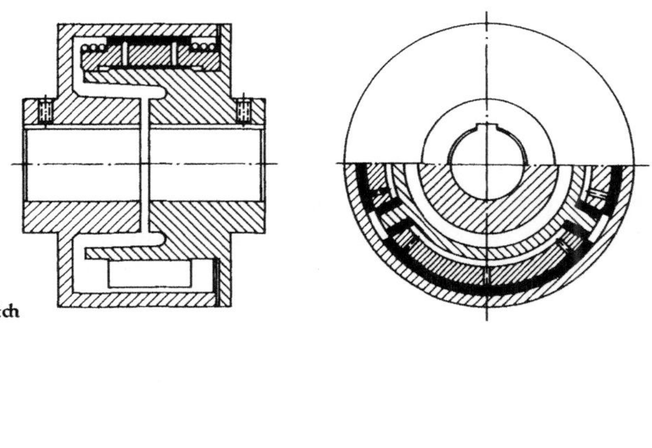

Fig. 268 Centrifugal clutch

ring of the driving part are pressed by centrifugal force against the outer ring of the driven part. If the centrifugal force has a predetermined value then the complete torque may be transmitted. Retracting springs are used to keep the clutch disengaged below the predetermined speed. The clutch is used where the torque of the starting motor is too low to drive a machine e. g. for internal combustion engines which cannot be started under load.

Overrunning or free wheeling clutch

An overrunning clutch, shown in Fig. 269, is a clutch which transmits torque in only one direction. It is used, if a shaft is driven by a main and an auxiliary engine and the auxiliary engine should be prevented from being driven by the main engine. Examples are: the bicycle coaster brake and the drive between an electric motor as main engine or a Diesel engine as auxiliary engine and an electric generator. If the speed of the inner race is greater than that of the outer race, then the rollers are jammed and transmit the torque on the outer race. If the speed of the outer race is greater than that of the inner race, then the outer race free wheels.

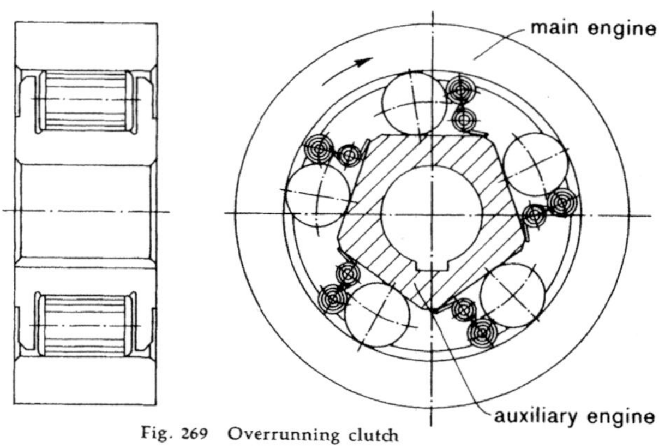

Fig. 269 Overrunning clutch

8.2.3.3. Brakes

A brake is a machine element the purpose of which is to stop or to slow down a moving part. An additional purpose of the brake is to hold a shaft in position at rest and to prevent an unwanted reversal of the direction of rotation. The major functional difference between a clutch and a brake is that a clutch connects a moving member to another moving member, whereas a brake connects a moving member to a stationary frame. The dissipation of the energy absorbed as heat is a greater problem for brakes than for clutches.

Shoe brake

For the shoe brake, represented in Fig. 270, a shoe is pressed against the surface of a rotating cylinder called a "brake drum". The tangential frictional force between the shoe and the brake drum produces the braking torque. A pivoted shoe gives a more uniform pressure distribution than a lever mounted shoe.

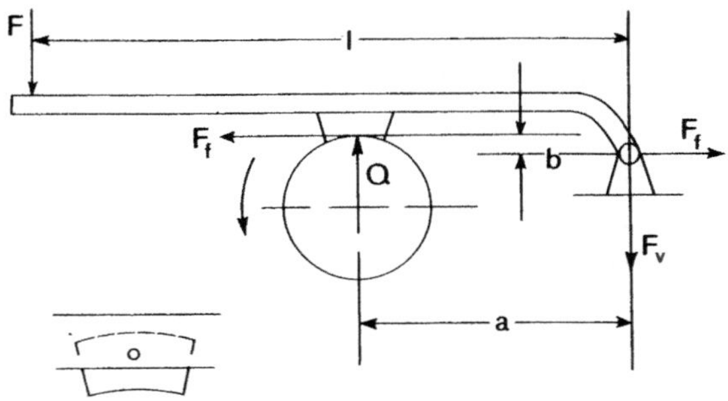

Fig. 270 Shoe brake

From the condition of equilibrium of all moments, the force F on the end of the lever may calculated thus:

$$F = \frac{Q}{l}(a - \mu\, b)\,.$$

From the condition of equilibrium of all vertical forces, the vertical force acting on the support may be calculated thus:

$$F_v = Q - F\,.$$

The total force acting on the support is:

$$F_s = \sqrt{F_v^2 + (Q\mu)^2}\,.$$

Band brake

The band brake, represented in Fig. 271, consists of a steel band lined on one side with leather, wooden blocks or an asbestos composition material.

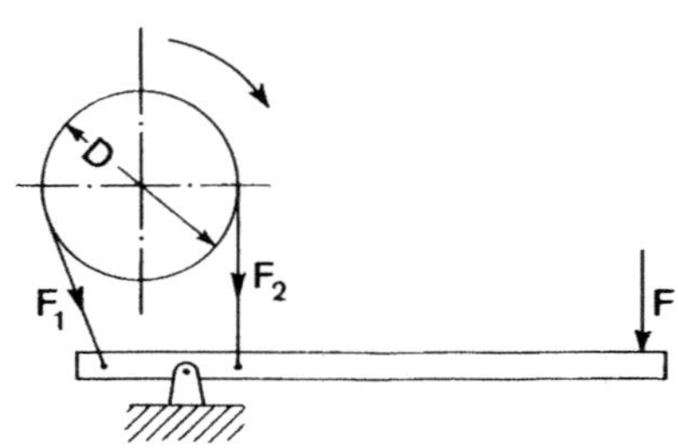

Fig. 271 Band brake

The torque taken up by the brake is

$$M_t = (F_1 - F_2)\,\frac{D}{2}\,.$$

The relationship between the forces acting on the band ends is given by the Eytelwein equation $F_1 = F_2\,e^{\mu\alpha}$

where μ = coefficient of friction between the brake drum and the band
and α = wrap angle of the band.

The difference $F_1 - F_2$ is equal to the frictional force between the brake drum and the band.

The force F on the end of the lever and the force F_s acting on the support of the lever may be calculated in a similar way to the shoe brake.

Ratchet mechanism

If a wheel, rotating in a clockwise direction, should be prevented from rotating in counter-clockwise direction, then a ratchet mechanism, shown in Fig. 272, may be used. If the wheel should also rotate in a counter clockwise direction, the pawl must be notched.

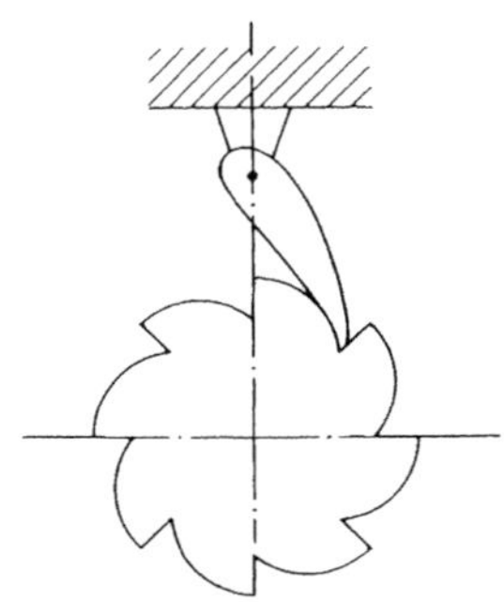

Fig. 272
Ratched mechanism

8.2.4. Belt Drive

Mechanical power is generally transmitted by means of belts, chains or gears. Each of these types of mechanical drives has specific features that often dictates its selection in a particular situation. In many cases, however, the overlapping of characteristics makes two or more of the drives competitive and the final choice will be based upon considerations such as costs, maintenance, reliability, availability, etc. A belt drive has several useful characteristics. It can be used to connect widely or closely spaced shafts. It does not require precise alignment of the shafts and pulleys. It is relatively quiet in operation. Under excessive load, slip takes place at the belt drive in the same manner as at a slip clutch. One important limitation of its use is that it is not a positive drive, slip and creep will occur, and it cannot be used therefore, where either a positive timing or an exact velocity ratio is necessary. There are two types of belt drives: the flat belt drive and the Vee-belt drive.

8.2.4.1. Flat Belt Drive

Belt materials

Originally only leather was used for making belts. Nowadays belts of woven cotton, of perlon and nylon and other materials are also used. Belts must have high strength and good adhesion and must be able to endure high speeds. Because there are no materials which possess all these qualities, belts of combined materials are used, thus the nylon chrome leather (Extremultus) belt and the polyvinylchloride (P V C) perlon belt.

Leather belt

Leather belts are used singly for thicknesses between 4 and 7 mm. For greater thicknesses double belts are used. Normally for single belts the flesh side is in contact with the pulley. For double belts the flesh sides are in contact. The ultimate strenght is less than 500 kp/cm^2. The maximum speed is 30 m/s, the maximum gear ratio is for open drive 1 : 5 and the coefficient of friction between the belt and the pulley is about 0.25. The pulleys may be flat or cambered. The belts are joined by one of the various fastening devices, e. g. by wire spirals with an inserted pin.

Extremultus belt

There are three types of extremultus belt, shown in Fig. 273. The first type consists of an inner layer of nylon, with one outside layer of chrome leather and the other outside layer of textile fabric. The second type consists of an internal layer of nylon and both outside layers of chrome leather. The third type consists of a layer of nylon and a layer of chrome leather. The nylon gives high strength and the chrome leather a good adhesion to the pulley. The side containing the chrome leather must always be in contact with the pulley. The ultimate strength of the belt is about 2000 kp/cm^2, the maximum speed 50 m/s, the maximum gear ratio 1 : 20 and the coefficient of friction between the belt and the pulley can be up to 0.8. The pulleys are slightly cambered. The belt is joined by gluing.

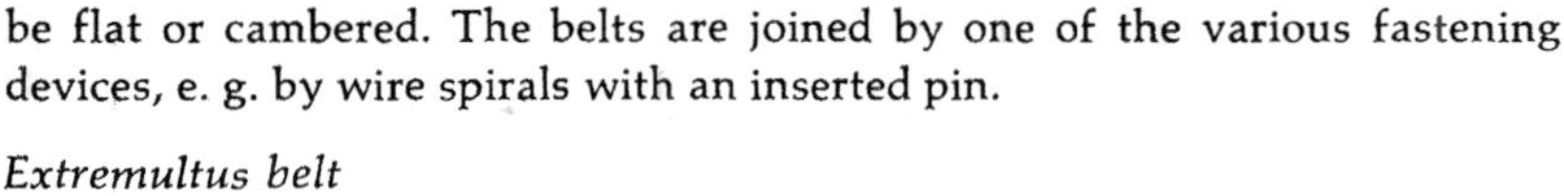

Fig. 273 Types of Extremultus belt

P V C belt

The P V C belt consists of perlon fabric which is covered on both sides with a layer of polyvinylchloride. The perlon fabric gives the strength and the layer of polyvinylchloride the adhesion to the belt. The ultimate strength of the belt is about 1700 kp/cm^2, the maximum speed 90 m/s, the maximum gear ratio 1 : 20 and the coefficient of friction between the belt and the pulley about 0.8. The working temperature of the belt should not exceed 60° C. The pulleys may be flat or cambered. The belt is joined by gluing.

Types of belt drive

Open drive

The leather belt drive of the open type should lie either horizontally or inclined. It should never be vertical. The tight portion of the belt should be below the slack portion, so that the angle of contact of the belt on both sides is increased. The distance between the pulley centres should be 2.5 to 3.5 times the sum of the pulley diameters. The PVC and the Extremultus belt drive of the open type may be horizontal, vertical or inclined. Any distance between the pulley centres may be chosen (Fig. 274).

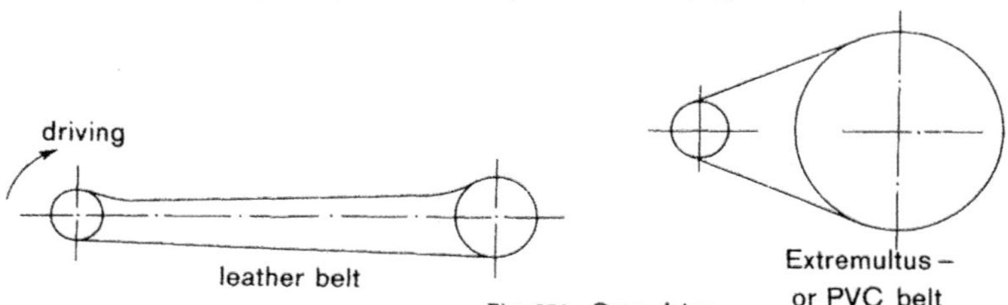

Fig. 274 Open drive

Open drive with belt tightener

Belt tighteners, shown in Fig. 275, are only used for leather belts to increase the angle of contact of the belt on the two pulleys. The belt tightener must always be arranged on the slack side of the belt if a belt tightener is employed. The distance between the pulley centres can be reduced. The maximum gear ratio is 1 : 16.

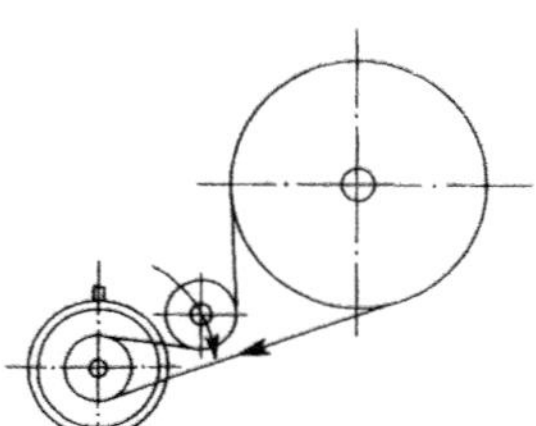

Fig. 275 Open drive with belt tightener

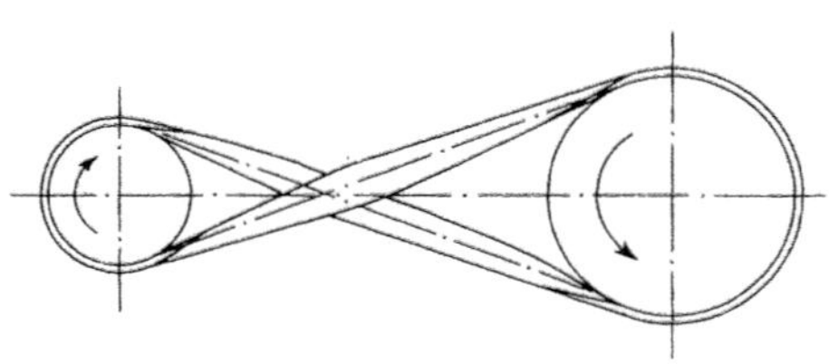

Fig. 276 Crossed belt drive

Crossed belt drive

The crossed belt drive, shown in Fig. 276, is used if the two pulleys should rotate in opposite direction. A disadvantage of the crossed belt drive is that wear occurs by the belt sides rubbing together.

Quarter turn belt drive

For the quarter turn belt drive, represented in Fig. 277, the pulleys must be arranged so that the centre line of the belt running on the driving pulley coincides with the centre line of the driving pulley. The pulleys must have flat rims. The direction of rotation cannot be changed.

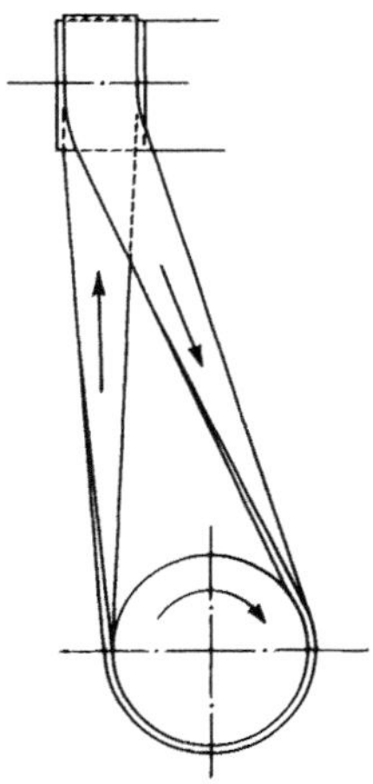

Fig. 277 Quarter turn belt drive

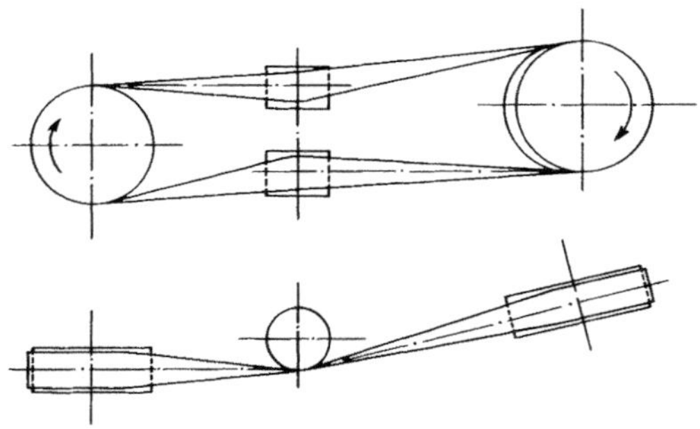

Fig. 278 Belt drive with guide pulleys

Belt drive with guide pulleys

If for a turn belt drive the direction of rotation is to be changed, then guide pulleys must be arranged (Fig. 278).

Pulleys

Pulleys are usually made of cast iron. They may be made either in one piece as solid pulleys or of two pieces as split pulleys. The rim is fixed to the hub either by a web for small pulleys or by arms for large pulleys. The rim may be flat or cambered. A cambered rim keeps the belt better in its place. There are fixed pulleys keyed to the shaft and loose pulleys which are allowed to rotate freely on the shaft. Loose pulleys are fitted with bushes or with rolling contact bearings. Motor pulleys have flat rims (Fig. 279).

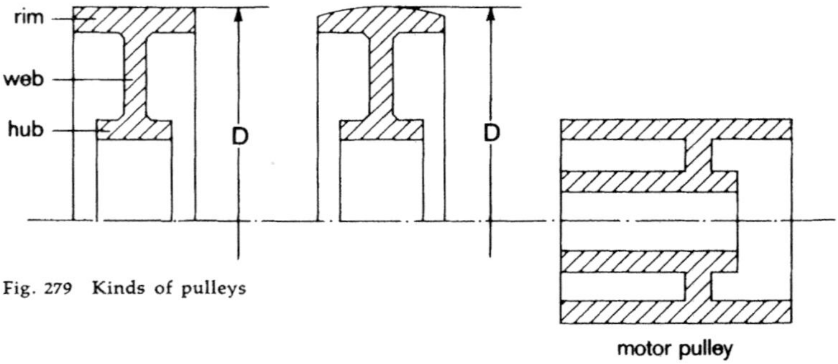

Fig. 279 Kinds of pulleys

Calculations for a flat belt drive

The median length of the belt (Fig. 280), may be calculated thus:

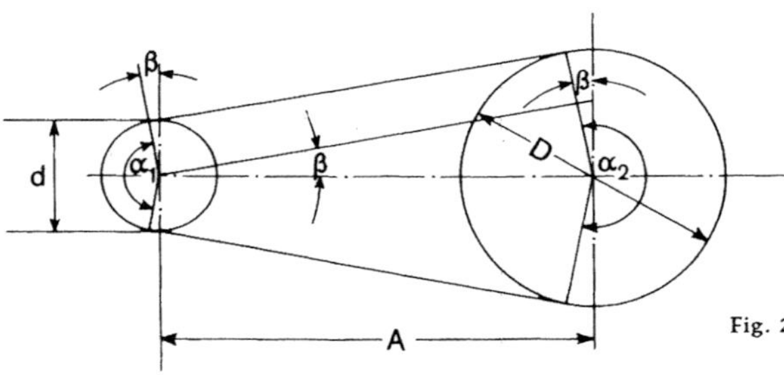

Fig. 280

$$L = \frac{\pi}{2}(D + d) + 2\,A\cos\beta + \frac{(D - d)}{2}\beta_{\text{radians}}$$

since $\beta^\circ = \frac{\pi}{180}\beta_{\text{radians}}$

$$L = \frac{\pi}{2}(D + d) + 2\,A\cos\beta + (D - d)\,\beta^\circ\,\frac{\pi}{180}. \tag{157}$$

The angle β may be calculated from $\sin\beta = \frac{D - d}{2\,A}$.

The required width of the belt may be calculated from

$$b = \frac{c\,P}{P_1\,f} \tag{158}$$

where b = width of belt (dm)
c = service factor

P = power transmitted by the belt (PS)
P_1 = power per dm belt width (PS/dm)
f = correction factor for angle of contact.

The service factor c, the power per dm of belt width P_1 and the correction factor f must be taken from tables of belt drives. The correction factor f is for the wrap angle $\alpha = 180°$, $f = 1$, for $\alpha < 180°$ less than 1 and for $\alpha > 180°$ greater than 1.

SESPA belt drive

If a belt drive is at rest the two belt forces are equal. If the belt is loaded, the belt force F_1 in the tight portion increases and the belt force F_2 in the slack portion is reduced by the same amount. The sum of the two forces is the same for each loading. If the force F_2 is zero then the belt begins to slide and the load transmission is stopped. For the SESPA belt drive the forces in the belt are adapted to the loading. The ratio of the belt forces remains constant and the force F_2 never becomes zero. There are two types of SESPA belt drive, demonstrated in Fig. 281. For the first type the housing of the motor is eccentrically supported, and for the secound type the torque of the motor shaft is transmitted by two gear wheels which gives the same effect as the arrangement of the first type. The motor moment produces a reaction moment in the eccentrically supported motor pulley which causes stretching of the belt.

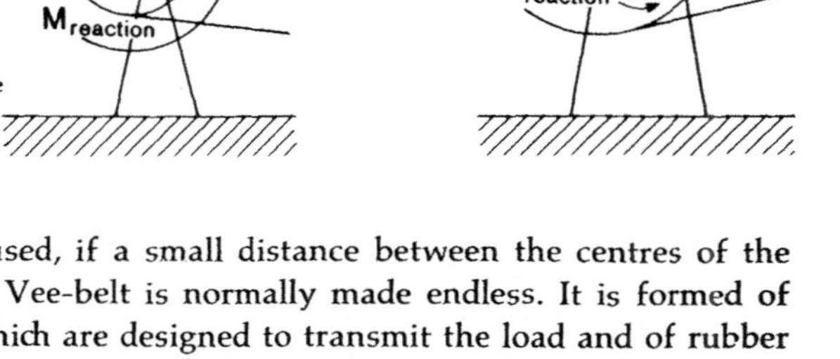

Fig. 281

Two types of SESPA belt drive

8.2.4.2. Vee-Belt Drive

The Vee-belt drive is used, if a small distance between the centres of the pulleys is desired. The Vee-belt is normally made endless. It is formed of layers of cord fabric which are designed to transmit the load and of rubber which gives the belt the necessary stiffnes. It is covered by a rubber casing which is designed to protect the belt from abrasion and chemical damage (Fig. 282).

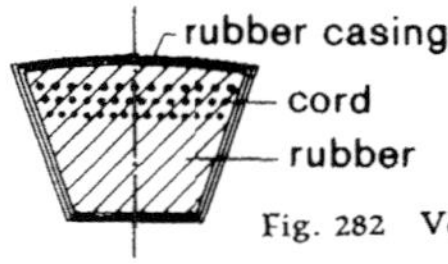

Fig. 282 Vee-belt

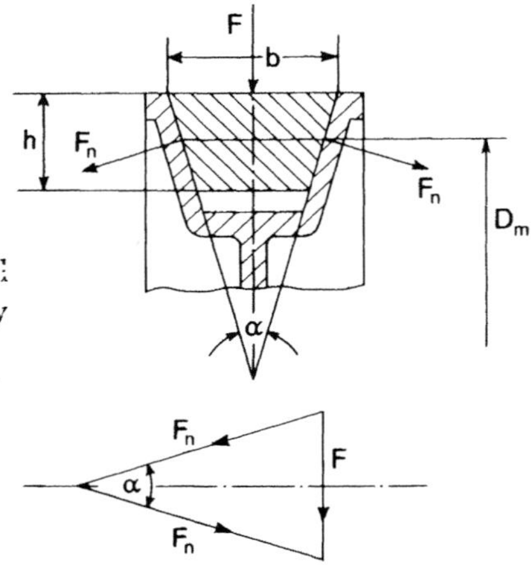

Fig. 283 Forces acting on a Vee-belt

As shown in Fig. 283, the tangential force acting on a Vee-belt pulley may be calculated from:

$$F_f = 2\,F_n\,\mu\,,$$

and since $F_n = \dfrac{F}{2\sin(\alpha/2)}$,

this gives $F_f = \dfrac{F\,\mu}{\sin(\alpha/_1)}$.

The tangential force on the pulley of a flat belt drive is $F_f = F\,\mu$.

By comparing the two formulae we see that for the same pressure force (F) the transmitting power is greater for a Vee-belt drive than for an equivalent flat belt drive.

The maximum gear ratio is 1 : 10. The usual distance between the pulley centres is $A = D_m + h$, where D_m = mean pulley diameter and h = thickness of the Vee-belt. The maximum speed of the belt is 28 m/s.

The width b, the thickness h of the belt and the belt lengths are standardized and can be obtained from tables. The profile angle is normally $\alpha = 38°$.

Calculations for Vee-belt drive

The size of the belt profile is dependent on the smaller pulley diameter and may be found from tables. After choosing the size of the belt profile the required number of belts may be calculated from the following formula:

$$z = \frac{c\,P}{P_1\,f} \tag{159}$$

where P = power transmitted by the belt drive (PS)
c = service factor
P_1 = power transmitted by one belt
f = correction factor for angle of contact.

The service factor c, the power transmitted by one belt and the correction factor f must be taken from tables of Vee-belt drives.

8.2.5. Chain Drive

The various kinds of chains are divided into two groups: the round iron chain used for cranes and lifting tackles and the gear chain used for various purposes in transmitting power, if a positive drive is desired.

8.2.5.1. Round Iron Chain

The chain links are produced by bending and electrically welding of round iron rods. For calculation the links are asumed to be stressed in tension only. Bending and reduction of strength at the welding points is allowed for by choosing a low value value of σ_{all}. The load transmitted by a chain link is calculated from formula (67) thus:

$$F = 2\,\frac{4}{d^2\,\pi}\,\sigma_{z\,all}$$

where d = diameter of the chain link, $\sigma_{z\,all}$ is choosen between 400 and 600 kp/cm². The chain wheels on lifting tackles may be made with or without teeth, as seen in Fig. 284.

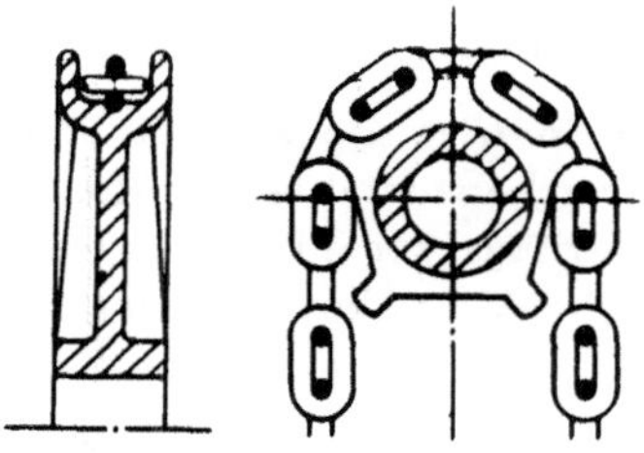

Fig. 284
Round iron chain driving
a wheel with and without teeth

8.2.5.2. Gear Chains

Gear chains have no slip, but they do have a certain elasticity. The distance between centres of the chain wheels may be small or large. A major advantage of the chain drive is that it is possible to drive several units at the same time. Chains must be lubricated. Up to a speed of 2.5 m/s chains are lubricated by grease or by oil which is provided from an oil can or a brush. For speeds over 2.5 m/s gravity feed provided by dropping or douching devices or splash lubrication is used.

Roller chain

The roller chain is extensively used for driving machinery. It may be used also for high speeds. The gear ratio may be up to 1 : 8.The components of the chain are made of steel which is case hardened and treated to resist corrosion. The chain wheels are also made of steel and their teeth are case hardened. The rollers are mounted on bushes through which pass the link connecting pins. These carry the inner and the outer plates. The pins have their ends pressed over or riveted in order to secure the links (Fig. 285). The pins of the end links are made with recesses in which a spring clip fits. If the spring clip is removed, the chain may be connected or disconnected. The roller chains are made as single strand chains or as multi-strand chains.

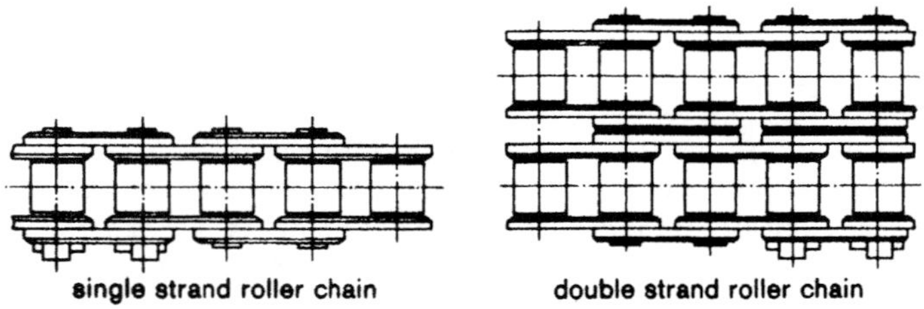

Fig. 285

Types of roller chains

The chain drive should be arranged so that the centres of the chain wheels lie either horizontal or inclined. For a vertical drive the chain may be lifted by centrifugal force from the teeth of the wheels. The stretched part of the chain should be above the loose part, otherwise clamping of the chain may occur. If three or more wheels are to be connected or if the direction of rotation is to be changed, then guide wheels must be provided. A long chain should be provided with tightening devices to keep the chain tight. Chain drives without a tightening device must be adjustable i. e. one wheel must be able to be displaced (Fig. 286).

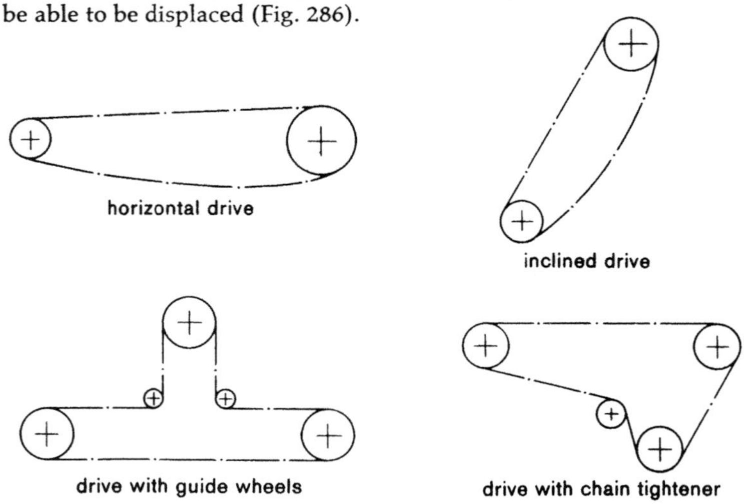

Fig. 286 Types of roller chains

Calculations for roller chain drive

The chain pitch is selected so that the drive runs noiselessly and the service life of the chain is as high as possible. It may be selected from tables in a roller chain catalogue. The distance between the chain wheels must be chosen so that the angle of contact is at least 120°. Normally this distance is 30 to 50 x the chain pitch. The number of teeth should be for small wheels, $z_1 =$ 18 to 25, the minimum number of teeth is $z_{1\,min} = 15$.

The number of link strands is calculated from

$$b = \frac{c\,P}{P_1}. \tag{160}$$

where P = power transmitted by the chain drive
P_1 = power of one link strand
c = service factor.

The service factor c and the power of one link strand P_1 may be found from tables in a chain catalogue.

Morse rocker joint chain

The morse rocker joint chain, shown in Fig. 287, eliminates the sliding friction of the pins and replaces it by a rolling friction. Instead of an ordinary pin bearing a rocking bearing is provided at each joint. It consists of two pins, the seat pin and the rocker pin. The seat pin has a plane surface against which the edge of the rocker pin rocks.

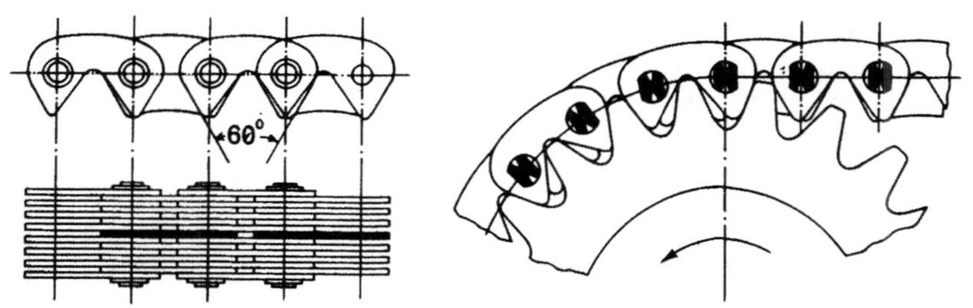

Fig. 287 Morse rocker joint chain

8.2.6. Friction Drive with Rubber Wheels

A friction drive with rubber wheels is used for transmitting small and medium powers at high speed. The hubs of the friction wheels are made of cast iron or steel. The rubber rings are vulcanized on the hubs (Fig. 288) The temperature of the friction wheels should not exceed 65 to 70 °C. The coefficient of friction is 0.6 to 0.8. The diameters of the friction wheels are the same as the diameters of the pulleys of the belt drive. The motor housing has on one end a pinned support, and on the other end springs which press the friction wheel mounted on the motor shaft against the driven wheel. The driven wheel may be made of cast iron or steel.

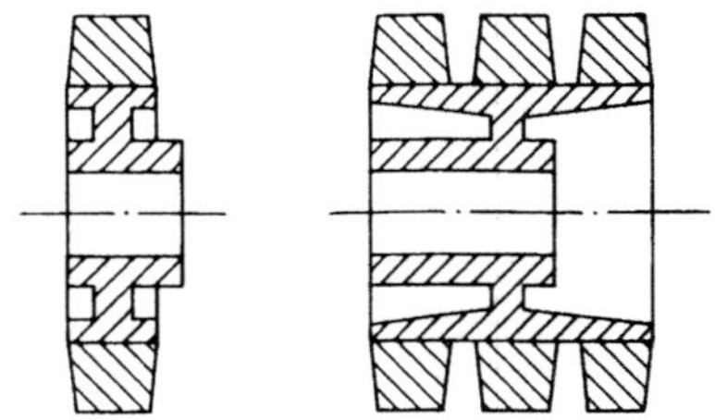

Fig. 288 Rubber wheels

Calculations for the friction gear

In Fig. 289 the acting forces on a friction gear are demonstrated. The pressure force F_a between the friction wheel and the driven wheel is composed of the pressure force F_{a1} which is adjusted automatically and the pressure force F_{a2} which is produced by a spring. The pressure force is calculated from:

$$F_a = F_u \frac{(\tan\alpha - \mu)\, c + \mu}{\mu \tan\alpha} \tag{161}$$

where F_u = tangential force
α = 0.075 (mean value)
c = service factor
(taken from tables).

The spring force F is calculated from the condition of equilibrium of all moments acting on the motor with respect to the swivelling axis on which the motor is supported thus:

$$F = \frac{G\,g + F_a\,p - F_u\,u}{f},$$

where G = weight of motor.

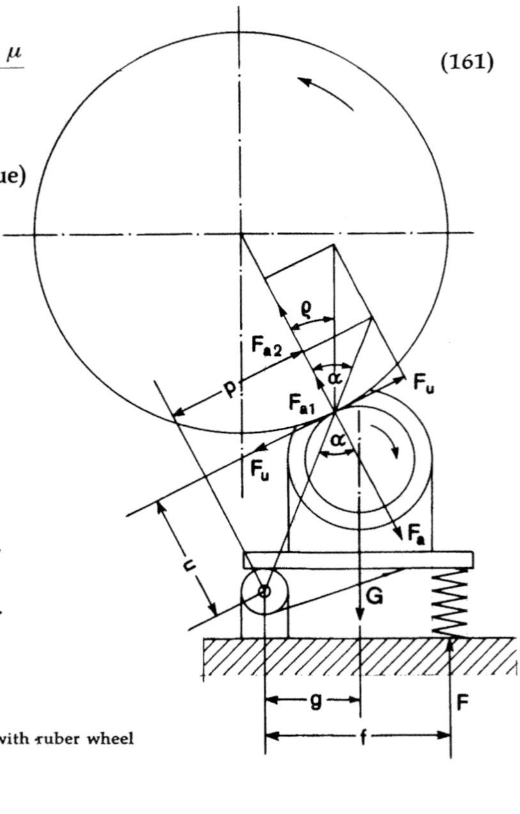

Fig. 289
Forces acting on a friction gear drive with ruber wheel

The width of a friction ring may be calculated from:

$$b = \frac{F_a}{p_{all}\, z} \tag{163}$$

where z = number of rubbing rings on the wheel
and p_{all} = specific pressure between the rubbing rings and the driven wheel.

8.2.7. Gear Drive

The gear drive is a positive drive. It is used for transmitting power, if a precise timing is desired. In comparison with belt and chain drives, gear drives are most compact and can operate at higher speeds. On the other hand gear drives require more attention to lubrication, cleanliness and shaft alignment than belt and chain drives and usually operate in a closed case with provision for lubrication.

Gear wheels may be made by a non-machining process e. g. by casting or drop forging or by a machining process e. g. by milling. For cast gears normally cast iron is used. Cast gear wheels may only be employed for low speeds. Normally steel gears are used which are milled, hardened and ground.

8.2.7.1. Forms of Gear Teeth

Types, Terms and Symbols

A gear train consists at least of three members: Two gear wheels and a connecting part (Fig. 290). Each of the three members may be driving. Usually the connecting part is stationary and one of the two wheels is driving. If one wheel is stationary and the connecting part is rotating, then the gear train is called a "planetary gear train".

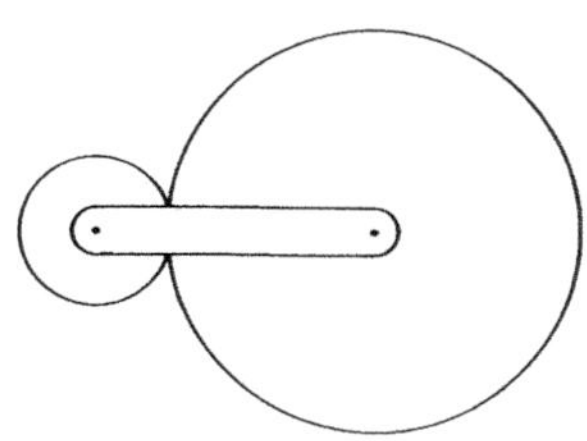

Fig. 290

Gears are classified according to the position of the axes of the two gears meshing; firstly in gears the axes of which are parallel, secondly in gears the axes of which are intersecting and thirdly in gears the axes of which are crossing. In the first two types the axes coplanar i. e. the axes are in one plane. In the third type the axes are noncoplanar i. e. the axes lie in different planes. Gears having parallel axes are called "spur gears" (Fig. 291). The smaller gear gear wheel is often known as a "pinion". A pinion is usually of solid block construction, whereas a bigger gear wheel is composed of a hub, a rim on which the teeth are formed and a web or arms which connect the hub with the rim. The maximum gear ratio is 1:8. The efficiency is from 96 to 98%. A rack may be considered as a spur gear of in infinite radius (Fig. 292). Gears having intersecting axes are called "bevel gears" (Fig. 293).

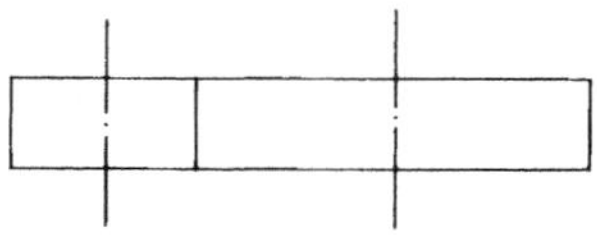

Fig. 291 Spur gears

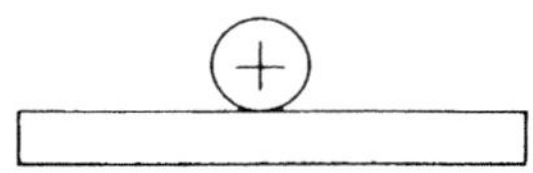

Fig. 292 Rack- and pinion gear

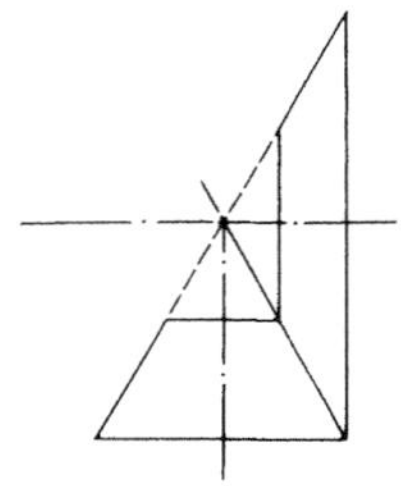

Fig. 293 Bevel gear

Their axes are normally at right angles, but other angles may be used. The maximum gear ratio is 1:6. The efficiency is about 98%. In Fig. 294, three types having crossing axes are represented; skew gears having straight teeth and hypoid gears having curved teeth, screw gears and worm qear. A worm qear may be considered as a very long screw gear. The maximum gear ratio for screw gears is 1:5, the maximum gear ratio for worm and worm gear is 1 : 100. The efficiency in both types is from 95 to 40%.

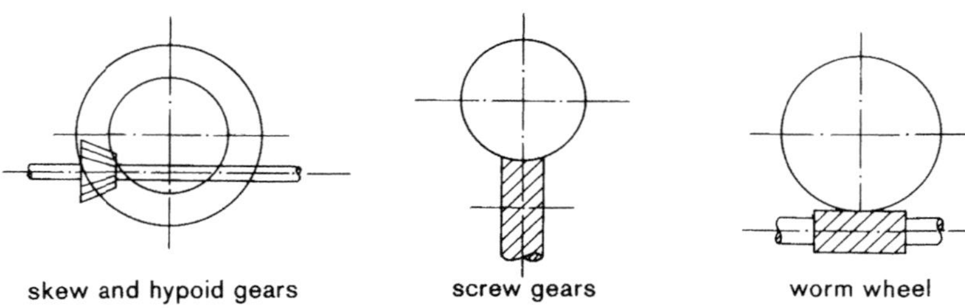

Fig. 294

The above gears are called "external gears". If a pinion is in mesh with an annular ring, then the gears are called "internal gears". For the latter ones both wheels have the same direction of rotation (Fig. 295).

For the tooth profiles the involute and cycloidal curves (cycloid, epicycloid and hypocycloid) are used. The involute (Fig. 296) is generated, if a straight line rolls on a circle. The epicycloid (Fig. 297) is generated, if a circle rolls outside another circle, and it is used for the curve of the face o a tooth. The hypocycloid (Fig. 297) is generated, if a circle rolls inside another circle, and is used for the curve of the flank of a tooth. The cycloid (Fig. 298) is generated, if a circle rolls on a straight line, and it is used for the curves of the teeth of a rack. The cycloidal curves are only used in special cases, since their production is more difficult than the production of the involute. Also, the involute has other advantages e. g. it is not sensible to small changes of the distance between the axes of gears. The involute has been standardized for gears.

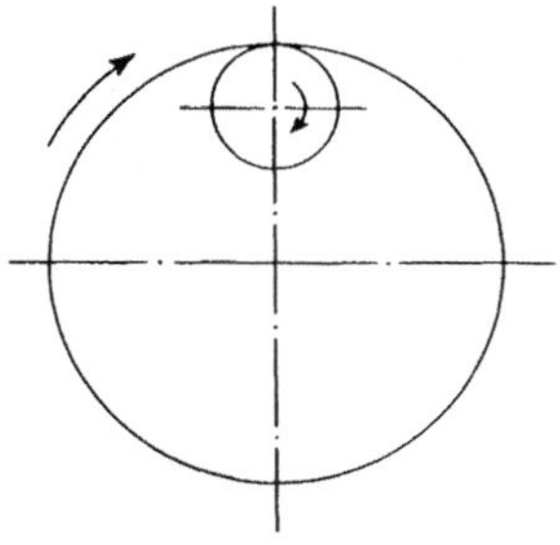

Fig. 295
Pinion and internal gear wheel

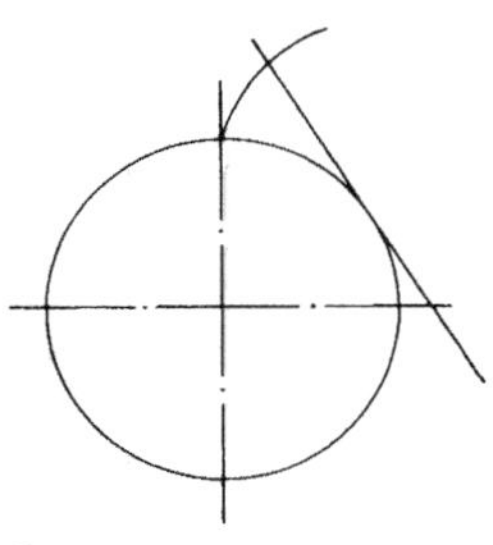

Fig. 296
Generating of an involute

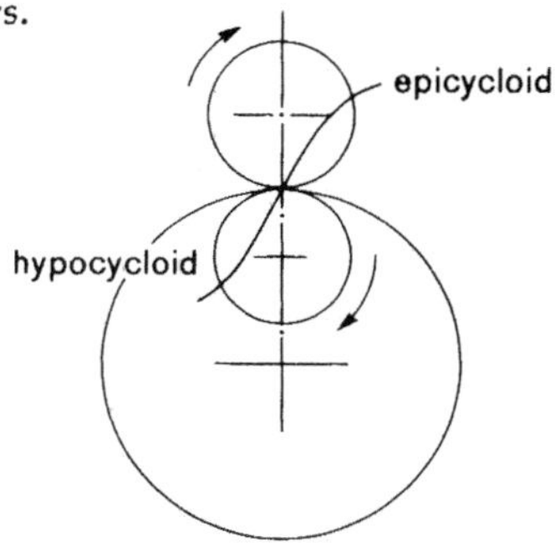

Fig. 297 Generating of an epicycloid and an hypocycloid

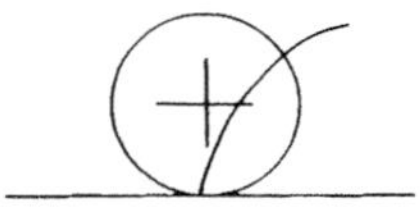

Fig. 298
Generating of a cycloid

Gear wheels have different forms of teeth. The most common types of gear wheels, represented in Fig. 299, are: the straight spur and the straight bevel wheel, the helical spur and the helical bevel wheel, the herring bone spur wheel and the spiral bevel wheel.

In helical spur wheels a side pressure is exerted which causes a tendency to push each wheel out of mesh sideways. The side pressure may be compensated for, if herring bone spur gears are used.

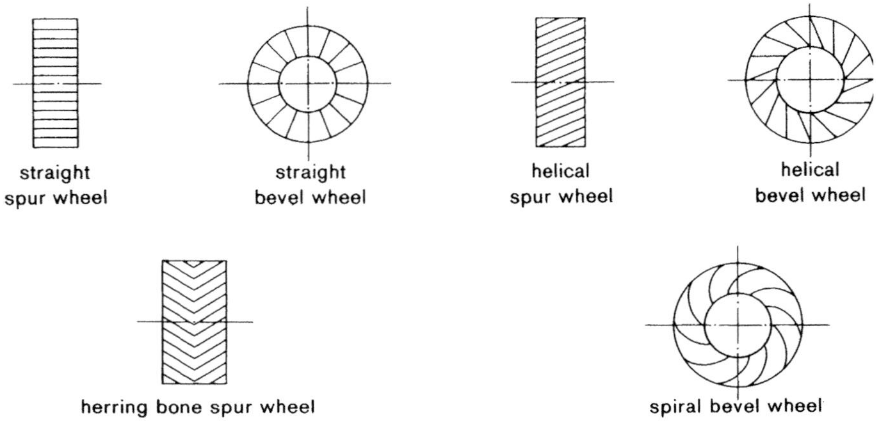

Fig. 299 Forms of teeth on a gear wheel

Terms for gears

The following terms for gears are in common use (Fig. 300).

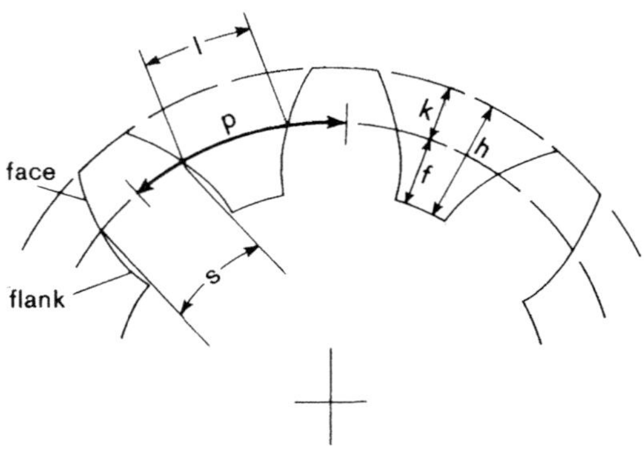

Fig. 300

pitch circle = circle on which the wheels are rolling
addendum circle = circle passing through the outer ends of the teeth
dedendum-or root circle = circle passing through the bottom of the tooth spaces

tooth profile = curve forming the tooth
tooth face = portion of the lateral tooth surface which is outside the pitch circle
tooth flank = portion of the lateral tooth surface which is inside the pitch circle

tooth depth (h)
addendum (k)
dedendum (f)
tooth width (b)
tooth thickness (s)
tooth space (l)
pitch (p) = distance from the centre of one tooth to the centre of the next tooth, measured on the pitch circle
pitch point (C) = point of contact of the pitch circles of two gears which are in mesh

$$h = f + k\,, \quad s + l = p\,, \quad f - k = c \text{ (clearance)}\,.$$

The diameter of the pitch circle is calculated from:

$$d_0\,\pi = z\,p \quad d_0 = \frac{z\,p}{p} \quad \text{or} \quad d_0 = z\,m$$

where z = number of teeth

and $m = \dfrac{p}{\pi}$ = module.

The pitch diameter is a simple number, if the module is a simple number. The module is standardized from 0.3 to 75 mm and may be found from tables. The dimensions of a tooth are taken in relation to the module. For standardized gears the following relationships are valid:

$$k = m\,, \quad f = 1.2\,\text{m}\,, \quad h = 2.2\,\text{m}\,, \quad c = f - k = 0.2\,\text{m}$$

$$s = l = \frac{p}{2}\,.$$

Fundamental law of gear tooth action

The tooth profiles must be made such that the velocity ratio of two gear wheels is constant at all times,

i. e. ω_1/ω_2 = constant.

At the point of contact (P) of two meshing teeth the velocity of gear wheel 1 is designed with v_1 and the velocity of gear wheel 2 with v_2. The velocities v_1 and v_2 are resolved into the components s_1 and s_2 normal to the tooth profiles and into the components p_1 and p_2 tangential to the tooth profiles (Fig. 301). Since the tooth profiles must make permanent contact with each other, the component s_1 must be equal to the component s_2.

$$s_1 = v_1 \cos \alpha_1 = l_1 \, \omega_1 \frac{\varrho_1}{l_1} = \omega_1 \, \varrho_1 \, ,$$

$$s_2 = v_2 \cos \alpha_2 = l_2 \, \omega_2 \frac{\varrho_2}{l_2} = \omega_2 \, \varrho_2 \, .$$

Now, $s_1 = s_2 \, , \quad \omega_1 \, \varrho_1 = \omega_2 \, \varrho_2$

or $\dfrac{\omega_1}{\omega_2} = \dfrac{\varrho_2}{\varrho_1} = \dfrac{r_2}{r_1} \, .$

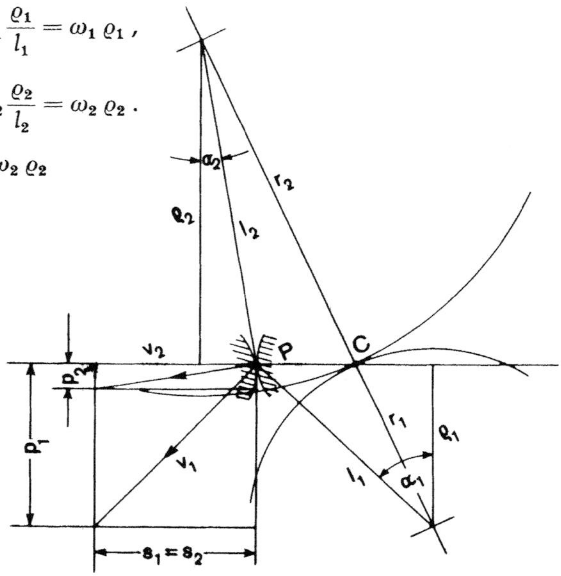

Fig. 301

The velocity ratio remains constant, if r_2/r_1 remains constant or, if the point of intersection (C) of a line normal to the point of contact of the tooth profiles and the connecting line of the centres of the gear wheels is always at the same place. In other words: the velocity ratio remains constant, if the normal to the tooth curves at all points of contact passes through the pitch point. As the components p_1 and p_2 of the velocities v_1 and v_2 have different values at all points of contact except the pitch point, the tooth faces and tooth flanks are sliding on each other. All curves, the normals of which intersect the pitch circle or at least are tangential to it, may be used as the profile of a tooth. The profile of the tooth of the mating gear may be found by means of the fundamental law of gear tooth action. If the involute or the cycloid (only for racks) or the epicycloid or the hypocycloid are taken as tooth profiles, the profile of the tooth of the mating gear is also an involute or an epicycloid or hypocycloid, respectively.

Involute spur gears

External gears

The involute is a curve traced by a point on a line as the line rolls on a circle (Fig. 302). The line is called a "generating line" and the circle a "base circle". The generating line is a tangent to the base circles of two meshing gears. The angle between the generating line and the line perpendicular to the connecting line of the centres of the gears is called the "angle of obliquity" (α). Within the base circle the tooth profile is made radial.

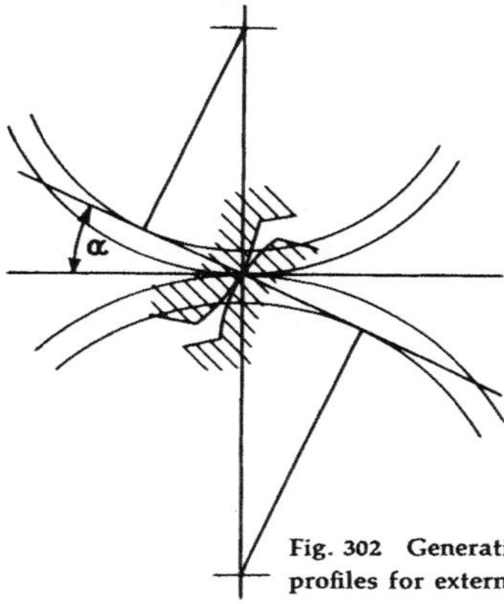

Fig. 302 Generating of involute tooth profiles for external gears

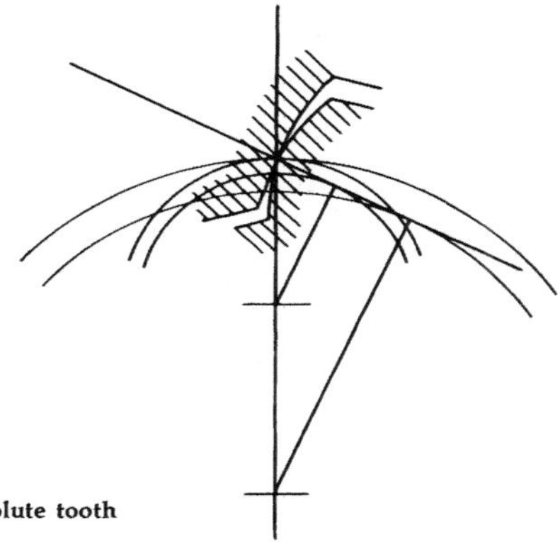

Fig. 303 Generating of involute tooth profiles for internal gears

Pinion and rack

The rack may be considered as a gear wheel with an infinite number of teeth. The pitch circle and the base circle are straight lines. The involute is also a straight line, perpendicular to the generating line (Fig. 304).

Internal gears

The involute of the annular ring is drawn in a similar way to the involute of the pinion. Now, the material is on the other side of the curve (Fig. 303).

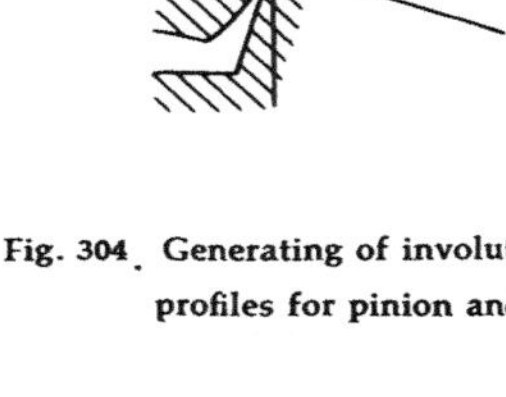

Fig. 304. Generating of involute tooth profiles for pinion and rack

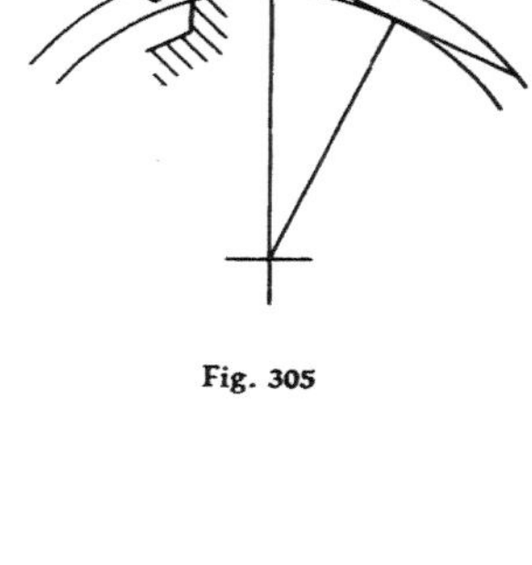

Fig. 305

The normal to any point of the involute is tangential to the base circle. At any point of contact of two teeth the normals to the two tooth profiles form one line which is tangential to the base circle of both gears. All points of contact lie on this line which is at the same time the generating line. So the path of contact and the generating line form one line. Since this line passes through the pitch point, the fundamental law of gear tooth action is fulfilled (Fig. 305).

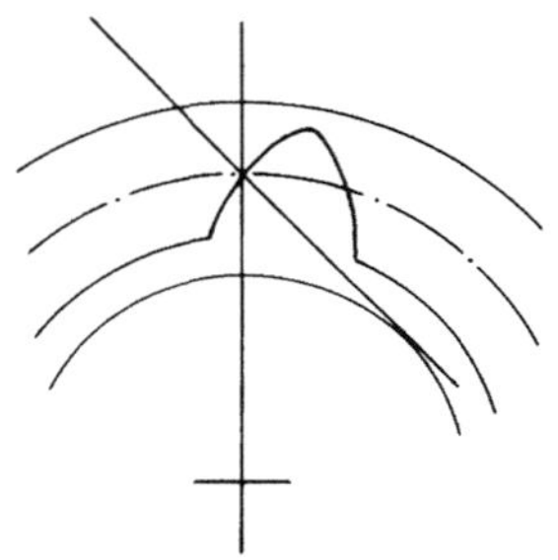

Fig. 306

The angle of obliquity must be greater than 0°, since otherwise a part of the working flank is not an involute, but it must not be too great or the face will not extend to the addendum circle (Fig. 306). The angle of obliquity is standardized. For gears, $\alpha = 20°$; for pinion and rack, $\alpha = 15°$.

The tooth profile of the driving gear is independent of the magnitude of the driven gear. Thus a gear may work with a gear of any size or of any number of teeth, if the pitch or the module and the angle of obliquity are the same.

Length of path of contact and length of arc of contact

The points of intersection between the path of contact and the addendum circles of two meshing gear wheels give the limits for the contact of two tooth profiles (Fig. 307).

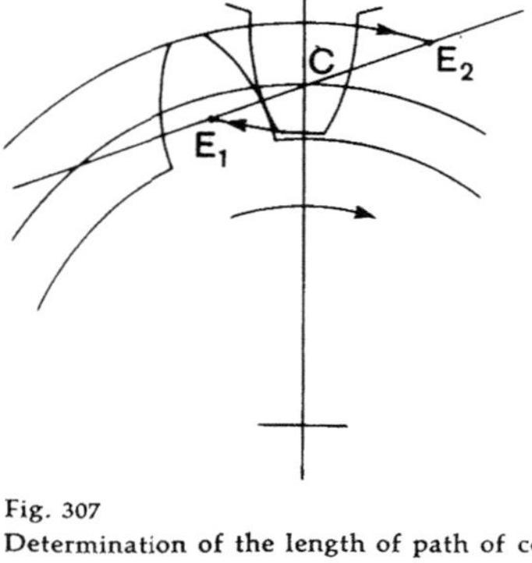

Fig. 307
Determination of the length of path of contact

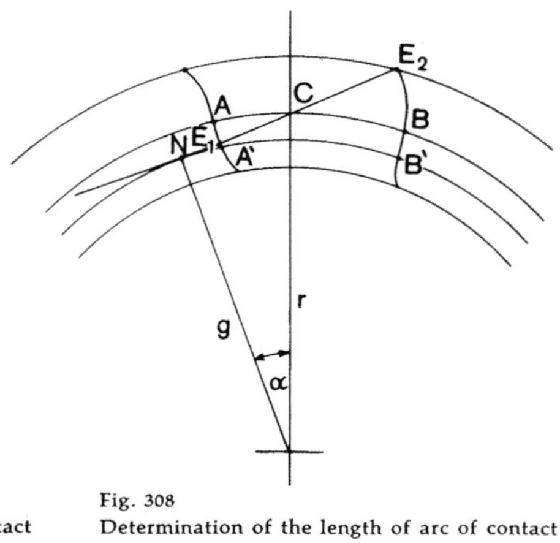

Fig. 308
Determination of the length of arc of contact

As shown in Fig. 308, the length of arc of contact $\overset{\frown}{AB}$ measured on the pitch circle is given by:

$$\overset{\frown}{AB} = \overset{\frown}{A'B'}\,\frac{r}{g} = \frac{\overset{\frown}{A'B'}}{\cos\alpha},$$

where $\overset{\frown}{A'B'}$ = length of arc of contact measured on the base circle.

Since $\overset{\frown}{A'B'} = E_1 E_2$ because the line $E_1 E_2$ rolls on the base circle, the length of arc of contact measured on the pitch circle becomes:

$$\overset{\frown}{AB} = \frac{E_1 E_2}{\cos\alpha}. \tag{164}$$

Since at least one tooth must be in mesh, the arc of contact must be at least equal to the pitch. The ratio of arc of contact to the pitch must be at least equal to 1. Normally this ratio is greater than 1.2.

Limit of the number of teeth

The tooth profile is formed by the cutting tool which rolls on the tooth profile like a rack on a pinion. If the end point E_1 of the path of contact lies outside NP, which is the case if the number of teeth is very small, then undercut on the tooth occurs, as shown in Fig. 309. The limit for teeth without undercut is when the points E_1 and N coincide (Fig. 310). For this case the number of teeth for standardized gears may be calculated thus:

$$k = m = NC\sin\alpha \quad \text{and since} \quad NC = r\sin\alpha,$$

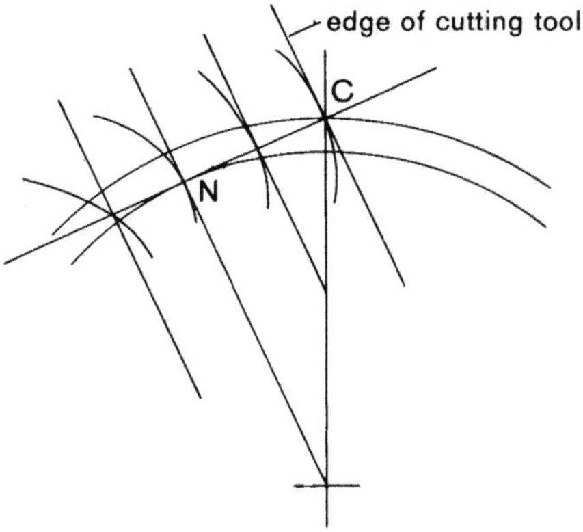

Fig. 309 Limit for undercut

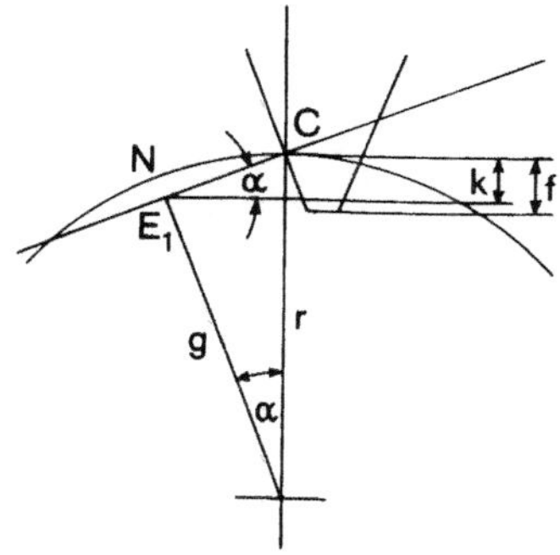

Fig. 310

this gives $m = r \sin^2 \alpha$.

$$\frac{m\, z_{lim}}{2} \text{ substituted for } r \quad \text{gives} \quad m = \frac{m\, z_{lim}}{2} \sin^2 \alpha ,$$

whence the limiting number of teeth without undercut follows as

$$z_{lim} = \frac{2}{\sin^2 \alpha} . \tag{165}$$

For $\alpha = 20°$, $z_{lim} = 17$; for $\alpha = 15°$, $z_{lim} = 30$.
If undercut in the non-working flank is allowed, then the limit of the number of teeth may be reduced. In this case the limit of the number of teeth becomes for

$$\alpha = 20° , \quad z_{lim}' = 14 \quad \text{and for} \quad \alpha = 15° , \quad z_{lim}' = 25 .$$

Spur gears with corrected teeth

Undercut for gears having smaller numbers of teeth than the limit number may be eliminated, if the rack cutter is displaced away from the centre of the gear when cutting the teeth. Since the distance between gears produced in this way is increased, this method of cutting teeth is called the "extended centre distance system". As gears generated in this manner have some advantages e. g. the roots of teeth get thicker or the pressure on the surfaces of the teeth is reduced, this method of cutting teeth is also employed for gears having greater number of teeth than the limiting number up to $z = 50$.

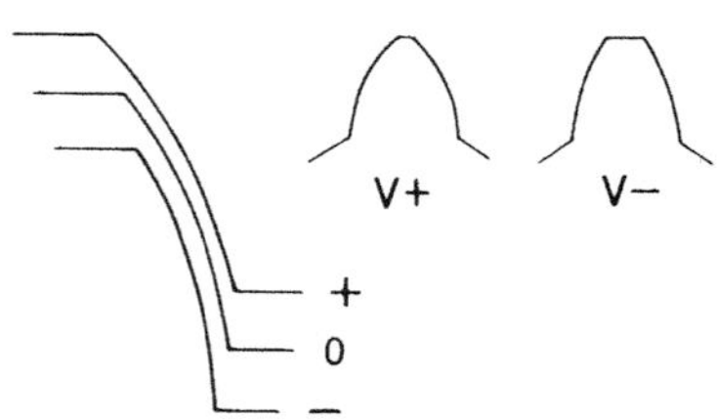

Fig. 311
Positive and negativ tooth correction

The tooth profile for gears having corrected teeth is formed by the same involute as for gears having normal teeth. Thus the base circle remains the same. It the cutter is displaced away from the centre of the gear, then the outer part of the involute is used for the tooth profile. This is called "positive tooth correction". If the cutter is displaced in the direction of the centre of the gear, then the inner part of the involute is used for the tooth profile. This is called "negative tooth correction" (Fig. 311).

Gears having teeth without correction are called: standard gears
Gears having positive tooth correction are called: $V +$ gears
Gears having negative tooth correction are called: $V -$ gears.

The amount of displacement of the cutter is taken in relation to the module. $v = x\,m$, where x is called "factor of tooth correction".

As Fig. 312 shows, the angle of obliquity, the rolling circles and the contact ratio change, if gears having corrected teeth are used. The radii of the base circles remain the same.

The velocity ratio is: $\frac{r_1}{r_2} = \frac{g_1 \cos\alpha}{\cos\alpha \cdot g_2} = \frac{g_2}{g_1}$.

Since the velocity ratio is only dependent on the ratio of the base circles which remains the same, it does not change.

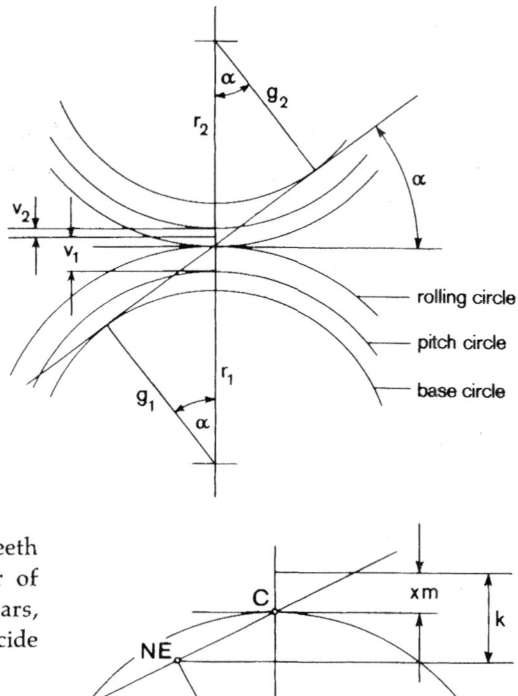

Fig. 312

Limit of the number of teeth

Gears having corrected teeth have their limiting number of teeth, if, as for standard gears, the points E_1 and N coincide (Fig. 313).

Thus

$$m = \frac{m\,z}{2}\sin^2\alpha + m\,x_{\text{lim}}$$

since $\frac{2}{\sin^2\alpha} = z_{\text{lim}}$

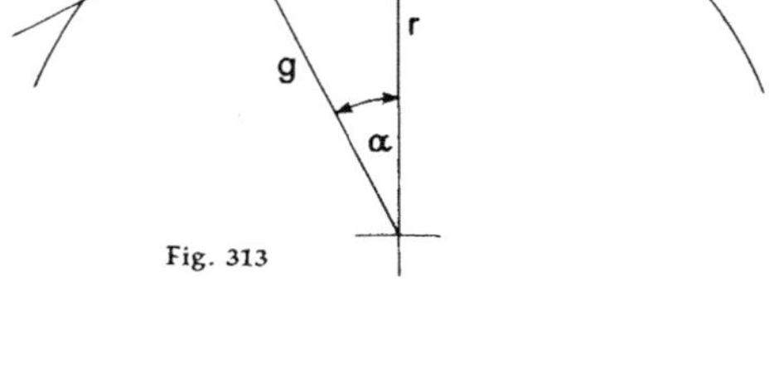

Fig. 313

the equation becomes:

$$1 = \frac{z}{z_{\text{lim}}} + x_{\text{lim}} \qquad \text{or} \qquad x_{\text{lim}} = \frac{z_{\text{lim}} - z}{z_{\text{lim}}}. \tag{166}$$

x_{lim} is the factor of tooth correction at which the number of teeth becomes the limiting number.

For $\alpha = 20^\circ$, $x_{\text{lim}} = \dfrac{17 - z}{17}$ and $x_{\text{lim}}' = \dfrac{14 - z}{17}$.

For $\alpha = 15^\circ$, $x_{\text{lim}} = \dfrac{30 - z}{30}$ and $x_{\text{lim}}' = \dfrac{25 - z}{30}$.

Each number of teeth may have a limiting number up to the minimum limiting number which is for $\alpha = 20^\circ$, $z_{\text{min}} = 7$ and for $\alpha = 15^\circ$, $z_{\text{min}} = 8$.

Example

Determine for a gear having 8 teeth and for a gear having 20 teeth the factors of tooth correction at which the given number of teeth become a limiting number ($\alpha = 20^\circ$).

$$x_{\text{lim}} = \frac{17 - 8}{17} = 0.53 \qquad x_{\text{lim}}' = \frac{14 - 8}{17} = 0.354$$

$$x_{\text{lim}} = \frac{17 - 20}{17} = -0.177 \qquad x_{\text{lim}}' = \frac{14 - 20}{17} = -0.354.$$

Bevel gears

Bevel gears are used to connect shafts intersecting at any angle. Normally the angle of intersection (γ) is 90°. There are two special types of bevel

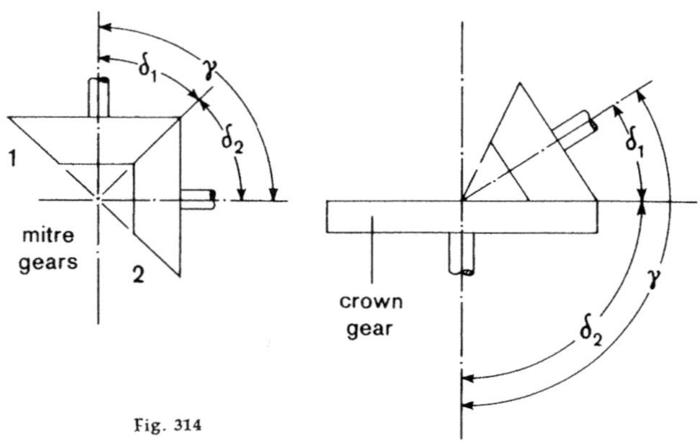

Fig. 314

gears: the mitre gear which has a pitch angle (δ) of 45° and the crown gear which has a pitch angle of 90°, represented in Fig. 314. The crown gear corresponds to a rack in spur gearing. For the pitch circle of a bevel gear, the pitch circle of the greater base of the cones is taken. The corresponding module is standardized. Bevel gears operate only as matched pairs and are not interchangeable as spur gears with straight gears.

For determining the limiting number of teeth, the bevel gear is replaced by an equivalent spur gear with the pitch radius $r_e = r_a/\cos\delta$, (Fig. 315). The limiting number of teeth may be found as follows:

$$2\, r_e\, \pi = z_e\, p$$

$$2\, r_a\, \pi = z\, p$$

by division of the two equations:

$$\frac{z_e}{z} = \frac{r_e}{r_a} = \frac{1}{\cos\delta}$$

or $$z_e = \frac{z}{\cos\delta}$$

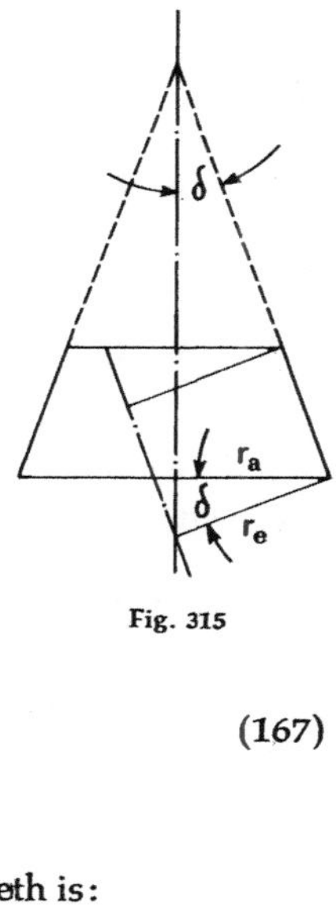

Fig. 315

where z = number of teeth on the bevel wheel
and z_e = number of teeth on the equivalent spur gear.

The limiting number of teeth is:

$$z_{lim} = z_e \cos\delta \tag{167}$$

for $\alpha = 20°$, $z_{lim} = 17\cos\delta$ and $z_{lim}' = 14\cos\delta$.

The factor of tooth correction for a limiting number of teeth is:

$$x_{lim} = \frac{z_{e\,lim} - z_e}{z_{e\,lim}}. \tag{168}$$

For $\alpha = 20°$, $x_{lim} = \frac{17 - z/\cos\delta}{17}$ and $x_{lim}' = \frac{14 - z/\cos\delta}{17}$.

Helical spur gears

A helical spur gear is a stepped spur gear with an infinite number of steps. The contact between the mating teeth is gradual starting at one end and moves along the teeth, so that at any instant the line of contact extends diagonally across the tooth surface. Helical spur gears rotate more noiselessly than straight spur gears. A disadvantage of helical spur gears is that the loss by friction is greater than for straight spur gears.

The following terms are in common use (Fig. 316).

α = helix angle
β = angle between helix and axis
p_s = circular pitch
p_n = normal pitch
b = width of the gear
s = overlap.

The following relationships hold:

$$\cos\beta = \frac{p_n}{p_s}, \quad \tan\beta = \frac{s}{b}.$$

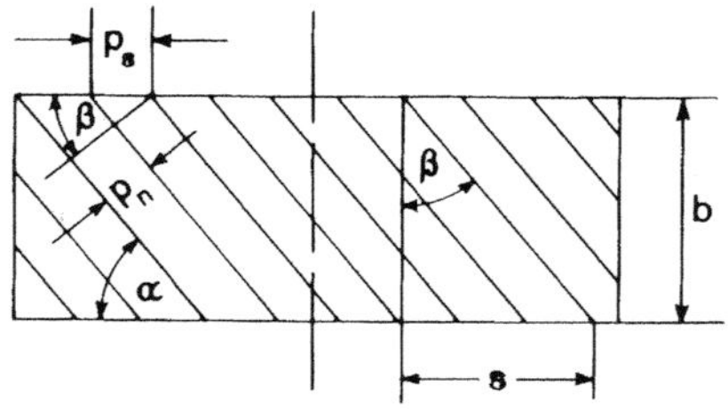

Fig. 316

The contact ratio is given by:

$$\varepsilon = \frac{(E_1 E_2/\cos\alpha + s)}{p_s}. \tag{169}$$

For determining the limiting number of teeth, the helical gear, similarly to the bevel gear, is replaced by an equivalent spur gear, the radius of which is equal to the major radius of curvature (r_n) of an ellipse formed by the section of the pitch cylinder on a plane normal to the helix (Fig. 317).

$$r_n = a\cot\alpha = \frac{a^2}{r}.$$

Now, since $a = \dfrac{r}{\cos\beta}$,

$$r_n = \frac{r}{\cos^2\beta}.$$

$$\frac{2\,r_n\,\pi}{2\,r\,\pi} = \frac{z_n\,p_n}{z\,p_s}.$$

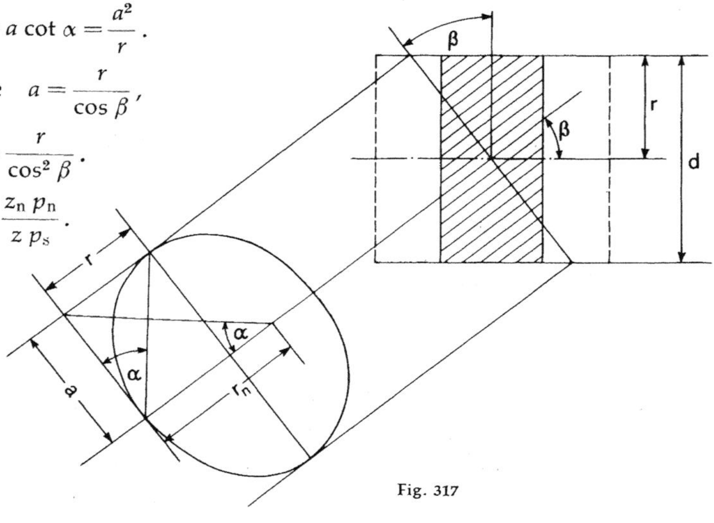

Fig. 317

It follows that the limiting number of teeth for the equivalent spur gears is: $z_n = \frac{r_n\, z\, p_s}{r\, p_n} = \frac{r\, z}{\cos^2\beta\, r\cos\beta}$ or $z_n = \frac{z}{\cos^3\beta}$.

The circular pitch diameter is:

$$d = \frac{z\, p_s}{\pi} = \frac{z\, p_n}{\pi\cos\beta} \quad \text{or} \quad d = \frac{m_n\, z}{\cos\beta}. \tag{170}$$

The normal module is standardized.

The limiting number of teeth is

$$z_{lim} = z_n \cos^3\beta\,. \tag{171}$$

For $\alpha = 20^\circ$, $z_{lim} = 17\cos^3\beta$ and $z_{lim}' = 14\cos^3\beta$.

The factor of tooth correction for the limiting number of teeth is:

$$x_{lim} = \frac{z_{n\,lim} - z_n}{z_{n\,lim}}. \tag{172}$$

For $\alpha = 20^\circ$ $x_{lim} = \frac{17 - z/\cos^3\beta}{17}$ and $x_{lim}' = \frac{14 - z/\cos^3\beta}{17}$.

Screw gears

The form of teeth for screw gears is the same as for helical spur gears. These gears work at an angle of γ which is normally 90°. The surfaces of the teeth of one gear slide along the surfaces of the teeth of the other gear. The efficiency is very poor so that screw gears may be used only for relatively light duty service e. g. for speedometer,cable drives or for an oil pump or distributor drive for motor cars.

Fig. 318 shows screw gears meshing together at an angle of γ. The velocity ratio and the pitch diameters are calculated as follows. The tangential velocities v_1 and v_2 are resolved into their normal and tangential components. The normal components must be equal.

$$v_{n\,1} = v_{n\,2} \quad \text{or} \quad r_1\,\omega_1\cos\beta_1 = r_2\,\omega_2\cos\beta_2$$

$$i_{12} = \frac{\omega_1}{\omega_2} = \frac{d_2\cos\beta_2}{d_1\cos\beta_1} \ \text{(The velocity ratio).} \tag{173}$$

Also

$$d_1 = \frac{p_{s\,1}\, z_1}{\pi} = \frac{p_n\, z_1}{\cos \beta_1\, \pi}$$

or

$$d_1 = \frac{m_n\, z_1}{\cos \beta_1} \quad \text{and} \quad d_2 = \frac{m_n\, z_2}{\cos \beta_2} \tag{174}$$

where z_1 and z_2 = number of teeth on the two gears, respectively.

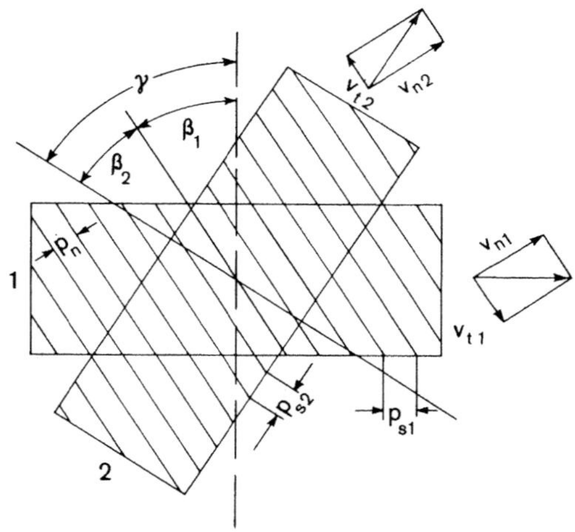

Fig. 318

Example

Two screw gears with the same pitch diameter located at right angles have a velocity ratio $i_{12} = 2:1$. Find the angles β_1 and β_2 of the two gears.

$$i_{12} = \frac{d_2 \cos \beta_2}{d_1 \cos \beta_1} = \frac{\cos (90^\circ - \beta_1)}{\cos \beta_1} = \frac{\sin \beta_1}{\cos \beta_1} = \tan \beta_1 = 2$$

$$\beta_1 = 63.5^\circ \quad \beta_2 = 26.5^\circ\,.$$

Worm gearing

The axes of the worm and the worm wheel are usually at right angles. The worm is made of steel usually case hardened and ground. The worm wheel is made for small speeds of cast iron, for higher speeds of cast iron with a rim mostly made of phosphor bronze which is pressed or screwed on the gear.

There are two types of worms, shown in Fig. 319, the spiral and the involute, respectively. In the first type the teeth of the worm in a section normal to the helix are straight sided, in the second type the teeth of the worm in a section normal to the helix have involute profiles.

The velocity ratio is $i = \frac{\omega_s}{\omega_r} = \frac{z_r}{z_s}$

where z_r = number of teeth on the worm wheel
and z_s = number of threads on the worm.

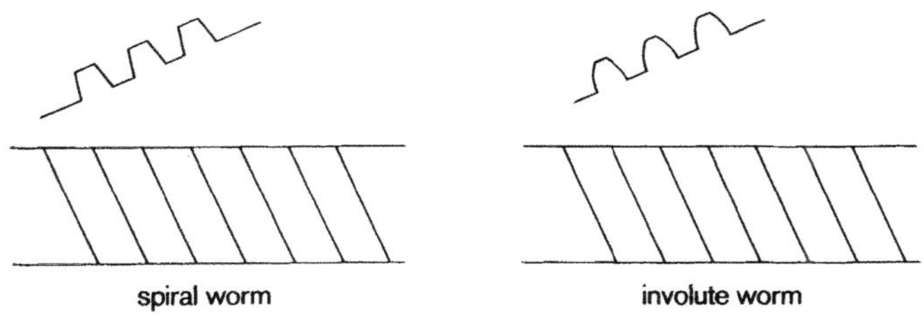

Fig. 319

The pitch diameter of the worm is calculated from:

$$\tan \alpha = \frac{z_s\, p}{d_s\, \pi} = \frac{z_s\, m}{d_s}$$

whence

$$d_s = \frac{z_s\, m}{\tan \alpha}. \tag{175}$$

8.2.7.2. Strength of Gears

Gears usually fail by fracturing one or more teeth or by excessive wear of the contact surfaces. For the first type of failure a loaded tooth may be considered as a cantilever beam under fatigue load. As can be seen from Fig. 320, the crack starts at the fillet on the tension side of the tooth, slowly progresses inward and downward and then, as the tooth becomes much weaker because of the reduced cross section, rapidly breaks through the remainder of the tooth. The second type of failure is caused by the wear of the gear teeth. The wear may occur as scoring, scratching or pitting. This type of failure also occurs for other direct mechanisms such as cams, ball and roller bearings, etc. For slow speed gears the acting

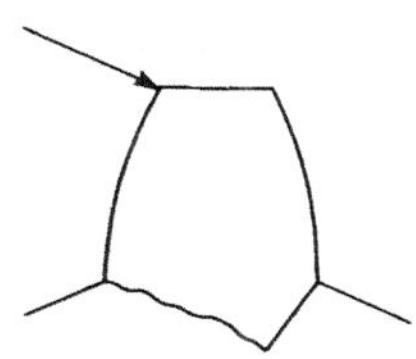
Fig. 320 Fracture of a tooth

force may be considered as a static load. For gears rotating at high speed the impact between the mating teeth gives an additional load. The impacts are due to the errors in the tooth profiles and in spacing the tooth profiles and due to the slight change in spacing caused by the deflexion of the teeth under load. The increased load is considered as a function of the accuracy of the gears and the velocity with which the teeth come together. The increased load is allowed for in the allowable specific pressure in the surfaces of contact or in the allowable bending stress in the roots of the teeth.

Straight spur gears

Wear strength

The calculation is based on the Hertz equation for the maximum pressure p_H occuring at the centre of the surfaces of contact of two convex cylinders.

$$p_H = \sqrt{0.35 \frac{F_n E}{2 b} \frac{\varrho_1 + \varrho_2}{\varrho_1 \varrho_2}}$$

where F_n = normal force acting on the pitch circle

E = modulus of elasticity; for unlike materials $\frac{1}{E} = \frac{1}{E_1} + \frac{1}{E_2}$

ϱ_1 and ϱ_2 = radii of curvature on the pitch circle, respectively

b = width of tooth.

The radii of curvature and the normal force may be calculated, from Fig. 321, thus:

$$\varrho_1 = r_1 \sin \alpha, \quad \varrho_2 = r_2 \sin \alpha$$

$$F_n = \frac{F_u}{\cos \alpha} = \frac{M_t}{r \cos \alpha}$$

these values inserted in the above equation gives:

$$p_H = \sqrt{0.35 \frac{M_t E 4}{b d^2 \sin 2 \alpha} \frac{i+1}{i}}$$

where $i = r_2/r_1$

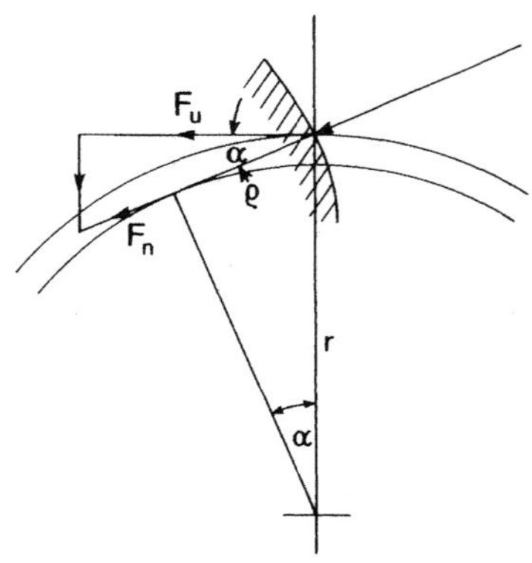

Fig. 321

$m z$ substituted for d and ψm substituted for b gives for $\alpha = 20°$

$$p_H = 1.475 \sqrt{\frac{M_t\, E}{\psi\, m^3\, z^2}\,\frac{i+1}{i}} \tag{176}$$

where M_t = torque transmitted by the gear
m = module
z = number of teeth
ψ = b/m.

The more accurate the manufacturing of the teeth, the wider the teeth may be made. For cast gears, $\psi = 5$ to 6, for machined gears, when overhang mounted $\psi = 10$ and when supported on both sides, $\psi = 15$ to 20.

The allowable specific pressure may be obtained for different materials from curves. The values in the curves are given for a certain service life, normally for 25 000 hours. If another service life is desirable, then the values for the specific pressure must be multiplied with a factor which can be taken from a curve of service life.

Beam strength

The normal force F_n acting on the upper end of the tooth is resolved into the radial component F_c and the tangential component F_b, as shown in Fig. 322.

The radial force produces a compressive stress and the tangential force a bending stress. Failure occurs on the tensile side of the tooth. The compressive stress which reduces the tensile stress may be neglected in comparison with the bending stress.

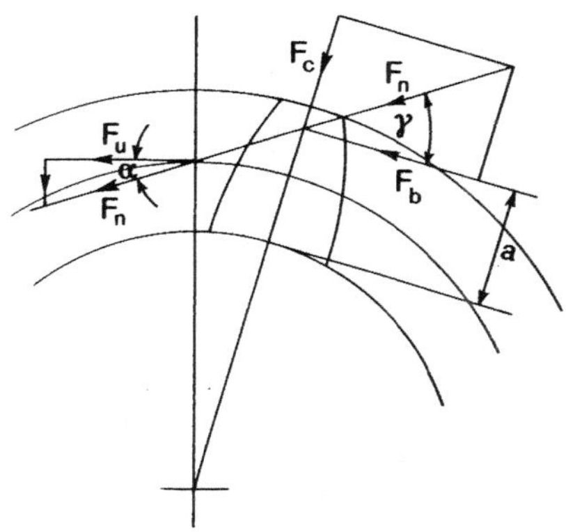

Fig. 322

The maximum value of the bending stress is:

$$\sigma_{b\,max} = \frac{F_n \cos \gamma\, a\, 6}{b\, s_1^2}.$$

Substituting $F_u/\cos\alpha$ for F_n, $x\,m$ for s_1 and $y\,m$ for a gives:

$$\sigma_{b\,max} = \frac{F_u}{b\,m}\,\frac{\cos\gamma\, y\, 6}{\cos\alpha\, x^2} \quad \text{or} \quad \sigma_{b\,max} = \frac{2\,M_t}{b\,d\,m}\,q \tag{177}$$

q is called the "tooth factor". It is dependent upon the number of teeth and may be found from tables.

The force acting on the shaft is:

$$F_n = \frac{F_u}{\cos \alpha}.$$

Straight bevel gears

For determining its strength the bevel gear is replaced by an equivalent spur gear with the pitch radius

$$r_{em} = r_m/\cos \delta \quad \text{(Fig. 323).}$$

The calculation for the wear strength is made in a similar way as for straight spur gears. The maximum pressure between the tooth surfaces is for $\alpha = 20°$

$$p_H = 1.475 \sqrt{\frac{M_t E}{\psi m_m^3 z^2} \sqrt{\frac{i^2+1}{i^2}}}$$

where $m_m = \dfrac{d_m}{z}$ and $\psi = \dfrac{b}{m_n}$

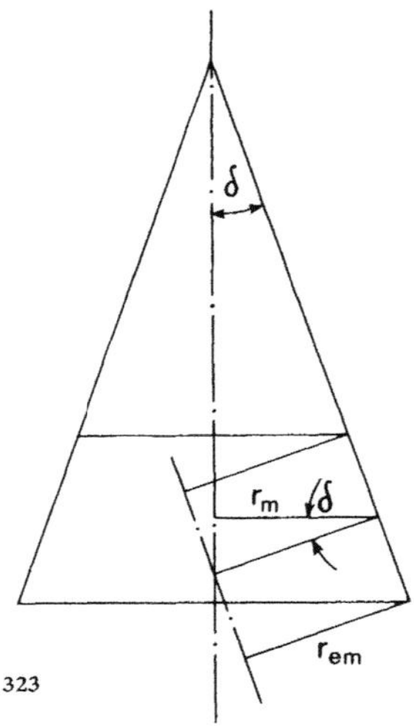

Fig. 323

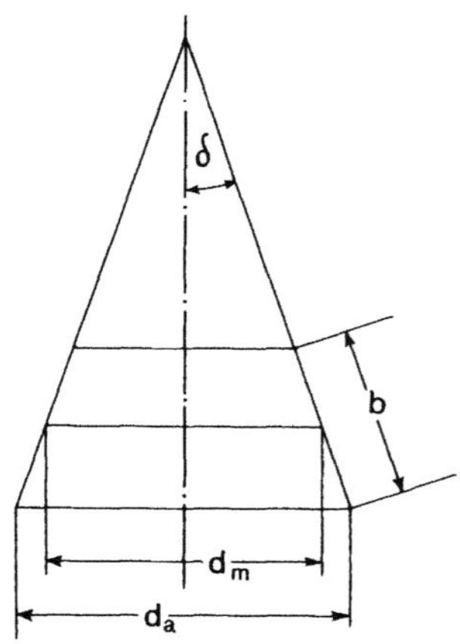

Fig. 324

The diameter of the base is calculated, as can be seen in Fig. 324, thus: $d_o = d\,m + b \sin \delta$.

The basic module, which is standardized, is: $m = \dfrac{d_a}{z}$.

For calculation of the beam strength formula (177) may be used. The values for the tooth factor remain the same, only d_m and m_m must be inserted for d and m.

The axial and radial forces acting in bevel gears, set at an angle of 90°, may be calculated, as seen in Fig. 325, thus:

$$F_{a1} = F_{r2} Q \sin \delta_1 = $$
$$= F_u \tan \alpha \sin \delta_1$$

$$F_{a2} = F_{r1} = Q \cos \delta_1 =$$
$$= F_u \tan \alpha \cos \delta_1 \, .$$

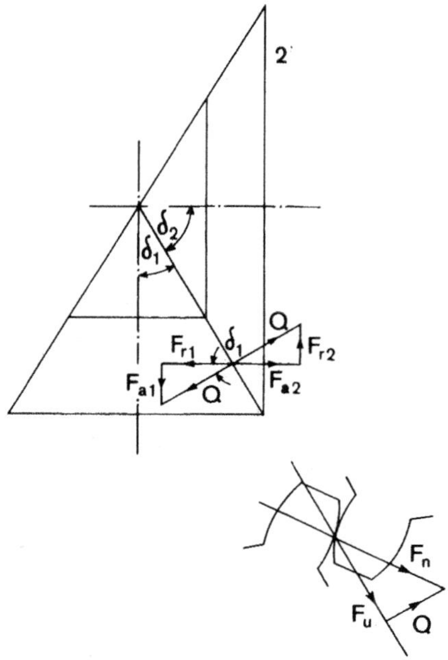

Fig. 325 Forces acting on bevel gears

Helical spur gears

For determining its strength the helical spur gear is replaced by an equivalent straight spur gear with the pitch radius

$$r_n = r/\cos^2 \beta \, .$$

The wear strength is calculated in similar way as for straight spur gears. The maximum pressure between the tooth surface is for $\alpha_n = 20°$

$$p_H = 1.475 \sqrt{\frac{M_t \, E \cos^4 \beta}{m_n^3 \, z^2} \frac{i+1}{i}} \qquad (181)$$

where $\psi = b/m_n$.

The beam strength may be calculated from formula (177). The values for the tooth factors q remain the same, only m_n must be inserted for m.

The radial and axial forces acting on helical spur gears may be calculated, as seen in Fig. 326, thus:

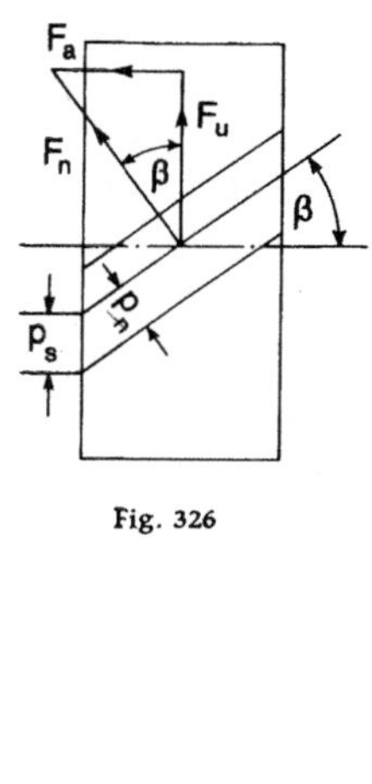

Fig. 326

$$F_r = \frac{F_u}{\cos \beta} \tan \alpha_n \text{ and } F_a = F_u \tan \beta \, . \qquad (182)$$

Worm gearing

In Fig. 327 the forces acting on a worm gearing are shown.

$$F_{uw} = F_{ag} = \frac{2\,M_{tw}}{d_w}$$

$$F_{aw} = F_{ug} = \frac{F_{uw}}{\tan(\gamma + \varrho)} \qquad (183)$$

$$F_{rw} = F_{rg} = \frac{F_{uw} \cos \varrho \tan \alpha_n}{\sin(\gamma + \varrho)} \,.$$

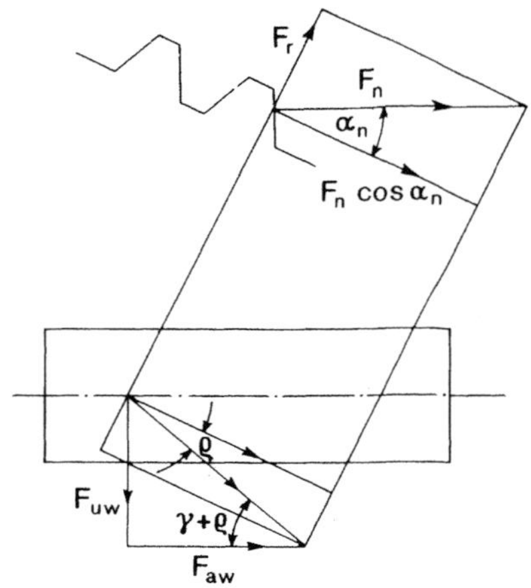

Fig. 327 Forces acting on worm gearing

The calculation of strength for the worm gear is based on a load factor. The module of the worm gear may be calculated from the following formula:

$$m = \sqrt[3]{\frac{2\,M_t\,\eta_w\,\eta_b}{z\,c\,f\,\psi\,\pi^2}} \qquad (184)$$

where M_t = twisting moment of the worm gear

η_w = efficiency of worm gearing $= \dfrac{\tan \gamma}{\tan(\gamma + \varrho)}$

η_b = efficiency of the worm bearing

z = number of teeth of the worm gear

c = load factor

f = tooth factor

$\psi = b/p_s = 1.5$ to 3.

The load factor c and the tooth factor f may be found from tables.

8.2.7.3. Cutting of Gear Teeth

The cutting of gear teeth is performed by forming cutters or by generating cutters. In form cutting, the tooth space takes the exact shape of the cutter. In generating, a tool having a shape different from the tooth profile is moved relative to the gear so that the proper tooth shape is obtained.

Gear teeth may be formed by milling, shaping or hobbing. They may be finished by shaving, burnishing, grinding or lapping. The oldest method of cutting gear teeth is that of milling by means of a form milling cutter. Theoretically it is necessary to use a different cutter for each gear having the same module, but the change in space is not too great, so that eight cutters may be used for a range of twelve teeth. Teeth may be generated with either a pinion cutter or a rack cutter. The cutter reciprocates, after each cutting stroke and the gear blank and the cutter roll slightly on their pitch circles. Hobbing is made with a tool which is shaped like a worm. The errors in tooth profiles may be diminished by finishing. Burnishing and shaving are used with gears which are not treated. Grinding and lapping are used for hardened gear teeth.

8.2.7.4. Gear Trains

A series of gears working together is called a gear train. Gear trains are normally used as speed changers. A gear train consisting of one pair of gears is called a "simple gear train" or a "single stage gear train". A gear train consisting of two, three or more gear pairs is called a "double, triple, etc. stage gear train". Gear trains consisting of only spur gears are called "spur gear trains", those which consist of only bevel gears "bevel gear trains" and those which consist of spur and bevel gears "spur bevel gear trains" (Fig. 328). Gear trains are normally provided with rolling contact bearings.

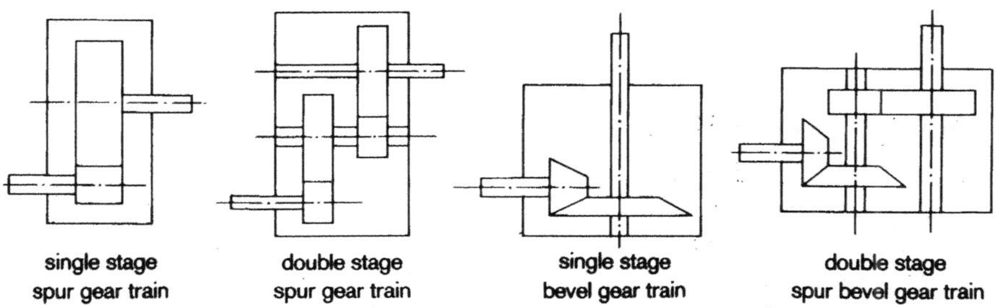

Fig. 328

A gear train may be mounted on an electric motor and form with the motor one unit. Such a unit is called a "geared motor". The speed ratio of geared motors varies from 1: 8 up to 1 : 20.

Planetary gear train

A planetary gear train includes a sun wheel, two or more planet wheels, a planet carrier and an internal gear wheel (Fig. 329). The motion of any part of this gear train is only determined, if the motion of two elements is known. For instance, if the sun wheel is driven and the internal gear wheel is stationary, the motion of the planet wheels and the planet carrier is known. For the case, when the sun wheel is driven and the internal gear wheel is stationary, the velocity ratio between the sun wheel and the planet carrier may be determined as follows.

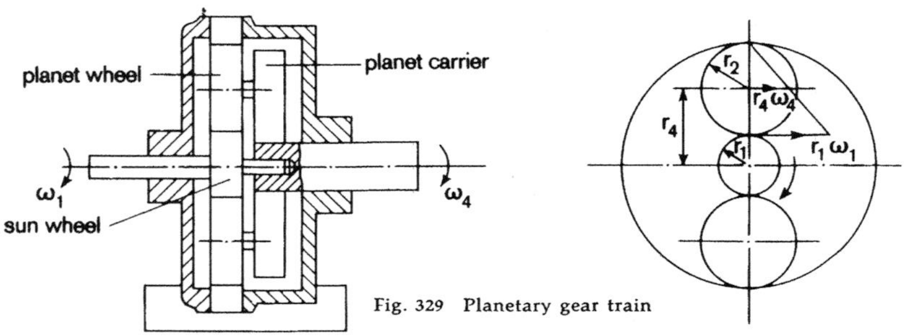

Fig. 329 Planetary gear train

The planet wheels are rotating about the point of contact with the internal gear wheel.

$$\frac{r_1\,\omega_1}{r_4\,\omega_4} = 2\,. \quad \text{i. e.}\ \frac{\omega_1}{\omega_4} = 2\,\frac{r_4}{r_1} \tag{185}$$

For $\dfrac{r_1}{r_2} = \dfrac{1}{2}$, $\dfrac{\omega_1}{\omega_4} = 2 \times 3 = 6$ since $r_4 = r_1 + r_2$.

8.3. Elements of a Slider Crank Mechanism

A slider crank mechanism is a mechanism which serves to convert reciprocating motion into rotary motion, such as in a reciprocating power engine, e. g. the internal combustion engine. Conversely it may be used to convert rotary motion into reciprocating motion, such as in a reciprocating working machine, e. g. the piston compressor. There are two types of reciprocating machine, the single and the double acting machine, shown in Fig. 330. In the first type the working fluid is on one side of the piston, in the second type the working fluid is on both sides of the piston. The mechanism of the first

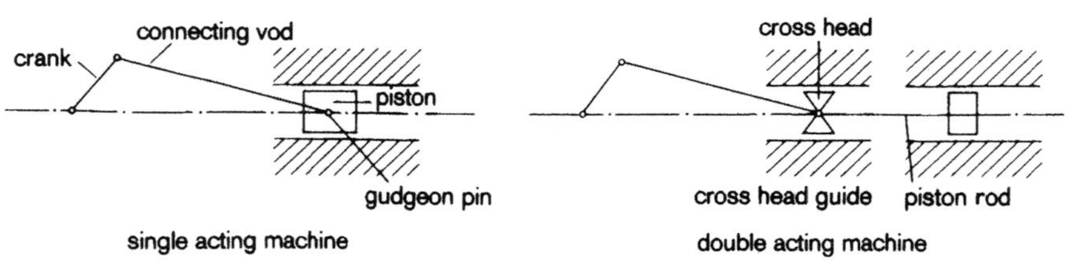

Fig. 330 Elements of a slider crank mechanism

type consists of the crank, the piston and the connecting rod; the mechanism of the second type has additionally a cross head and a piston rod. Examples for the first type are: the internal combustion engine and the single acting pump and compressor. Examples for the second type are: the steam engine, the double acting Diesel motor and the double acting pump and compressor.

8.3.1. Kinematics of the Slider Crank Mechanism

The displacement s, the velocity v and the acceleration a of the point C (piston) of the slider crank mechanism, seen in Fig. 331, may be calculated approximately, as follows:

$$s_C = r\left(1 + \frac{\lambda}{4} - \cos\alpha - \frac{\lambda}{4}\cos 2\alpha\right)$$

$$v_C = r\,\omega\left(\sin\alpha + \frac{\lambda}{2}\sin 2\alpha\right) \qquad (186)$$

$$a_C = r\,\omega^2\left(\cos\alpha + \lambda\cos 2\alpha\right)$$

where $\lambda = \frac{r}{l}$.

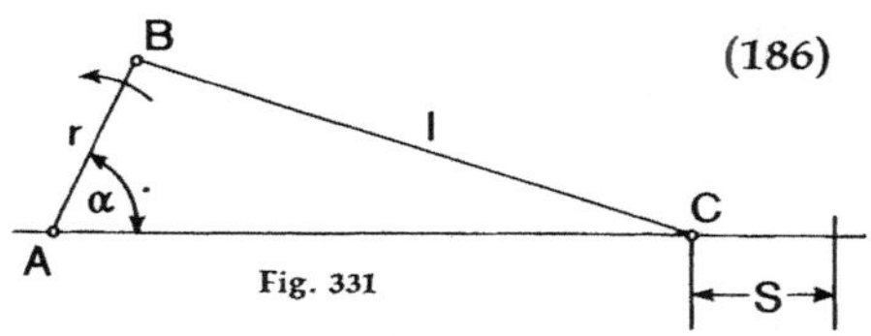

Fig. 331

Graphically the velocity of the piston may be determinated, as Fig. 332 shows, thus:

$$v_C = v_B \overset{\wedge}{+} v_{CB} \text{ (equation 39)}$$

where v_B = velocity of B

and v_{CB} = velocity of C relative to B

for $\omega = 1$, $v_B = r\,\omega = \overline{AB}$ and $v_{CB} = \overline{BD}$.

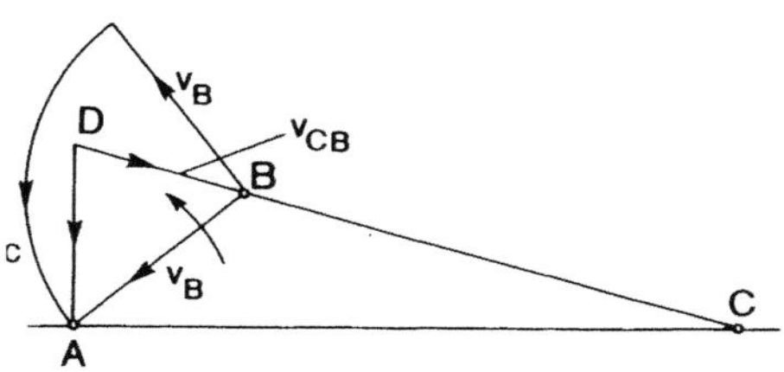

Fig. 332
Determination of the velocity of the piston

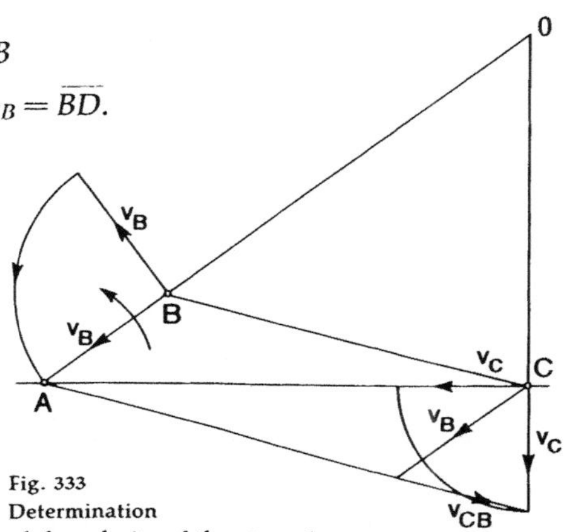

Fig. 333
Determination
of the velocity of the piston by means
of the instantaneous centre

The velocity may also be determinated by means of the instantaneous centre, as Fig. 333 shows.

The acceleration of the piston may be determined, as Fig. 334 shows, thus:

$$a_C = \overset{\wedge}{n_B} + \overset{\wedge}{n_{CB}} + t_{CB}$$

where n_B = normal acceleration of B
n_{CB} = normal acceleration of C relative to B
t_{CB} = tangential acceleration of C relative to B.

The velocity v_B is supposed to be uniform. The tangential acceleration t_B is therefore zero.

For $\omega = 1$, $a_B = n_B = r\,\omega^2 = \overline{AB}$.

Since $\Delta\, ABC \sim \Delta DBE$
and $\Delta\, ABD \sim \Delta\, EBF$,

$$\frac{\overline{BF}}{\overline{BD}} = \frac{\overline{BE}}{\overline{BA}} = \frac{\overline{BD}}{\overline{BC}}$$

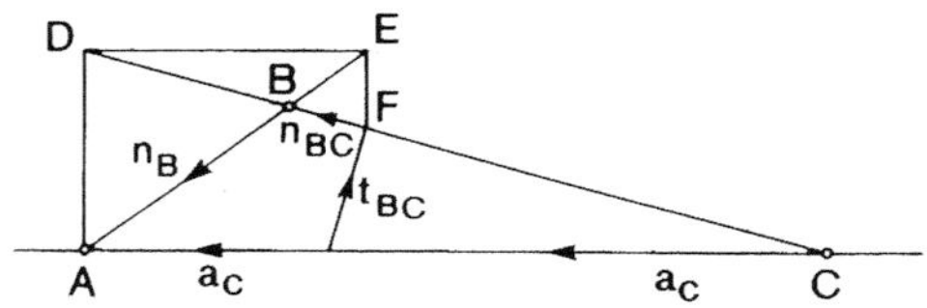

Fig. 334
Determination of the acceleration of the piston

hence

$$\overline{BF} = \frac{(\overline{BD})^2}{\overline{BC}} = \frac{v_{CB}^2}{\overline{BC}} = n_{CB} \text{ (equation 31).}$$

8.3.2. Forces acting on the slider crank mechanism

On the slider crank mechanism two kinds of forces are acting: the force of the working fluid and the forces of inertia of the moving masses. The force of the working fluid is acting on the crown of the piston and the forces of inertia are acting at the centres of gravity of the single parts of the slider crank mechanism.

The changing force of the working fluid may be taken from an indicator diagram. The forces of inertia of the single parts of the slider crank mechanism are found from the product of mass and acceleration of the parts. The masses of piston, piston rod and crosshead have reciprocating motion, the mass of the crankshaft rotary motion and the mass of the connecting rod partly rotary and partly reciprocating motion. For calculation, the mass m of the connecting rod is distributed, as Fig. 335 shows, in a rotary mass m_1 acting in the crank pin and in a reciprocating mass m_2 acting in the gudgeon pin. The mass distribution is determined statically. The mass portion m_1

is found as $m_1 = m\frac{b}{l}$ and the mass portion m_2 as $m_2 = m\frac{a}{l}$. The force of inertia acting on the crank is composed of the force of inertia produced by the mass of the crank and the portion of the mass of the connecting rod.

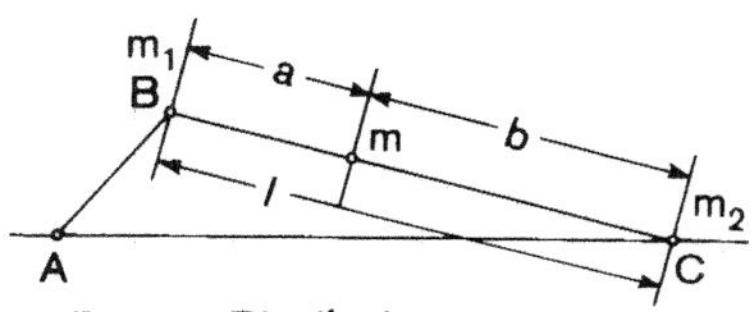

Fig. 335 Distribution of the mass of the connecting rod

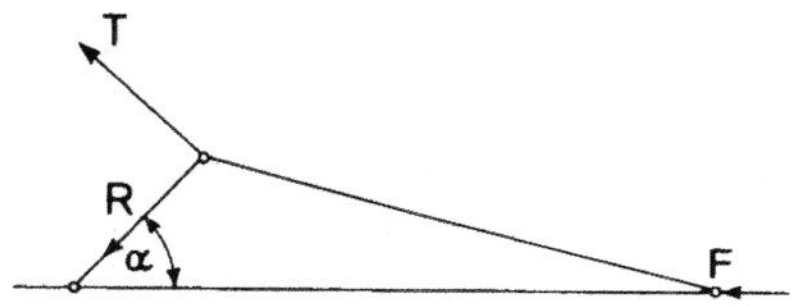

Fig. 336

The force of inertia acting on the crank may be compensated for by balance weights. The force F acting on the point C in Fig. 335 is composed of the force of the working fluid acting on the crown of the piston and the forces of inertia of the reciprocating masses i. e. the masses of the piston, the cross head, the piston rod and the reciprocating portion of the mass of the connecting rod. The force F is transmitted by the connecting rod on the crank pin. The force F_B acting on the crank pin is supposed to be resolved into a radial component R and a tangential component T, shown in Fig. 336.

The radial force is calculated approximately from

$$R = F\left(\cos\alpha - \frac{\lambda}{2} + \frac{\lambda}{2}\cos\alpha\right). \tag{187}$$

The tangential force is calculated approximately from

$$T = F\left(\sin\alpha + \frac{\lambda}{2}\sin\alpha\right). \tag{188}$$

The changing tangential force acting on the crank pin causes a variable angular velocity of the crank. This variable speed of the crank may be compensated for by using a flywheel. The calculation of the required size of the flywheel is based on a given coefficient of cyclic variation.

8.3.3. Parts of the Slider Crank Mechanism

Crankshaft

The crankshaft consists of the journals, the throws of the cranks and the coupling flange. The throw is formed by two webs and the crank pin. Normally each crank pin is situated between two journals. But sometimes, for small loads, two crank pins are between two journals (Fig. 337). For

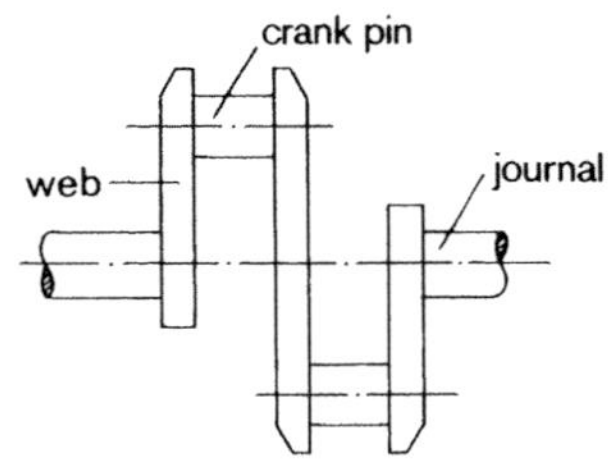

Fig. 337

single cylinder machines of horizontal construction an overhung crank may also be employed. For single cylinder machines balanced crankshafts or flywheels ar necessary for compensating for the free forces of inertia. In multicylinder machines the throws of the crankshaft are arranged so, that the forces of inertia are compensated. The crankshaft is normally forged of one piece, but for big machines the crankshafts are composed of separate parts. For lubricating the big end bearings of the connecting rod, oil holes are drilled in the crankshaft.

Piston

A piston is a cylindrical body which fits in a hollow cylinder in such a way that although free to slide in it, under the action of fluid pressure, as in a steam engine, or if acting against fluid pressure under the action of a force as in a pump, practically no escape of the fluid from one side of the piston to the other one takes place.

Pistons normally are made of cast iron, pistons for large machines are made of cast steel, and small pistons may also be made of aluminium. The construction of a piston is dependent upon the kind of machine for which the piston is used. Generally there are three types: the plunger piston for pumps, the piston for double acting heat engines and compressors and the piston for single acting heat engines and compressors (Fig. 338).

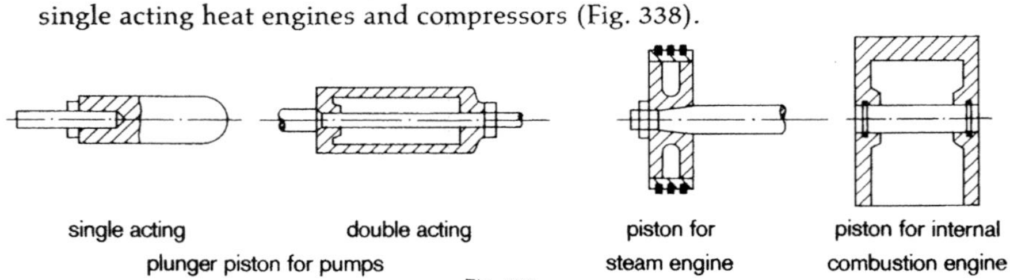

Fig. 338

Plunger pistons for pumps have no packing. The packing is fixed on the walls of cylinder. The pistons for heat engines and compressors have packing springs of rectangular cross section made of special cast iron to prevent the piston from leakage. The rings are machined to a diameter slightly larger than the cylinder bore and a gap is cut across the ring. The rings exert a constant pressure on the cylinder walls. Some pistons of internal combustion engines are fitted with a scraper ring which is mounted as the lowest ring in vertically constructed machines. The function of this ring is to clear surplus lubricating oil from the walls of the cylinder. This oil passes through small holes which are drilled in the cylinder skirt and return to the crank case. In the piston of internal combustion engines a hole is drilled which accomodates the gudgeon pin. On the ends of the gudgeon pin snap rings in grooves are located which prevent longitudinal movement of the pin.

Crosshead

The crosshead is that part of the slider crank mechanism which connects, for double acting machines, the ends of the piston and connecting rod. It consists, as seen in Fig. 339, of the crosshead body which is made of cast iron, cast steel or forged steel and the slide shoes which are cast or screwed on the body. In the body, the end of the piston rod and the journal for the small end bearing of the connecting rod are fastened. The guide surfaces may be made plane or round. In the normal construction the small end bearing of the connecting rod lies inside the crosshead, whereas for big machines the small end bearing of the connecting rod is shaped like a fork and works outside the crosshead.

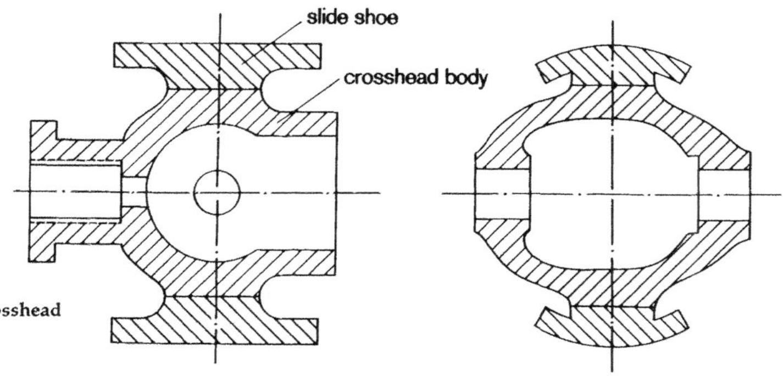

Fig. 339 Crosshead

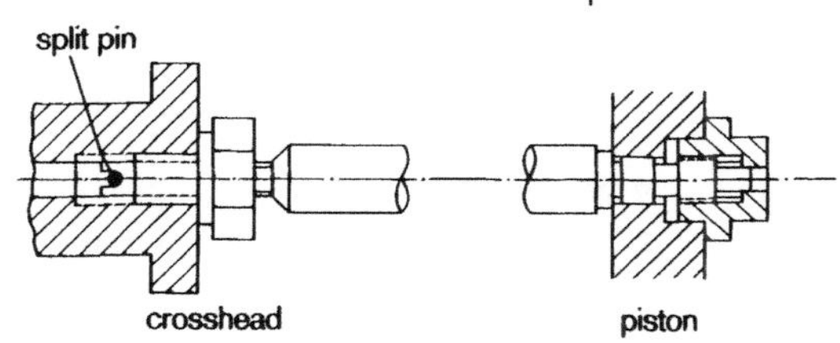

Fig. 340
Fixing of the piston rod
to the crosshead and to the piston

Piston rod

The piston rod connects together the piston and the crosshead. It passes through a stuffing box in the cover of the cylinder in which the piston works. The piston rod is made of steel having natural hardness or of case hardened steel. Fastening the piston rod to the piston and to the crosshead is made by screw connection, as shown in Fig. 340. The connection on the crosshead must be adjustable in order to be able to adjust the dead space in the cylinder.

Connecting rod

The connecting rod connects the crankshaft and the crosshead or the piston, respectively. The length of the connecting rod, measured from the centre of the crank pin to the centre of the gudgeon or crosshead pin varies from about 3.5 to 6 times the length of the crank according to the type of the machine. Connecting rods, normally, are drop forged and made of I-section. The big end which is on the crank side, represented in Fig. 341, is fitted with a split bearing, and the small end with a eye bearing.

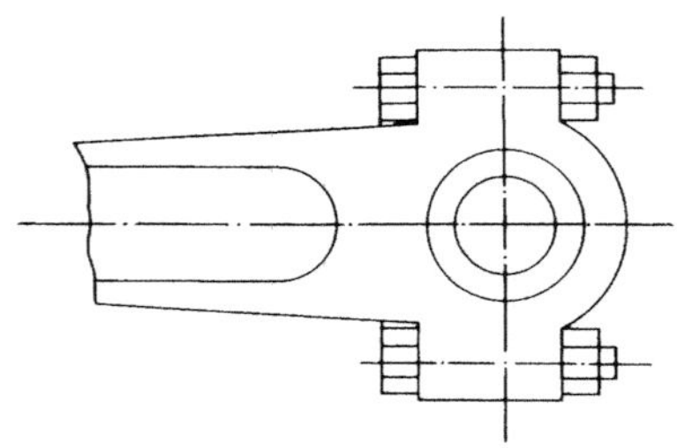

Fig. 341 Big end of a connecting rod

8.4. Pipes and valves

Pipes are used for conveyance of steam, water, gas, oil and other fluids. The material of pipes depends on varions points such as cost of production, nature of the fluid, pressure and velocity of the fluid, etc.

A pipe system generally consists of straight pipes, bends, elbows, pipe connections, expansion joints (for hot fluids) and valves.

The pipes are marked with different colours according to the kind of fluid which flows through them.

The following colours are used for the single fluids:

water	green	acid	orange
steam	red	alkaline	violet
air	blue	oil	brown
gas	yellow	vacuum	grey

The pipes are either completely painted or marked with stripes in the corresponding colour.

Pipes are designed for certain pressures. The nominal pressure is the pressure for which the pipes are marked. The working pressure is dependent upon the temperature of the fluid. Up to 120° C the working pressure is equal to the nominal pressure. For higher temperatures the working pressure is smaller than the nominal pressure. For each nominal pressure a proof pressure is stated which is about from 1.5 to 2 times the nominal pressure.

Nominal diameters which normally are equal to the internal diameter of pipes are standardized. Deviations between nominal and internal diameter result from the different thicknesses of walls, if the external diameter is kept constant.

8.4.1. Pipes

Pipes are made of cast iron (centrifugally cast) up to a nominal pressure of 16 kp/cm², of steel (seamless draught or rolled or for greater diameters welded) or of non-ferrous metal.

Calculations for pipes

Internal diameter of pipes

The internal diameter is found from:

$$Q = \frac{d^2 \pi c}{4} \quad \text{whence} \quad d = \sqrt{\frac{4\,Q}{\pi c}} \tag{189}$$

where Q = conveying capacity (m³/s)
and c = velocity of the fluid (m/s).

For steam the conveying capacity is given in to/h. The conveying capacity in m³/s is found from:

$$Q = \frac{m_t}{3.6 \times 10^6 \varrho}$$

where m_t = mass per time unit (to/h) and
ϱ = density (kg/m³).

Pressure reduction

The pressure reduction in pipe lines is calculated from:

$$p = \left(\lambda \frac{l}{d} + \sum \zeta\right) \frac{\gamma c^2}{2 g} \tag{190}$$

where l = length of the straight pipes in the pipe line
d = internal diameter of the pipes
λ = friction factor for pipes
ζ = loss coefficient for fittings and valves
γ = specific weight of the fluid
c = velocity of the fluid
g = acceleration due to gravity.

Thickness of steel pipes

The thickness of steel pipes may be calculated up to the temperature of 120 °C from following formula:

$$s = s_0 + c_1 + c_2 = \frac{d_e\, p}{2\, v\, \sigma_{all}} + c_1 + c_2 \tag{191}$$

where d_e = external diameter of pipe
p = pressure in the pipe
σ_{all} = allowable stress
v = efficiency of the joint (for seamless pipes, $v = 1$)
c_1 = factor for rust formation, normally 0.1
c_2 = additional factor (0.085 to 0.22) s_0 (cm)
s_0 = theoretical thickness.

8.4.2. Pipe joints

There are two types of joints: socket and flange joints.

Cast iron pipe joints

In Fig. 342, a socket joint and three types of flange joints for cast iron are represented. Socket joints are used for cast iron pipes conveying water or gas of low pressure, which are embedded in the ground. Socket joints have

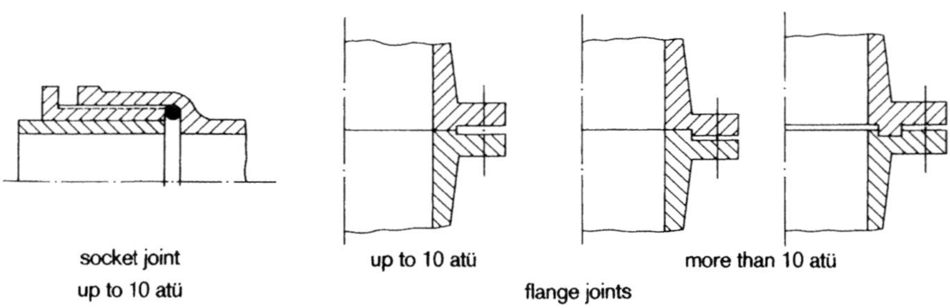

Fig. 342 Cast iron pipe joints

a certain flexibility allowing the pipe to accomodate itself to slight settlement of the earth. For work above ground cast iron pipes with flanges are used. Between the flanges sealing rings are inserted to form an effective seal.

Steel pipe joints

In Fig. 343 socket and in Fig. 344 flanged joints for steel pipes are demonstrated. The socket joint in steel pipes is used in a similar way to the socket joint in cast iron pipes for work below ground but for higher pressures. Steel pipes of small diameters and low pressures have sockets screwed on the pipes. Steel pipes with greater diameters and higher pressures have flange joints. The flanges may be screwed, rolled, riveted or welded on the pipe ends.

Socket joints

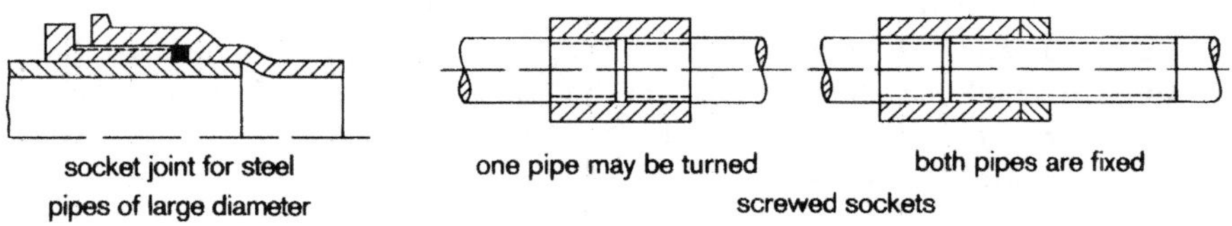

Fig. 343 Steel socket joints

flanged joints

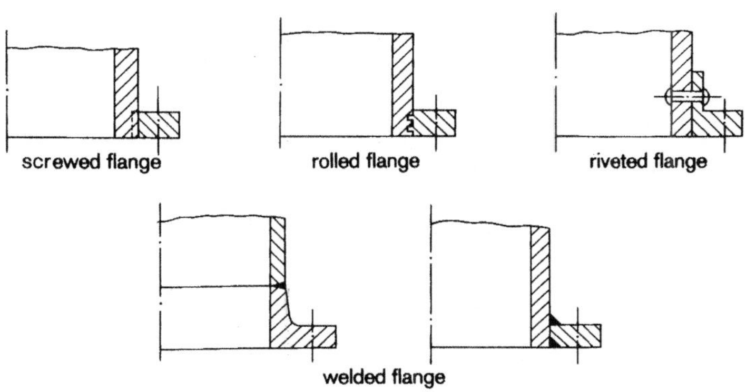

Fig. 344 Fixed steel flanged joints

Loose flange

Loose flanges, demonstrated in Fig. 345, are used for high pressure steam pipes and for joints of pipes of non-ferrous metals. For mounting pipes having joints with loose flanges the correct positions of the corresponding holes for the bolts must not be exactly observed.

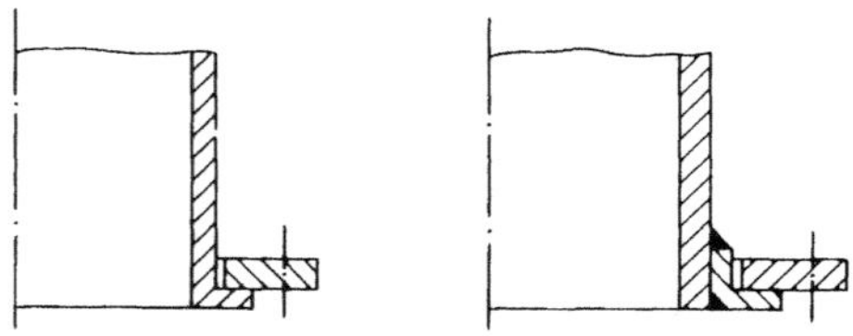

Fig. 345 Loose flanges

8.4.3. Fittings

Fittings are shaped joint parts which are used in pipe installations to form bends, branches, corners, etc. For pipes of small diameters fittings with internal or external screw threads on their ends are used. Fittings with external screw threads must have sockets to connect them up to the pipes; fittings with internal screw threads are connected by merely screwing the pipes into them. For pipes of large diameters, fittings with flanged ends are used. Fittings of small diameters are made of malleable cast iron, and those of large diameters are made of cast iron. In Fig. 346 some types of fittings of small diameters are demonstrated.

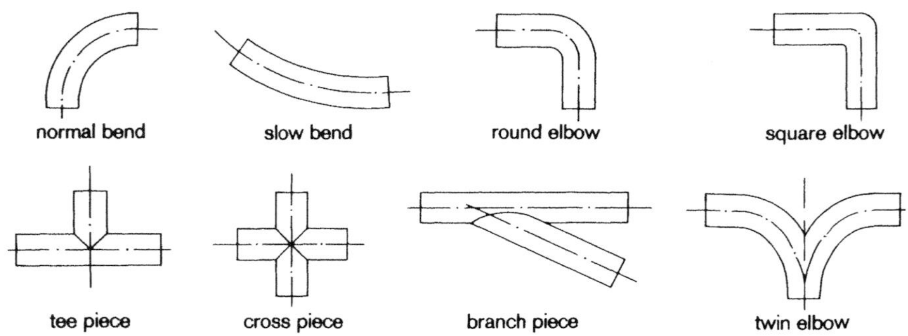

Fig. 346 Kinds of fittings

8.4.4. Expansion Joints

In steam or hot water pipe lines expansion joints must be inserted which accomodate alterations of length of the pipes due to the varying temperatures. Normally corrugated expansion joints made of alloyed steel or tombac are used. There are two types of corrugated expansion joints: the axial and the hinged expansion joint. The first one is stretched and inserted in the pipe line to reduce its compression. The second type is used for larger expansions.

The amount of thermal expansion may be calculated from:

$$\Delta l = \alpha_t \, l \, \Delta t$$

wher α_t = coefficient of thermal expansion
l = length of pipe line
t = change of temperature.

The coefficient of thermal expansion is a function of temperature. For calculations, an average value over a certain range of temperature is taken. In Fig. 347 two types of of corrugated axial expansion joints and in Fig. 348 two types of corrugated hinged expansion joints are demonstrated. Fig. 349 shows examples for mounting corrugated hinged expansion joints.

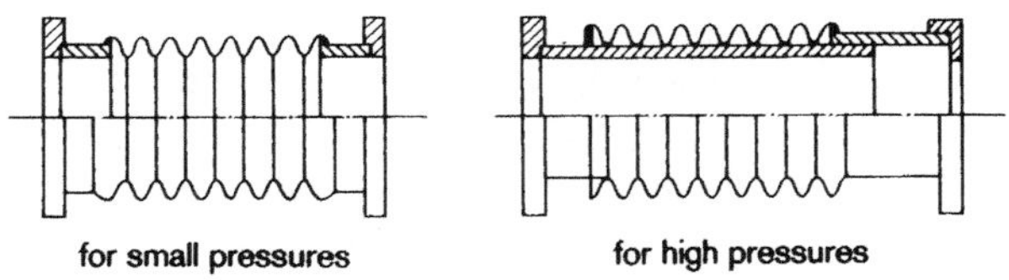

Fig. 347 Corrugated axial expansion joints

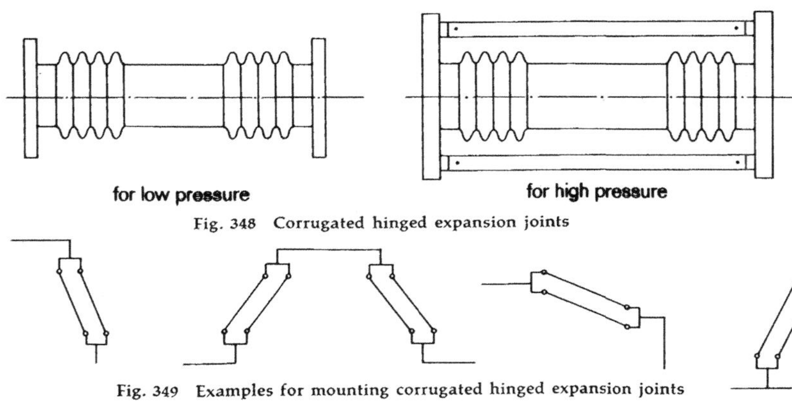

Fig. 348 Corrugated hinged expansion joints

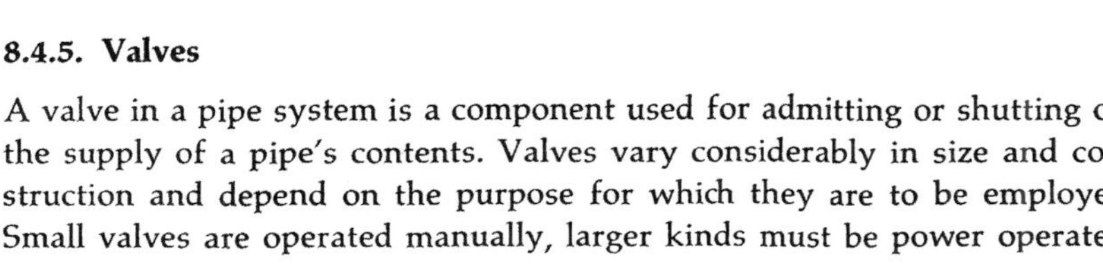

Fig. 349 Examples for mounting corrugated hinged expansion joints

8.4.5. Valves

A valve in a pipe system is a component used for admitting or shutting off the supply of a pipe's contents. Valves vary considerably in size and construction and depend on the purpose for which they are to be employed. Small valves are operated manually, larger kinds must be power operated.

Valves are made of cast iron, cast steel, forged steel and non-ferrous metals. Yellow and red brass are used to a large extent for the manufacture of valves used in cold water and domestic heating systems. Cast iron is employed for valves used for many industrial purposes and cast and forged steel for high pressure steam work.

Generally there are four types of valves used in a pipe system: the globe valve, the gate valve, the plug cock and the clack valve (Fig. 350).

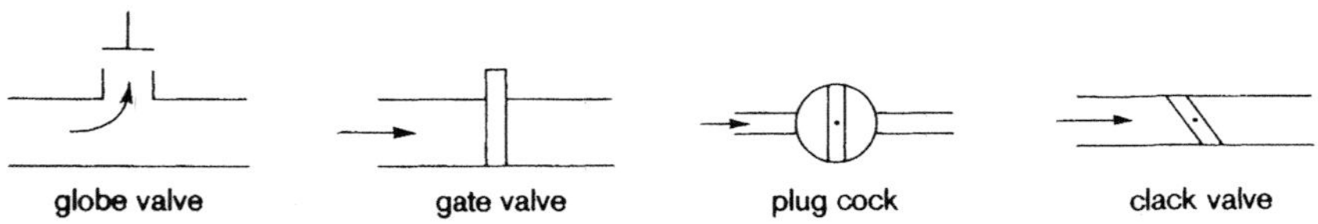

Fig. 350 Types of valves

In the construction of valves the following must be taken in consideration; shutting off must be tight and permanent, and the velocity and the direction of the fluid stream should be changed as little as possible. The gate valve and the plug cock fullfils this condition to a high degree. The sealing parts must be easily accessible for cleaning and changing.

8.4.5.1. Globe valves

There is a great variety of types of globe valves of which the most common are: the shut-off valve, the regulating valve, the non-return valve, the safety valve and the pressure-reducing valve.

Shut-off valve

The most commonly used globe valve is the shut-off valve. According to the shape of their housings shut-off valves are classified as straight-way or angle valves. The seat of the straight-way valve may be perpendicular or slanted to the axis of the pipe line. In Fig. 351 two types of housings of the straight-way valve and the housing of an angle valve are demonstrated.

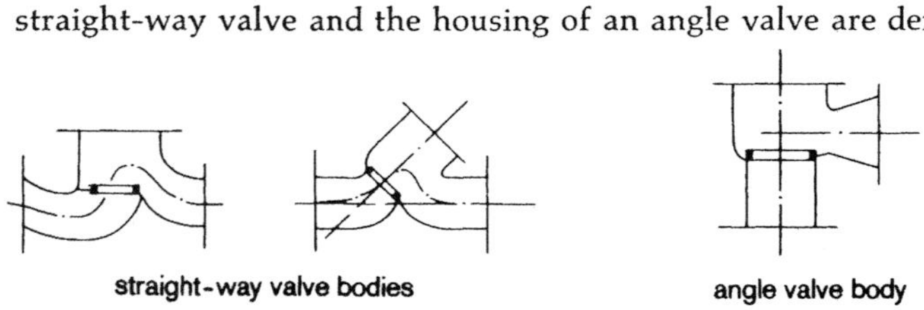

Fig. 351 Types of valve housings

As a spindle guide for smaller valves, epecially for domestic or industrial hot water systems a bonnet is used and for larger valves a guide frame is used

(Fig. 352). In the bonnet or guide frame, respectively, the packing is inserted. The packing is pressed against the spindle, and if the packing nut on the bonnet or the nuts of the bolts of the gland on the guide frame, respectively, are screwed in, this will give the sealing to the valve.

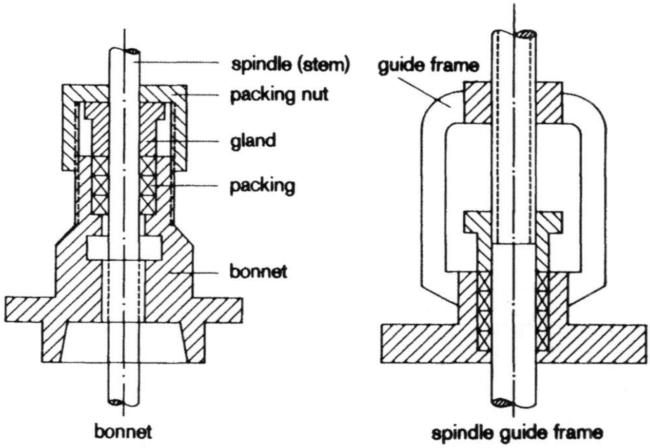

Fig. 352

On the lower end of the spindle the plunger is attached. Its sealing surface may be plane (valve disc) or conical (valve cone). By manipulation of the handle fixed on the upper end of the spindle, the spindle moves up and down and the plunger is lifted off or lowered on to its seating. Thus the orifice is opened or closed. The seating pressed into the housing is normally made of brass. In Fig. 353 two types of valve plungers are demonstrated.

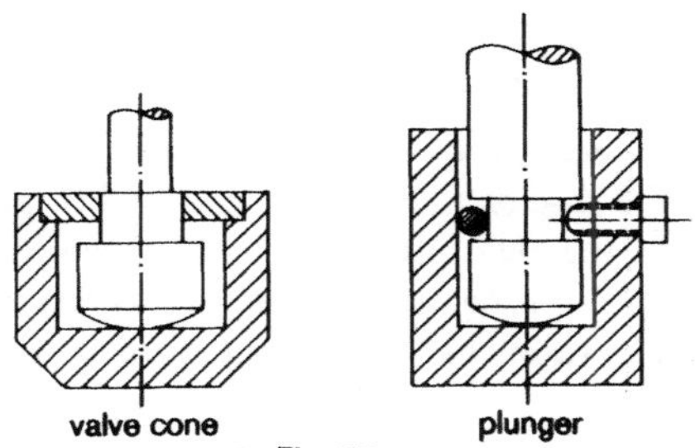

Fig. 353

Regulating valve

Regulating valves are used for regulating fluid quantity. The plungers of this type of valve have, on the sealing surface, an extended cone which is either shaped parabolic or made as a needle cone, as seen in Fig. 354. By lifting the plunger a gradual increasing of the orifice occurs, so that the rate of flow may be exactly regulated.

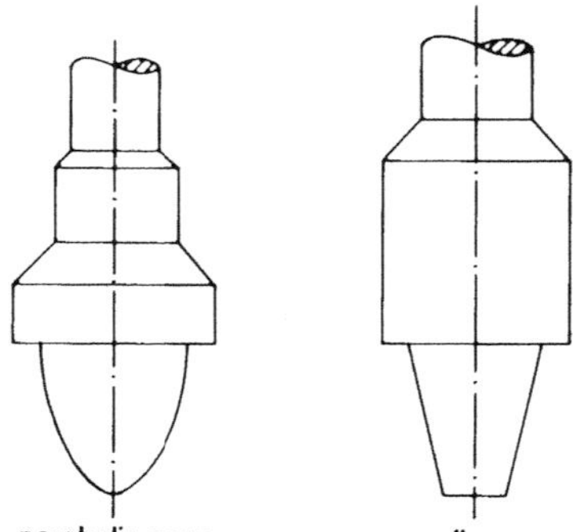

Fig. 354 Valve cones for regulating valves

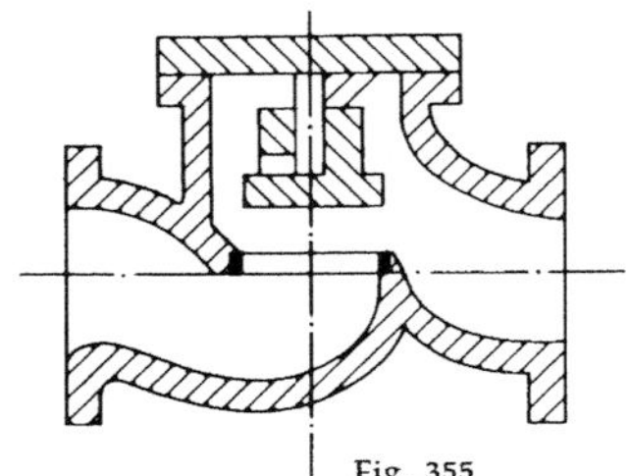

Fig. 355
Non-return valve

Non-return valve

Non-return valves, the most common type is demonstrated in Fig. 355, are used extensively for pipe lines in conjunction with pumping installations, especially, if the fluid is pumped up an incline. The purpose of this valve is to safeguard the pump, because if for any reason the pump stops working the column of liquid might run back and cause an impulse which might damage the pump mechanism.

Safety valve

Safety valves are used to safeguard boilers from overloading. The cone or the disc of the valve is retained in position on its seat against the pressure of the fluid either by a weighted lever or a spring, as shown in Fig. 356. Spring-type valves are usually given preference, because they withstand vibrations better. The pressure at which the valve should open can be regu-

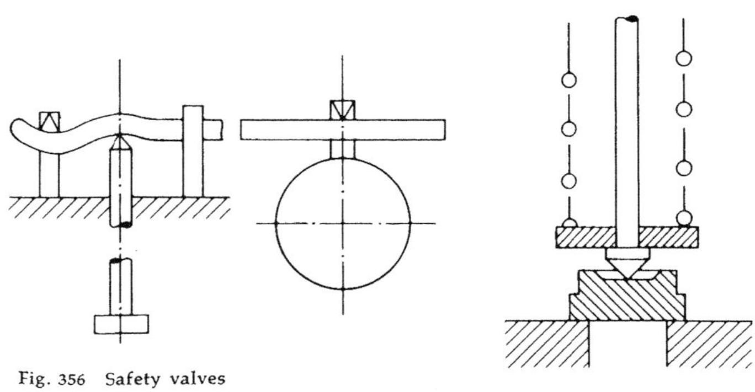

Fig. 356 Safety valves

lated by means of a screwed adjustable sleeve which is secured in any desired position by a lock nut. If the pressure rises above the adjusted amount, the force exerted by the fluid pressure lifts the cone or the disc and releases the fluid and thus prevents the boiler from bursting.

Pressure reducing valve

Pressure reducing valves, one simple type is shown in Fig. 357, are designed to reduce a high fluid pressure to a low fluid pressure. They must be constructed so that at varying high pressure the low pressure always remains constant. For this reason the plunger of the valve must be shaped in such a way that the high pressure acting on it is always compensated. Inside the valve is a jockey either made of rubber or of corrugated metal which is loaded on one side by the force of an adjustable spring and on the other side by the force of the fluid with the low pressure. If the fluid pressure has its adjusted value, then the spring force is in equilibrium with the fluid force. If the fluid pressure decreases then the valve opens, if the fluid pressure rises over the adjusted value then the valve shuts.

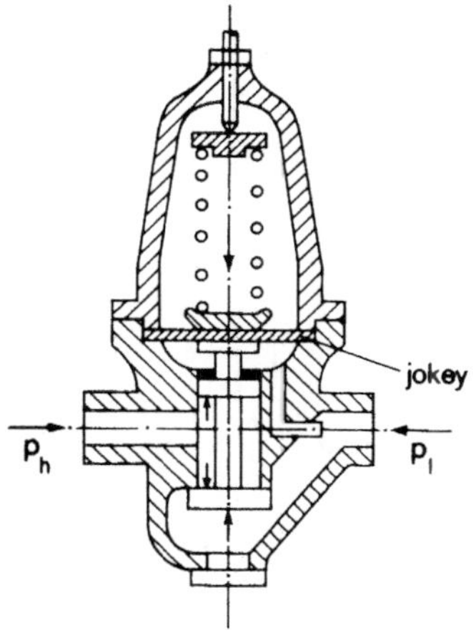

Fig. 357
Pressure reducing valve

8.4.5.2. Gate valves

Gate valves have full size straight way openings. The resistance of the fluid in these valves is very small, because the flow of the fluid has no change in direction or velocity. The gate may be either of tapered or of parallel-sided disc formation. Immediately below the gate a sediment chamber is formed and is fitted with a drain plug. After removing the drain plug any accumulated dirt or sediment which may be deposited from the fluid can easily be cleared.

Valves with a tapered gate have a non-rising spindle. A solid collar is formed on the spindle which is embedded in a casing on the bonnet so that the spindle cannot move in an axial direction. On the gate a nut is fixed in which the spindle is screwed in or out when turned and since the spindle cannot move in an axial direction the gate is raised or lowered (Fig. 358 (a)).

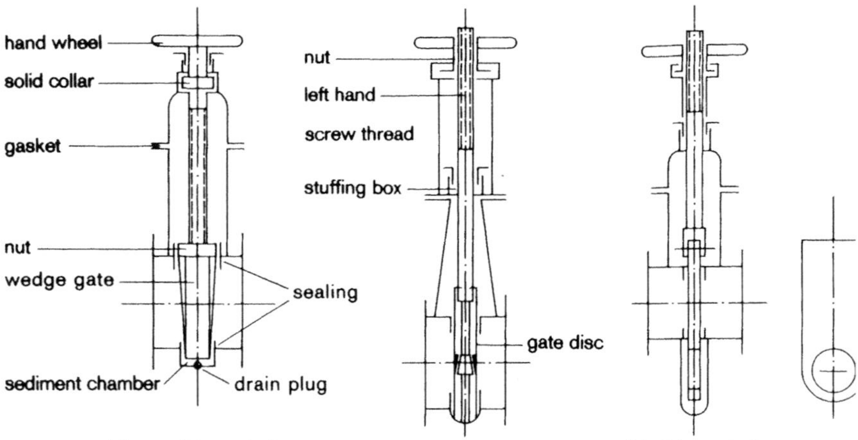

(a) non rising spindle (b) rising spindle

Fig. 358 Gate valves

For valves with parallel-sided discs the spindle raises or lowers when the handwheel is turned. The handwheel has inside a nut attached to it which is supported on the guide frame. If the handwheel is turned, the spindle with the gate is raised or lowered, whereas the handwheel does move in an axial direction. There are varying arangements for pressing the parallel discs on the seats. Two arrangements, seen in Fig. 358 (b), are in common use. For the first one the discs hang loosely on the spindle. If the gate reaches the bottom, the two discs are wedged against the seat by an adjusting wedge between them. For the second one a holed plate is moved between the seats which are pressed against the plate. If the plate is raised the fluid can flow through the hole in the plate, if the plate is lowered the flow of fluid is cut off.

8.4.5.3. Plug Cocks

Plug cocks are used in gas and water pipe systems of small diameters. They are made without or with stuffing boxes. Through the body centre at right angles to the line of flow a tapered hole is formed. Into this hole fits a similarly tapered plug which has a hole of almost rectangular formation. The top end of the plug has a square head to which a spanner can be applied to operate it. For the type without a stuffing box, seen in Fig. 359 (a), the diameter on the lower end of the plug is reduced and a square shoulder is formed from which a screw threaded end projects to fit a nut. A washer is fitted on the square shoulder to form a seal. For the type with a stuffing box, seen in Fig. 359 (b), the stuffing box is attached to the top of the body.

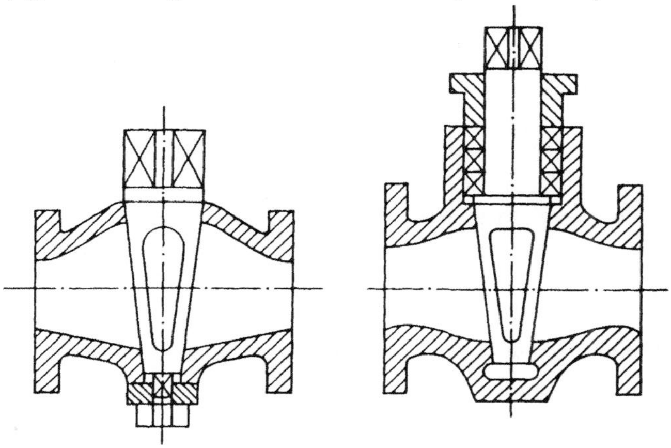

Fig. 359 Plug cocks

(a) plug cock without stuffing box (b) plug cock with stuffing box

8.4.5.4. Clack Valves

Clack valves are used as non-return valves in suction and pressure pipes of pumps. Closure occurs by the door's own weight or by an additional weight or by a spring. Clack valves are also employed as throttle valves for regulating the rate of flow, such as for smoke clack valves on furnaces or for the clack valves on carburettors of motor cars.

8.5. Springs

A spring is an elastic member used for many purposes in engineering. It is designed to give an elastic deformation under a given load. The forms of springs allow them to take external forces over considerable distances without losing their elasticity. Thus, for example, a helical spring made of steel wire can be expanded to twice its length without losing its elasticity. On the other hand, a bolt made of steel may be expanded only a small amount, about 0.004 of its length to remain within the elastic limit.

Springs are employed in machines and devices as follows:

1. to generate force as in clutches, brakes and valves or to generate force to keep cam followers in contact with cams.

2. to store energy as in clocks, door closing devices, etc.
3. to diminish shocks as in springs on wheels of vehicles, buffers on locomotives or railway carriages, etc.
4. to measure forces on scales.
5. to generate vibrations as in swing sieves or conversely to prevent sympathetic vibrations in vibration dampers.

Springs are made of special steel called "spring steel", because of its superior strength as well in normal as in elevated temperatures and of its relative freedom from creep or relaxation under load. But other metals, such as bronze are occasionally employed for special purposes. Sometimes non-metallic materials such as rubber, felt, various liquids and air are used where special properties such as high internal damping and low modulus of elasticity are desired.

Depending on the stresses arising in loaded springs they are classified into:

1. torsional stressed springs
2. flexural stressed springs
3. tension and compression stressed springs

Characteristic of a spring

The ratio of the force acting on a spring to the extension or contraction, respectively, of the spring is called the "spring stiffness".

$$c = dF/df$$

where dF = increment of force
and df = increment of extension or contraction, respectively.

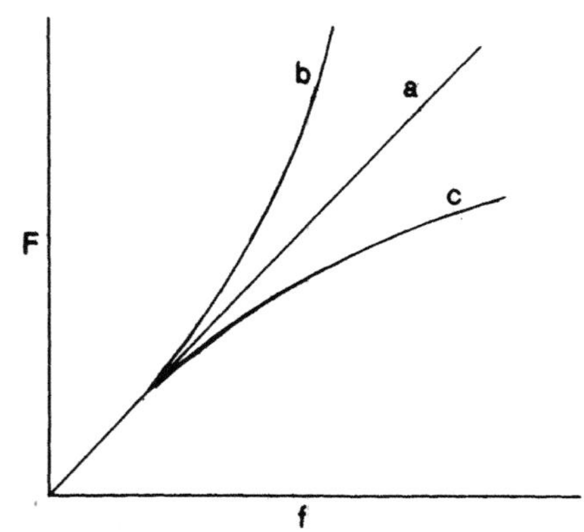

Fig. 360 Characteristics of springs

If the force on a spring is plotted as a function of its extension or contraction, respectively, on a graph, the characteristis of the spring is obtained (Fig. 360). For some springs this characteristic is a straight line (a). i. e.

$c = \text{const}$. For other springs this line is curved. An increasing characteristic (b) means that the spring becomes harder, if the stiffness of the spring is increasing. A decreasing characteristic (c) means that the spring becomes softer, if the stiffness of the spring is decreasing. According to its application, in some cases springs have an increasing characteristic, in other cases they have a decreasing characteristic.

Work done by a spring

The product of force and displacement of a spring is called the "spring energy".

$$A = \int F \, dx \, .$$

For springs having a constant spring stiffness the work done can be found from equation (50).

Efficiency and damping factor of a spring

If the friction in a spring is considered, the characteristic of a spring with an increasing load differs from the characteristic of a spring with a decreasing the load. In the graphical diagram, the characteristic of a spring having an increasing load is above the characteristic of a spring having a decreasing load.

The efficiency of a spring is defined as the ratio of the work output (W') to the work input (W) of the spring (Fig. 361).

$$\eta = \frac{W'}{W} = \frac{W - W_f}{W}$$

where $W_f = W - W' =$ work due to friction.

In vibration phenomena the damping factor δ is taken as,

$$\delta = \frac{W_f}{W + W'} \, .$$

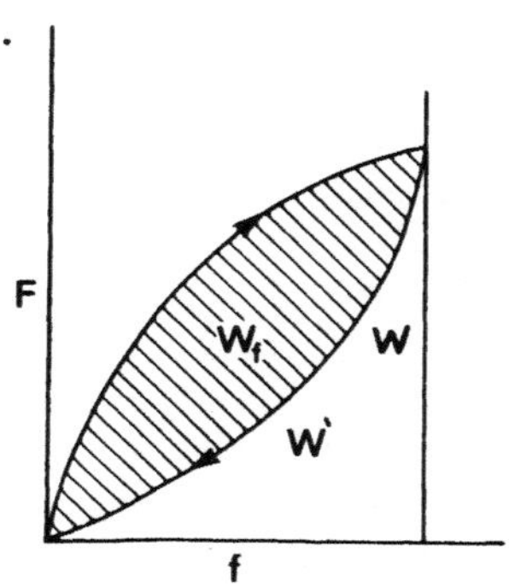

Fig. 361

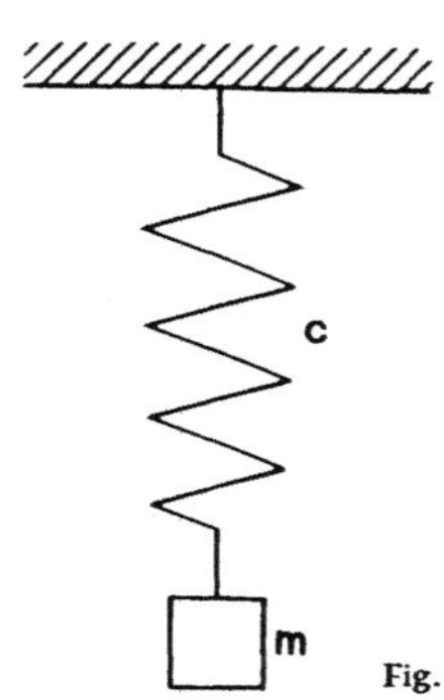

Fig. 362

The damping factor represents the ratio of the friction work W_f to the total work $W + W'$ which is done during loading and unloading of the spring.

Vibrations of springs

The natural frequency of a loaded spring, as shown in Fig. 362, is calculated from:

$$f_r = \frac{1}{2\pi}\sqrt{\frac{c}{m}}. \tag{192}$$

8.5.1. Torsional Stressed Springs

8.5.1.1. Helical Cylindrical Tension and Compression Springs

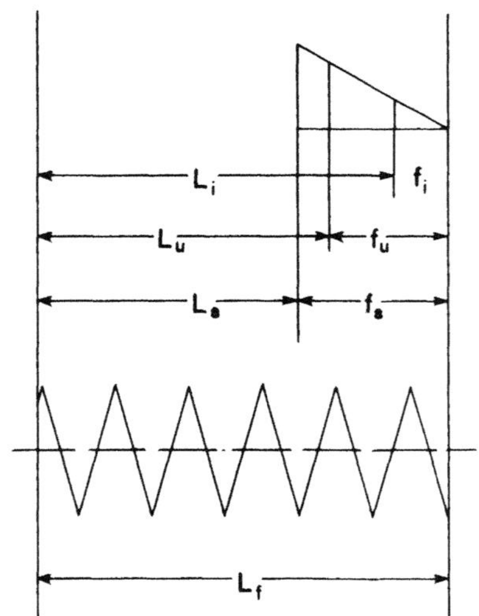

Fig. 363

A spring must possess sufficient strength, the desired elastic deformation and the desired characteristic. It should not buckle under load, if it is a compression spring and should not be in resonance with a cyclic disturbance. After the material and its preparation have been specified the only variables are the mean coil diameter D, the wire diameter d and the effective number of turns of the coil, z. Since the stress of the spring is independent of the effective number of turns, the coil diameter D and the wire diameter d are determined on the basis of the allowable stress. The first step should be the construction of the load deformation curve which is seen in Fig. 363 for a compressive spring, where L_f = free length, f_s = solid deformation, L_s = solid length, f_u = maximum useful deformation, L_u = minimum working length, f_i = initial deformation and L_i = maximum working length. The spring is usually designed to carry the specified maximum load for the allowable stress.

Helical cylindrical spring with round section

The shear stress distribution across a section of a wire is determined by superposition of the stresses due to the torsional moment and due to the shearing force. In Fig. 364, the curve 'a' represents the stress distribution, if the curvature of the wire is neglected. The effect of curvature is to change the stress distribution, (curve 'b'). The increase of the maximum stress in the wire section is allowed for by a factor k. For a tension or compression spring, as represented in Fig. 365, the torsional shearing stress τ_t and the deformation f is calculated from:

$$\tau_t = \frac{8\,F\,D\,k}{\pi\,d^3}, \qquad (193)$$

$$f = \frac{8\,z\,D^3\,F}{\pi\,G\,d^4} \qquad (194)$$

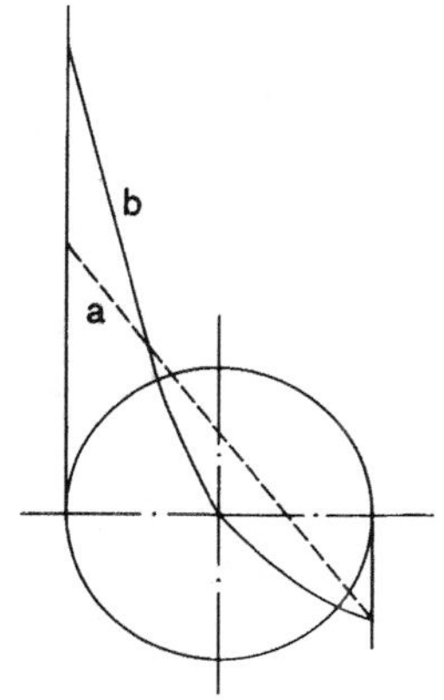

Fig. 364
Stress distribution on the cross section of a spring
tension spring
compression spring

where G = modulus of rigidity
and z = effective number of turns of the coil.

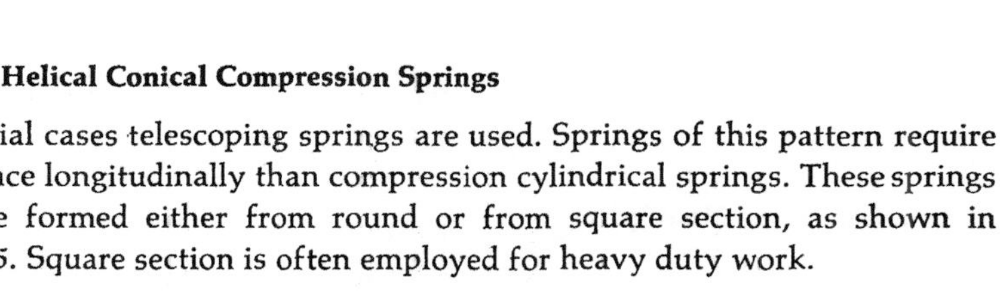

Fig. 365

8.5.1.2. Helical Conical Compression Springs

In special cases telescoping springs are used. Springs of this pattern require less space longitudinally than compression cylindrical springs. These springs may be formed either from round or from square section, as shown in Fig. 365. Square section is often employed for heavy duty work.

8.5.1.3. Torsion Bar

The torsion bar is widely used in hinges, brush holders, door locks etc. The torsional shearing stress τ_t and the angle of twist ψ is calculated from equation (82) or (84), respectively.

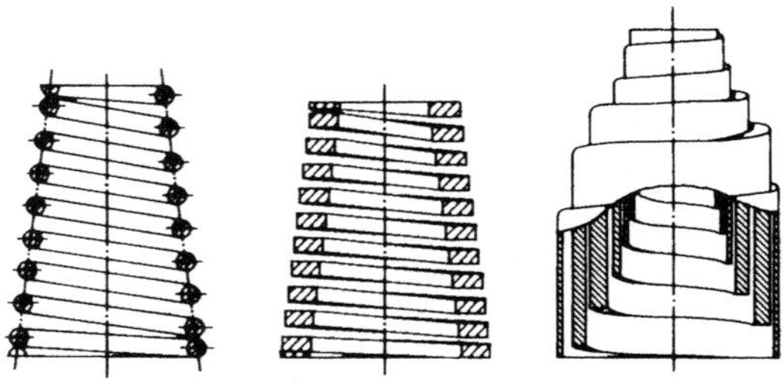

Fig. 366
Helical conical
compression spring

8.5.2. Flexural Stressed Springs

8.5.2.1. Flat Springs

Flat springs are extensively used in farm machinery and in the electrical industry.

The bending stress σ_b and the defexion f of *a* flat spring with constant rectangular section, as shown in Fig. 367, is calculated from:

$$\sigma_b = \frac{6\,F\,l}{b\,h^2} \quad (195), \qquad f = \frac{4\,F\,l^3}{E\,b\,h^3} \quad (196)$$

where E = modulus of elasticity

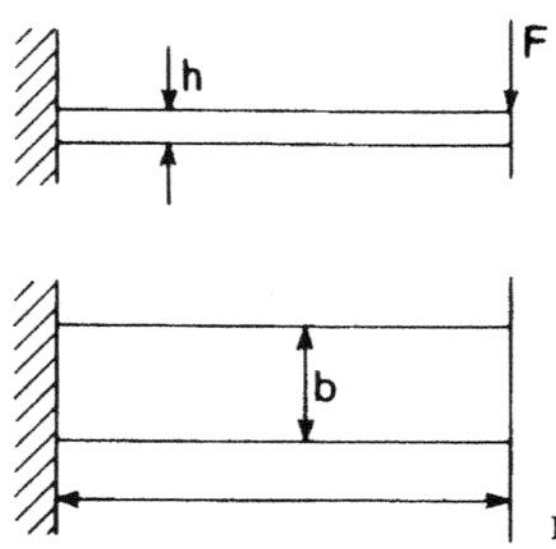

Fig. 367 Flat spring

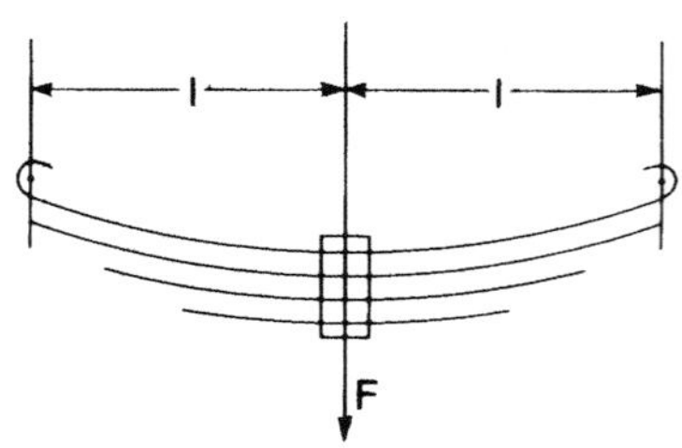

Fig. 368 Laminated plate spring

8.5.2.2. Laminated Plate Spring

The laminated plate spring, demonstrated in Fig. 368, is mainly employed in shock absorbing devices of automotive vehicles.

The bending stress may be calculated from:

$$\sigma_b = \frac{6\,F\,l}{z\,b\,h^2}\,. \qquad (197)$$

The deflexion may be calculated from the approximate formula:

$$f = \frac{12\,F\,l}{(2 + z'/z)\,E\,z\,b\,h^3} \tag{198}$$

where z = total number of leaves
z' = number of leaves, the lengths of which are up to the end of the top spring
h = thickness of one leaf
b = width of the spring.

8.5.2.3. Spiral Spring

Spiral springs are widely used in clocks and watches and in various instruments. They are loaded by torque.

For calculations, the ends of the spring are supposed to be fixed, as shown in Fig. 369. The bending stress σ_b and the angle of twist ψ is calculated from:

$$\sigma_b = \frac{6\,M_t}{b\,h^2} \tag{199},$$

$$\psi = \frac{12\,M_t\,l}{E\,b\,h^3} \tag{200}$$

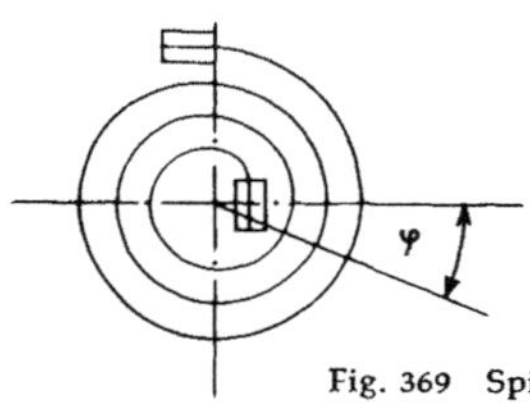

Fig. 369 Spiral spring

where l = length of spring
h = thickness of spring
b = width of spring.

8.5.2.4. Helical Cylindrical Torsion Spring

Helical cylindrical torsion springs represented in Fig. 370, are extensively used in motor brush holders, in counterbalances of machine tools, in spring motors etc. The bending stress σ_b and the angle of twist ψ of a helical cylindrical spring with round section may be calculated from:

$$\sigma_b = \frac{32\,F\,r}{\pi\,d^3} \tag{201},$$

$$\psi = \frac{64\,F\,l\,r}{E\,\pi\,d^3} \tag{202}$$

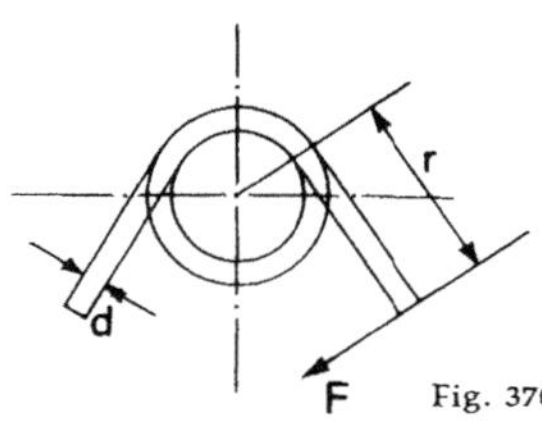

Fig. 370

where l = length of spring

8.5.3. Tension and Compression Stressed Springs

8.5.3.1. Ring Spring

A ring spring, demonstrated in Fig. 371, is composed of a set of specially profiled rings. When the end rings are loaded on the perimeter, the outer rings move over the inner ones. As a result the outer rings expand and the inner rings contract. Simultaneously the total height of the spring decreases. Ring springs are mainly used as buffers.

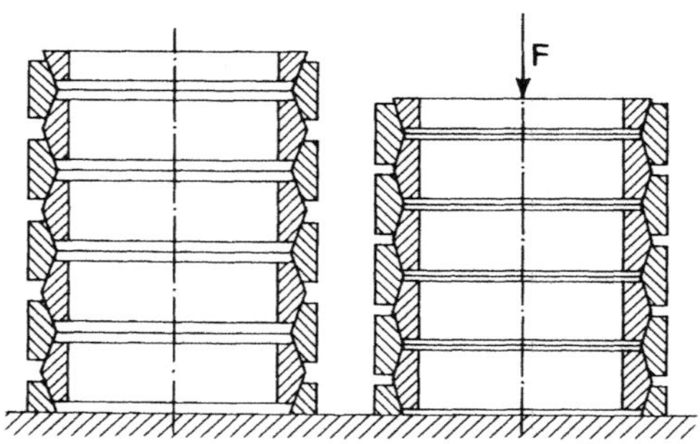

Fig. 371 Ring spring

8.5.3.2. Block Spring

Block compression springs made of rubber in the form of blocks of various shape are mainly used as shock absorbers to cushion shocks and to decrease vibrations.

English-German Vocabulary

1. Mathematics

engineering *Ingenieurswesen*
to comprise, to include *einschließen*
calculus *Rechnen*
to establish *aufstellen*
problem *Aufgabe*
appropriate *passend*
applicable *verwendbar*
range *Bereich*
applicability *Verwendbarkeit*
to derive *ableiten*
whole number *ganze Zahl*
even number *gerade Zahl*
odd number *ungerade Zahl*
common fraction *gemeiner Bruch*
decimal fraction *Dezimalbruch*
proper fraction *echter Bruch*
improper fraction *unechter Bruch*

1.1.1. bracket *Klammer*
term *Glied einer Gleichung*
sum *Summe*
to provide *versehen*
sign *Vorzeichen*
small 'a' *klein a*
capital 'A' *groß a*
fraction line *Bruchstrich*
numerator *Zähler*
denominator *Nenner*
dash *Strich*
round bracket *runde Klammer*
square bracket *eckige Klammer*
brace *geschweifte Klammer*

1.1.2. quantity *Rechengröße*

1.1.3. power *Potenz*
root *Wurzel*
to raise *erheben*
digit, figure *Ziffer*

1.1.4. equation *Gleichung*
derivation *Ableitung*
hence *daher*
value *Wert*
to calculate, to compute *berechnen*
to substitute *einsetzen*
to equate *gleichsetzen*
to miss *fehlen*

1.2. plane geometry *Planimetrie*
solid geometry *Stereometrie*
to deal with *behandeln*
plane figure *ebenes Gebilde*
solid figure or solid *räumliches Gebilde*
area *Fläche*
circumference *Umfang*
angle *Winkel*
acute *spitz*
obtuse *stumpf*
altitude *Höhe eines Dreiecks*
scalene *schiefwinklig*
right angled *rechtwinklig*
isosceles *gleichschenklig*
equilateral *gleichseitig*
trapezium *Trapez*
rectangle *Rechteck*
square *Quadrat*
circle *Kreis*
chord *Sehne*
diameter *Durchmesser*
sphere *Kugel*
right square prism *Quader*
cube *Würfel*
pyramid *Pyramide*
cone *Kegel*
frustum of cone *Kegelstumpf*

1.3. measurement *Messen*
sexagesimal system *Gradmaßsystem*
circular system *Bogenmaßsystem*
to subtend *umspannen*
conversion *Umrechnung*

ratio *Verhältnis*
opposite *gegenüberliegend*
adjacent *anliegend*
1.4. coordinate geometry *analytische Geometrie*
to determine *bestimmen*
straight line *gerade Linie*
to cut off *abschneiden*
intercept *Abschnitt*
conic section *Kegelschnitt*
to trace out *aufzeichnen*
perpendicular *senkrecht*
focus *Brennpunkt*
directrix *Leitlinie*
1.5.1. derivative *Ableitung*
slope *Neigung*
1.5.2. inverse *umgekehrt*
to bound *begrenzen*
to set up *aufstellen*

2. Mechanics

to deal with *handeln von*
engineering mechanics *technische Mechanik*
applied mechanics *angewandte Mechanik*
state of aggregation *Aggregatszustand*
collinear *auf derselben Geraden liegend*
coplanar *auf derselben Ebene liegend*
inertia *Trägheit*
2.1. rigid *starr*
to endure *erleiden*
distorsion *Verformung*
subsequent *folgend*
previous *vorhergehend*
displacement *Verschiebung*
velocity *Geschwindigkeit*
acceleration *Beschleunigung*
point of application *Angriffspunkt*
magnitude *Größe*
system of units *Maßsystem*
2.1.1. composition *Zusammensetzung*
resolution *Zerlegung*
to compose *zusammensetzen*
to resolve *zerlegen*
equilibrium *Gleichgewicht*
concurrent *zusammenlaufend*
plane *Ebene, eben*
space diagram *Lageplan*
force diagram *Kräfteplan*
to draw *zeichnen*
to lay off *auftragen*
point of intersection *Schnittpunk*
in turn *wieder*
order *Reihenfolge*
arbitrary *willkürlich*
arrow *Pfeil*
perpendicular *senkrecht*
theorem *Lehrsatz*
2.1.2. to displace *verschieben*
line of action *Wirkungslinie*
clockwise *im Uhrzeigersinn*
counter-clockwise *gegen Uhrzeigersinn*
link polygon or funicular polygon *Seileck*
link *Gelenk*
to join *verbinden*
with respect to or in regard to *mit Bezug auf*
couple *Kräftepaar*
to tend *versuchen*
stable *stabil*
unstable *labil*
neutral *indifferent*
to subject *unterwerfen*

to load *belasten*
load *Last*
edge *Kante*
2.1.4. structure *Tragwerk*
beam *Träger*
cantilever beam *eingespannter Träger, Kragträger*
hinge *Gelenk*
three hinged arc *Dreigelenk-bogenträger*
truss *Fachwerk*
concentrated load *Einzellast*
distributed load *Streckenlast*
support *Auflager*
to transmit *übertragen*
roller *Walze*
pin *Bolzen*
pinned support *drehbares Auflager*
built in or encastre *fest eingebaut*
shearing force *Scherkraft*
bending moment *Biegemoment*
reaction *Auflagekraft*
to assume *annehmen*
section *Schnitt*
either side *jeder Seite*
shear force diagram *Querkraft-diagramm*
uniformly varying *linear steigend*
to delineate *aufzeichnen*
intercept *Abstand*
derivative *Ableitung*
to state *festlegen, aufstellen*
agent *wirkende Kraft*
upper- lower chord *Ober-, Unter-gurt*
web member *Gitterstab*
frictionless *reibungslos*
stress *Spannung*
tensile stress *Zugspannung*
compressive stress *Druckspannung*
joint *Knoten*
adjacent *angrenzend*
panel *Feld*
hence *daher*
2.1.5. centre of gravity *Schwerpunkt*
centroid *Schwerpunkt einer Fläche*
particle *Massenteilchen*
to attract *anziehen*
plate *Platte, Scheibe*
trial *Versuch*
revolution *Umdrehung*
surface of revolution *Rotations-fläche*
solid of revolution *Rotations-körper*
path *Weg, Bahn*
2.1.6. friction *Reibung*
to occur *vorkommen*
amount *Betrag*
roughness *Rauhigkeit*
to rub *reiben*
inclined plane *schiefe Ebene*
to include *einschließen*
to impend *bevorstehen*
2.2. reference *Berücksichtigung*
to cause *verursachen*
2.2.1. to undergo *unternehmen*
2.2.1.1. to neglect *vernachlässigen*
rate *Anteil*
initial velocity *Anfangs-geschwindigkeit*
final velocity *Endgeschwindigkeit*
to rearrange *umformen*
above *obig*
rectilinear *geradlinig*
to decrease *abnehmen*
retardation *Verzögerung*
uniform *gleichförmig*
to overtake *überholen*
to project *abschleudern*
acceleration due to gravity *Erdbeschleunigung*
slight *geringfügig*
curvilinear *krummlinig*
instantaneous *augenblicklich*

centre of curvature *Krümmungsmittelpunkt*
angular velocity *Winkelgeschwindigkeit*
angular acceleration *Winkelbeschleunigung*
curvature *Krümmung*
projectile *Wurfkörper*
gravity *Schwerkraft*
to rifle *mit Zügen versehen*
barrel *Lauf*
trajectory *Wurfbahn*
range *Wurfweite*
elevation *Höhe*
to elapse *verstreichen*

2.2.1.2. bar *Stange*
link *Kettenglied*
linkage *Verkettung*
four-bar linkage *Kurbelviereck*
frame *Rahmen*
crank *Kurbel, Schwinge*
rod *Stab*
connecting rod *Koppel, Pleuelstange*
to attach *befestigen*
block *Block, Gleitstein*
to slide *gleiten*

2.2.2.1. unbalanced *nicht im Gleichgewicht befindlich*
to experience *erfahren*
inverse *umgekehrt*
implicit *eingeschlossen*
statement *Festsetzung*
cable *Schnur*
pulley *Rolle*

2.2.2.2. specific gravity *Wichte*

2.2.2.3. work *Arbeit*
power *Leistung*
helical spring *Schraubenfeder*
spring stiffness *Federkonstante*
to stretch *dehnen*
to compress *zusammendrücken*
increase *Zunahme*
decrease *Abnahme*
converse *umgekehrt*

2.2.2.4. linear momentum *Bewegungsgröße*
linear impulse *Impuls*
impact *Stoß*
angular momentum *Drall*
angular impulse *Moment des Antriebes*
increment *Zunahme*
to collide *zusammenstoßen*
collision *Zusammenstoß*
plane of contact *Berührungsfläche*
direct *gerade*
oblique *schief*
period of deformation *erste Stoßperiode*
restitution *Wiederherstellung*
period of restitution *zweite Stoßperiode*
separation *Entfernen*
approach *Annähern*

3. Strength of Materials

3.1. strength of materials *Festigkeitslehre*
stress *Spannung*
strain *Dehnung*
to fix *festlegen*
to confirm *bestätigen*
bending moment *Biegemoment*
twisting moment *Drehmoment*
face *Stirnseite*
extensometer *Dehnungsmesser*
specimen *Probestück*
wire *Draht*
photo elastic stress analysis *Spannungsoptik*
band *Streifen*
screen *Bildschirm*

fringe *Franse, Rand*
3.2. behaviour *Verhalten*
toughness *Zähigkeit*
brittleness *Sprödigkeit*
hardness *Härte*
flexure *Biegung*
rate *Art*
to strike *schlagen*
blow *Schlag*
to creep *kriechen*
creep test *Dauerstandversuch*
endurance *Ausdauer*
fatigue *Ermüdung*
fatigue test *Dauerprüfung*
3.2.1. circular *rund*
mild steel *Maschinenbaustahl*
release *Befreiung*
to coincide *zusammentreffen*
yield point *Streckgrenze*
to retain *zurückbehalten*
ultimate strength *Bruchfestigkeit*
to endure *aushalten*
brittle *spröde*
ductile *dehnbar*
to neck *sich einschnüren*
rupture *Bruch*
3.2.2. indentation *Eindrücken*
ball *Kugel*
standard test *genormte Prüfung*
to range *reichen*
alloy *Legierung*
increment *Zunahme*
soft *weich*
rebound *Rückprall*
tube *Röhre*
to drop *herunterwerfen*
3.2.3. shaft *Stange, Welle*
ball bearing *Kugellager*
disc *Scheibe*
to notch *einkerben*
anvil *Ambos*
3.2.4. plastic strain *bleibende Dehnung*
to plot *einzeichnen*
3.2.5. to fluctuate *schwanken, schwingen*
to pulsate *vibrieren, schwingen*
fluctuating stress *Schwingungsbeanspruchung*
completely reversed stress *Wechselbeanspruchung*
cycle *Folge*
endurance limit *Dauerfestigkeit*
failure *Bruch*
to alternate *abwechseln*
3.2.6. abrupt *plötzlich*
hole *Loch*
keyway *Keilnut*
groove *Hohlkehle*
screw *Schraube*
thread *Gewinde*
to arise *entstehen*
fatigue stress concentration factor *Kerbwirkungszahl*
3.3.2. vessel *Gefäß*
circumferential or hoop stress *Tangentialspannung*
3.3.3. average *mittlere*
3.4. to restrain *zurückhalten*
3.5.1. beam *Träger*
fibre *Faser*
3.5.3. stiffness *Steifheit*
to exeed *überschreiten*
3.6. to buckle *knicken*
slender *schlank*
column *Säule, senkrechter Träger*
strut *Verstrebungsstab*
slight *leicht*
due *herrührend*
accidental *axial*
radius of gyration *Trägheitsradius*
3.7. revolution *Umdrehung*
solid circular section *Kreisquerschnitt*
hollow circular section *Kreisringquerschnitt*
angle of twist *Verdrehungswinkel*
3.8. mutual *gegenseitig*

state of stresses *Spannungszustand*
theory of failure *Festigkeits-hypothese*
maximum shearing stress theory *Hauptschubspannungshypothese*
theory of elastic strain energy of distorsion *Gestaltänderungs-hypothese*
3.8.1. to convert *umformen*
3.8.2. to store *aufspeichern*
elastic energy of volume dilatation *Volumenänderungsarbeit*
elastic strain energy of distorsion *Gestaltänderungsarbeit*

4. Hydraulics

to deal with *handeln von*
significant *bedeutsam*
to concern with *sich befassen mit*
to govern *beherrschen*
behaviour *Verhalten*
strict *genau*
research *Untersuchung*
particle *Teilchen*
perfect liquid *vollkommene (reibungsfreie und unzusammen-drückbare) Flüssigkeit*
4.1.1. to withstand *widerstehen*
tension *Zug*
compression *Druck*
pressure head *Druckhöhe, Flüssigkeitshöhe*
mercury column *Quecksilberhöhe*
4.1.2. buoyancy *Auftrieb*
to submerge *untertauchen*
to immerse *eintauchen*
upward *aufwärts*
downward *abwärts*
buoyant force *Auftriebskraft*
submergence *Untertauchen*
4.1.3. reservoir *Behälter*
opening *Öffnung*
lock gate *Schleusentor*
valve *Ventil*
to exert *ausüben*
location *Lage*
centre of pressure *Angriffspunkt der Druckkraft*
width *Breite*
convenient *passend*
scale *Maßstab*
to hatch *schraffieren*
4.1.4. to float *schwimmen*
floatation *Schwimmen*
to displace *verdrängen*
to rise *heben*
infinite *unendlich viel*
caisson *Schwimmkasten*
pontoon *Ponton*
stable *stabil*
to restore *wiederherstellen*
couple *Kräftepaar*
unstable *labil*
disturbance *Störung*
neutral *indifferent*
metacentre *Metazentrum*
metacentric height *metazentrische Höhe*
war ship *Kriegsschiff*
4.2.1. to improve *verbessern*
4.2.1.1. to make up *zusammensetzen*
to trace *ausfindig machen*
instant *Augenblick*
tubular *röhrenförmig*
to bound *begrenzen*
to behave *sich verhalten*

steady *stationär*
flow pattern *Strömungsbild*
path *Weg*
to specify *einzeln festlegen*
4.2.1.2. orifice *Ausflußöffnung*
partition *Trennwand*
edge *Kante*
tap *Hahn*
discharge *Ausfluß*
jet *Strahl*
4.2.1.3. theorem *Lehrsatz*
pressure head *Druckhöhe*
velocity head *Geschwindigkeitshöhe*
elevation head *Ortshöhe*
to modify *abändern*
subsequent *folgend*
preceding *vorhergehend*
interpretation *Auslegung*
lost energy *Verlustenergy*
4.2.1.4. rack and pinion *Zahnstange mit Ritzel*
pointer *Zeiger*
stagnation pressure *Staudruck*
rate of flow *Durchflußmenge*
4.2.1.5. channel *Kanal*
guide vane *Leitschaufel an Turbine*
runner blade *Laufschaufel an Turbine*
diffusion vane *Leitschaufel an Zentrifugalpumpe*
impeller blade *Laufschaufel an Zentrifugalpumpe*
4.2.2. viscous *zäh*
4.2.2.1. viscosity *Zähigkeit*
real *wirklich*
to display *entfalten*
stationary *ruhend*
layer *Lage, Schicht*
conduit *Leitung*
uniform *gleichmäßig*
adjacent *anliegend, benachbart*
slip *Schlupf*
dy apart *für sich*
property *Eigenschaft*
boundary *Begrenzung*
efflux *Ausfluß*
to empty *leeren*
to suspend *aufhängen*
torsion wire *Torsionsdraht*
torsion spring *Torsionsfeder*
clearance *Spiel*
4.2.2.2. to exhibit *aufweisen, zeigen*
to fulfill *erfüllen*
to superimpose *überlagern*
dye *Farbe*
to adjust *regulieren*
filament *Faden*
to maintain *beibehalten*
4.2.2.3. perimeter *Umfang*
to designate *bezeichnen*
to wet *benetzen*
smooth *glatt*
rough *rauh*
sheet iron *Stahlblech*
to draw bright *tief ziehen*
concrete *Beton*
bend *Krümmer*
elbow *Kniestück*
4.2.2.4. artificial *künstlich*
sewerage tunnel *Abwasserkanal*
semicircle *Halbkreis*

5. Thermodynamics

5.1. transition, transfer *Übergang*
storage *Speicherung*
to store *speichern*
transformation *Umformung*

conduction *Leitung*
radiation *Strahlung*
heat of fusion *Schmelzwärme*
5.2.1. equation of state *Zustandsgleichung*
to result *sich ergeben*
5.2.2. cycle *Kreisprozeß*
to raise *heben*
simultaneous *gleichzeitig*
device *Vorrichtung*
steam boiler *Dampfkessel*
centrifugal compressor *Turbokompressor*
steady *stetig*
reversible *umkehrbar*
to radiate *abstrahlen*
useful work *Nutzarbeit*
to put in *hineinstecken*
to put out *herausnehmen*
delivery *Lieferung, Füllung*
exhaust *Ausschub*
5.2.3. to retain *zurückhalten*
energy reception *Energieaufnahme*
energy rejection *Energieabgabe*
to add *zuführen*
to reject *abführen*
to jacket *umhüllen*
piston *Kolben*
to intercool *zwischenkühlen*
to reciprocate *hin und hergehen*
state path *Zustandskurve*
5.2.4. irreversible *nicht umkehrbar*
to throttle *drosseln*
mixture *Mischung*
5.2.5. to perform *ausführen*
plant *Anlage*
internal combustion engine *Verbrennungskraftmaschine*
to hatch *schraffieren*
to supply *liefern*
efficiency *Wirkungsgrad*
coefficient of performance *Wirkungsgrad einer Kältemaschine*
charge *Ladung*
escape *Entweichen*
5.2.6. to furnish *liefern*
incentive *Antrieb*
gradient *Gefälle*
5.3. vaporisation *Verdampfung*
latent heat of fusion *Schmelzwärme*
latent heat of vaporation *Verdampfungswärme*
stage *Stadium*
to saturate *sättigen*
wet steam *Naßdampf*
saturated steam *Sattdampf*
superheated steam *Heißdampf*
dryness fraction *Dampfgehalt*
liquefaction *Verflüssigung*
to transfer *übertragen*
locus *geometrischer Ort*
to plot *einzeichnen*
5.4. source *Ursprung*
combustible *verbrennbar*
fuel *Brennstoff*
oxygen *Sauerstoff*
to release *entlassen*
5.4.1. pit coal *Steinkohle*
lignite *Braunkohle*
fuel oil *Heizöl*
producer gas *Generatorgas*
carbon *Kohlenstoff*
hydrogen *Wasserstoff*
hydrocarbon *Kohlenwasserstoff*
sulphur *Schwefel*
nitrogen *Stickstoff*
moisture *Feuchtigkeit*
impurity *Verunreinigung*
percentage *Prozentsatz*
constituent *wesentlich*
carbon monoxide *Kohlenmonoxyd*
carbon dioxide *Kohlendioxyd*
boiler furnace *Kesselfeuerung*
5.4.3. to quote *angeben*
calorific value *Heizwert*

higher or gross c. v. *oberer Heizwert*
lower or net c. v. *unterer Heizwert*
to cool back *zurückkühlen*

6. Engineering Materials

copper *Kupfer*
tin *Zinn*
alloy *Legierung*
wood *Holz*
stone *Stein*
brick *Ziegel*
concrete *Beton*
rubber *Gummi*
leather *Leder*
plastic *Kunstharz*
efficient *wirksam*
strength *Festigkeit*
stiffness *Steifheit*
ductility *Dehnbarkeit, Ziehbarkeit*
malleability *Schmiedbarkeit*
toughness *Zähigkeit*
machinability *Bearbeitbarkeit (spanabhebend)*
to standardize *normen*
6.1.1. to manufacture *herstellen*
pig iron *Roheisen*
ore *Erz*
silicon *Silizium*
hematite *Roteisenstein*
limonite or hydrated hematite *Brauneisenstein*
siderite or spatic ore *Spateisenstein*
magnetite *Magneteisenstein*
to mine *fördern*
flux *Zuschlag*
gangue *Gangart*
fusible *schmelzbar*
acid *sauer*
limestone *Kalkstein*
basic *basisch*
slag *Schlacke*
to float *schwimmen*
to smelt or to melt *schmelzen*
molten *geschmolzen*
blast furnace *Hochofen*
feature *Hauptteil*
steel shell *Stahlmantel*
to line *auskleiden*
fire brick *feuerfester Ziegel*
to withstand *widerstehen*
skip hoist *Kübelaufzug*
charge *Beschickung*
hopper *Einfülltrichter*
bell *Glocke*
blast pipe *Druckluftrohr*
nozzle or tuyere *Düse*
tapping notch *Abstichöffnung*
blast *Strom*
to draw off or to tap off *ablassen*
to cast *gießen*
bar *Stange*
pig *Massel*
cinder notch *Schlackenöffnung*
nitrogen *Stickstoff*
cast iron *Gußeisen*
6.1.2.1. puddling process *Puddelverfahren*
to refine *raffinieren*
wrought iron *Schweißstahl*
approach *Annäherung*
carbon *Kohlenstoff*
open- hearth process *Flammofenfrischen*
reverberatory furnace *Flammofen*
fire box *Feuerraum*
bridge wall *Feuerbrücke*
hearth *Schmelzbecken*
sloping *geneigt*
stack *Rauchschacht*
to feed *zuführen*

melting stage *Schmelzvorgang*
to stir *zerschlagen*
hoe *Hake*
pasty *teigig*
to stick *sinken*
mill scale *Oxydationsmittel*
to purify *reinigen*
lump *Stück*
to squeeze *quetschen*
bloom *Luppe*
to roll *walzen*
fibrous *fasrig*

6.1.2.2. vessel *Behälter*
to pierce *durchsetzen*
to tip *kippen*
to drop *fallen, abnehmen*
ladle *Gießpfanne*
ingot mould *Kokille*
to calcine *verkalken*

6.1.2.3. scrap *Schrott*
shallow *flach*
refractory *feuerfest*
chamber *Kammer*
checkerwork *Gitterwerk*
exhaust gases *Abgase*
previous *vorhergehend*
producer gas *Generatorgas*
incandescent *weißglühend*

6.1.2.4. crucible process *Tiegelverfahren*
crucible *Schmelztiegel*
ingredients *Zusatzstoffe*
charcoal *Holzkohle*
tungsten *Wolfram*
fire clay *feuerfester Ton*
to discard *wegwerfen*

6.1.2.5. constituent *Bestandteil*
stainless *rostfrei*
high speed steel *Schnellstahl*
coil *Windung*
alternating current *Wechselstrom*
tool *Werkzeug*
die *Gesenk*
ball bearing *Kugellager*

6.1.3.1. foundry cupola *Kupolofen*
to light *anzünden*
layer *Lage, Schicht*
mould *Gußform*
moist *feucht*
over and over again *immer mehr*
centrifugal mould *Schleudergußform*

6.1.3.2. malleable cast iron *Temperguß*
to anneal *glühen*
to malleabilize *tempern*
black hearth process *Schwarzgußverfahren*
white hearth process *Weißgußverfahren*
grain *Korn*
to embed *einbetten*
screw spanner *Schraubenschlüssel*
spring hanger *Federhaken*
differential gear *Differentialgetriebe*
steering gear *Lenkeinrichtung, Steuermechanismus*

6.1.3.3. shrinkage *Schwinden*
cavity *Lunker*
to forge *schmieden*

6.1.3.4. solidification *Erstarren*
to strip *abziehen*
soaking furnace *Glühofen*
hexagon *Sechskant*
tee section *T-Profil*
channel section *U-Profil*
I-section *Doppel T-Profil*
H-section *Breitflanschprofil*
rod *Stange*
drop forgings *Gesenkschmiedestücke*
axle *Achse*
shaft *Welle*
shear blade *Schermesser*
chisel *Meisel*
punch *Lochstempel*
razor *Rasiermesser*

gauge *Meßwerkzeug*
6.1.4. to solidify *erstarren*
to normalize *vergüten*
wear *Verschleiß*
to plunge *eintauchen*
to quench *abschrecken*
to perform *durchführen*
suitable *geeignet*
to temper *anlassen*
case hardening *Einsatzhärten*
carbonaceous *kohlenstoffhaltig*
hoof clipping *Hufschnitzel*
bone dust *Knochenmehl*
to carburize *kohlen*
core *Kern*
case *Härtetiefe*
cam *Nocken*
wrist pin or gudgeon pin *Kolbenbolzen*
to dissociate *trennen*
6.2.1. lead *Blei*
chalcocite *Kupferglanz*
chalcopyrite *Kupferkies*
blister copper *Rohkupfer*
blister *Luftblase*
cooking vat *Kochtopf*
cassiterite *Zinnstein*
agent *Mittel*
fluor spar *Flußspat*
redistillation *wiederholte Destillation*
die casting *Spritzguß*
tubing *Rohrleitung*
galena *Bleiglanz*
to extrude *kalt pressen*
to sheath *armieren*
to dissolve *auflösen*
to deposit *ablagern*
to drain out *herausziehen*
6.2.2. brass *Messing*
solder *Lot*
pump lining *Pumpenleitung*
moisture *Feuchtigkeit*
step *Lagerschale*
bearing *Lager*
plumbing *Klempnerarbeit*
6.3. concrete *Beton*
plastics *Kunstharz*
decay *Fäulnis*
coniferous tree *Nadelbaum*
spruce *Fichte*
fir *Tanne*
pine *Föhre, Kiefer*
larch *Lärche*
deciduous tree *Laubbaum*
hornbeam *Weißbuche*
red beech *Rotbuche*
elm *Ulme*
ash *Esche*
poplar *Pappel*
maple *Ahorn*
timber *Bauholz*
grain *Maserung*
interior finish *Innenverkleidung*
cabinet work *Möbelschreinerei*
to cement *zusammenkitten, zusammenkleben*
gypsum *Gips*
lime *Kalk*
gravel *Kies*
foundation *Fundament*
arch *Bogen, Gewölbe*
reinforced concrete *Stahlbeton*
footing *Sockel*
coherence *Bindekraft*
thermoplastics *thermoplastisches Kunstharz*
thermosetting plastics *duroplastisches Kunstharz*
binder *Bindemasse*
filler *Füllmasse*
to shred *zerschneiden*
plasticizer *Flußmittel*
catalyst *Katalysator*
housing *Gehäuse*
to insulate *isolieren*
gear wheel *Zahnrad*
handle *Handgriff*

7. Technical Drawing

to draw *zeichnen*
to design *konstruieren*
drawing *Zeichnung*
design *Entwurf*
to record *aufzeichnen*
specifications *Angaben*
designing engineer or designer *Konstrukteur*
to carry out *ausführen*
draughtsman *Zeichner*
Standards Institution *Normenausschuß*
standards specifications *Normen*
to issue *herausgeben*
to adopt *annehmen*
to adapt *anpassen*

7.1. drawing equipment *Zeichengerät*
drawing board *Reißbrett*
spruce *Fichte*
pine *Kiefer*
to warp *sich werfen*
strip *Streifen*
to glue *leimen*
edge *Kante*
ebony *Ebenholz*
to insert *einsetzen*
T-square *Reißschiene*
blade *Schiene*
stock *Querholz*
to varnish *firnissen*
moisture *Feuchtigkeit*
parallel rule attachement *Reißschienenparallelführung*
set-square *Zeichendreieck*
scale *Lineal*
box wood *Buchsbaumholz*
compasses *Zirkel*
yoke *Joch*
to adjust *feineinstellen*
socket joint *Muffenverbindung*
divider *Stechzirkel*
to purchase *beziehen, kaufen*
velvet *Samt*
spring bow for pencil *Bogenzirkel für Bleistift*
spring bow for pen *Bogenzirkel für Feder*
spring bow dividers *Bogenstechzirkel*
ruling pen *Ziehfeder*
clay *Ton*
protractor *Winkelmesser*
to set oft *ansetzen*
French curves *Kurvenlineal*
adjustable *einstellbar*
engineer's triangle *Zeichenschablone*
drawing ink or China ink *Tusche*
rubber *Radiergummi*
erasing knife *Radiermesser*
thumbtack *Reißnagel*
draughting tape *Klebstreifen*

7.2. perspective projection *Zentralprojektion*
oblique projection *schiefe Parallelprojektion*
orthographic projection *gerade Parallelprojektion*
projector *Projektionsstrahl*
to converge *zusammenlaufen*
front view or front elevation *Vorderansicht, Aufriß*
side view or side elevation *Seitenansicht, Seitenriß*
top view or plan from above *Draufsicht, Grundriß*
bottom view or plan from below *Untersicht*
back view *Rückansicht*

7.3. pictorial drawing *perspektivisches Zeichnen*
to visualize *sich vorstellen*

to foreshorten *verkürzen*
7.4. chain line *Strichpunktlinie*
broken line *Strichlinie*
root lines of bolts *Kernlinien von Schrauben*
root circle *Fußkreis*
dimension line *Maßlinie*
extension line *Maßhilfslinie*
to hatch *schraffieren*
pitch circle *Teilkreis*
7.5. to frame *einrahmen*
title strip *Schriftfeld*
parts list *Stückliste*
working drawing *Werkzeichnung*
assembly drawing *Zusammen-stellungszeichnung*
detail drawing *Teilzeichnung*
alteration strip *Änderungsfeld*
to enter *eintragen*
7.6. upright *aufrecht*
section lines *Schraffurlinien*
simultaneously *gleichzeitig*
numeral *Zahlzeichen*
machining symbol *Bearbeitungs-zeichen*
to cut apart *wegschneiden*
spacing *Abstand*
adjacent *anliegend*
to stagger *versetzen*
rib *Rippe*
shaft *Welle*
ball *Kugel*
rod *Stange*
rivet *Niet*
key *Längskeil*
cotter *Querkeil*
7.7. convenient *passend*
arrow head *Pfeilspitze*
numeral or dimension figure *Maßzahl*
bar *Stange*
taper *Kegel- und Verjüngungs-verhältnis*
slope *Neigung, Schrägung*
incline *Neigung, Steigung*
alteration *Änderung*
craftman *Facharbeiter*
to plane *hobeln*
to shape *hobeln mit der Shapingmaschine*
to mill *fräsen*
to turn *drehen*
to drill *bohren*
to ream *ausreiben*
to bore *ausbohren*
to machine *bearbeiten*
face *Stirnfläche*
7.8. evenness *Glätte*
roughness number *Rauhigkeits-zahl*
to cast *gießen*
to roll *walzen*
to forge *schmieden*
to grind *schleifen*
to seal *dichten*
to finish *fein bearbeiten*
to superfinish *feinst bearbeiten*
to polish *polieren*
to scrape *schaben*
to lap *läppen*
to hone *honen*
to anneal *glühen*
to normalize *vergüten*
7.9. fit *Passung*
interchangeability *Austausch-barkeit*
accuracy *Genauigkeit*
to match or to fit *passen*
gauge *Meßlehre*
caliper gauge *Rachenlehre*
plug gauge *Meßdorn*
go end *Gutseite*
no-go end *Ausschußseite*
plane fit *Flachpassung*
round fit *Rundpassung*
clearance fit *Spielpassung*

interference fit *Preßpassung*
transition fit *Übergangspassung*
clearance *Spiel*
interference *Übermaß*
bush or bushing *Buchse*
coupling *feste Kupplung*
clutch *Schaltkupplung*
gear wheel *Zahnrad*
basic hole system or hole basis system *System der Einheitsbohrung*
basic shaft system or shaft basis system *System der Einheitswelle*
to range *reichen*
range *Bereich*
reference line *Bezugslinie*

8. Machine Elements

definite *bestimmt*
capable *fähig*
to perform *verrichten, leisten*
useful work *Nutzarbeit*
power engine *Kraftmaschine*
to convert *umwandeln*
working machine *Arbeitsmaschine*
to dam up *stauen*
to store *speichern*
reciprocating steam engine *Kolbendampfmaschine*
internal combustion engine *Verbrennungskraftmaschine*
pump *Pumpe*
machine tool *Werkzeugmaschine*
concrete mixer *Betonmischer*
efficiency *Wirkungsgrad*
power output *gewonnene Leistung*
power input *aufgewendete Leistung*
joining element *Verbindungselement*
drive *Antrieb*
slider crank mechanism *Kurbeltrieb*
pipe *Rohr*
valve *Rohrschalter*
spring *Feder*
8.1. soldering *Löten, Weichlöten*
brazing *Hartlöten mit Messinglot*
welding *Schweißen*
alternatively *andrerseits*
rivet *Niet*
pin *Stift, Bolzen*
bolt *Schraube*
key *Längskeil*
detachable *lösbar*
tight *dicht, fest*
8.1.1. steel structure *Stahlkonstruktion*
building frame *Rahmenkonstruktion*
roof truss *Dachfachwerk*
to match *passen*
to punch *stanzen*
to drill *Bohren*
to carry out *ausführen*
to shrink *schrumpfen*
shank *Schaft*
to grip *greifen*
brass *Messing*
snap head *Halbrundkopf*
countersunk *versenkt*
beam *Träger*
girder *Hauptträger*
lap riveting *Überlappungsnietung*
butt riveting *Laschennietung*
to overlap *überlappen*
alignment *Fluchtrichtung*
to caulk, to fuller *verstemmen*
edge *Kante*

leak proof *wasserdicht*
blunt chisel *stumpfer Meisel*
to seal *abdichten*
centroid *Schwerpunkt einer Fläche*
8.1.2. to bond *verbinden*
parent metal *Grundmetall*
intimate *innig*
fusion welding *Schmelzschweißung*
torch *Brenner*
rod *Stange*
dead mild steel, mild steel } *see 6.1.3.4.*
direct current *Gleichstrom*
alternating current *Wechselstrom*
coating *Umhüllung*
to wrap *einhüllen*
fluxing material *Schutzmaterial*
union melt process *Unterpulververfahren*
inert gas shielded arc welding *Schutzgasschweißen*
tip *Spitze*
to switch *schalten*
tungsten *Wolfram*
butt welding *Stumpfschweißen*
spot welding *Punktschweißen*
seam welding *Nahtschweißen*
grip, jaw *Greifzange*
roller *Walze*
flame cutting *Brennschneiden*
spot *Fleck, Platz*
slot *Schlitz*
severance *Trennung*
groove weld *Stumpfnaht*
fillet weld *Kehlnaht*
arched *überwölbt*
butt joint *Stumpfstoß*
8.1.3. grooved pin *Kerbstift*
drive stud *Kerbnagel*
groove *Kerbfurche*
to groove *einkerben*
to swage *hineindrücken*
locking fit *fester Sitz*
locking force *Sperrkraft*
to ream *ausreiben*
to mill *fräsen*
to tap *Innengewinde schneiden*
taper *Schrägung*
taper pin, cotter pin *konischer Stift*
lock washer *Spannplatte*
full length taper gr. pin *Kegelkerbstift*
half length taper gr. pin *Paßkerbstift*
full length parallel gr. pin *Zylinderkerbstift*
half length reverse taper gr. pin *Steckkerbstift*
half length centre gr. pin *Knebelkerbstift*
round head dr. stud *Halbrundkerbnagel*
flat head dr. stud *Senkkerbnagel*
pan head dr. stud *Flachkopfkerbnagel*
truss head dr. stud *Flachkerbnagel*
shoulder head dr. stud *Ansatzkerbnagel*
bevel wheel *Kegelrad*
handle of a valve *Handgriff eines Ventils*
axle of a pulley *Achse einer Riemenscheibe*
decorative escutcheon *Zierscheibe*
pipe clamp *Rohrschelle*
8.1.4. clamp connection *Klemmverbindung*
hub *Nabe*
8.1.5. shrunk ring *Schrumpfring*
fly wheel *Schwungrad*
interference fit *Preßpassung*
8.1.6. key *Längskeil*
cotter *Querkeil*
coupling *feste Kupplung*

clutch *Schaltkupplung*
keyway *Keilnut*
clearance *Spiel*
sunk key *normaler Längskeil*
flat end plain key *Treibkeil*
round end plain key *Einlegekeil*
gib head key *Nasenkeil*
flat key *Flachkeil*
saddle key *Hohlkeil*
feather key *Paß- und Gleitfeder*
bearing *Lager*
recess *Aussparung*
splined shaft *Keilwelle*
8.1.6.2. socket *Muffe, Gabel*
8.1.7. bolt, screw *Schraube*
power screw *Bewegungsschraube*
stud *Stiftschraube*
to dismantle *lösen*
emergency *Dringlichkeit*
to lubricate *schmieren*
to fit *ausstatten*
washer *Unterlegscheibe*
lock nut *Gegenmutter*
8.1.7.1. helix *Schraubenlinie*
thread *Gewinde*
single thread *eingängiges Gewinde*
double thread *zweigängiges Gewinde*
multiple thread *mehrgängiges Gewinde*
notation *Bezeichnung*
major diameter *Außendurchmesser*
minor diameter *Innendurchmesser*
pitch *Steigung*
lead *Gewindeabstand*
pitch diameter *Flankendurchmesser*
depth of engagement *Tragtiefe*
helix angle *Steigungswinkel*
thread angle *Flankenwinkel*
fastening bolt *Befestigungsschraube*
coarse thread *Grobgewinde*
crest of thread *Gewindespitze*
root of thread *Gewindegrund*
ACME thread *Trapezgewinde*
to disengage *lösen*
round or knuckle thread *Rundgewinde*
environment *Umgebung*
buttress thread *Sägengewinde*
8.1.7.2. inadmissible *unzulässig*
hexagon head bolt *Sechskantschraube*
square head bolt *Vierkantschraube*
cup head bolt *Halbrundschraube*
countersunk head bolt *Versenkschraube*
recessed *vertieft*
recess *Vertiefung*
to screw on *anziehen*
timber *Bauholz*
neck *Hals*
to sink *versenken*
foundation bolt *Steinschraube*
jagged *zackig*
elastic bolt *Dehnschraube*
cap screw *Kopfschraube*
socket head screw *Innensechskantschraube*
cheese head screw *Zylinderschraube*
countersunk head screw *Versenkschraube*
rounded countersunk head screw *Linsensenkschraube*
fillister head screw *Zylinderschraube*
set screw *längerer Gewindestift*
cone point *Kegelspitze*
dog point *Ansatzspitze*
grub screw *kurzer Gewindestift*
8.1.7.3. locking device *Sicherungsvorrichtung*
to work loose *locker werden*

to attach *befestigen*
positive device *formschlüssige Vorrichtung*
friction device *kraftschlüssige Vorrichtung*
split pin *Splint*
spring washer *Federring, federnde Scheibe*
castle nut *Kronenmutter*
tooth washer *Zahnscheibe*
elastic stop nut *selbstsichernde Mutter*
to tighten *anziehen*
spanner *Schraubenschlüssel*
screw driver *Schraubenzieher*

8.1.7.4. strain *Dehnung*
to prestress *vorspannen*
bolt scale *Schraubenkennziffer*
plate scale *Plattenkennziffer*
to cancel out *aufheben*
hollow circular cross section *Kreisringquerschnitt*
yield stress *Streckgrenze*
endurance limit *Dauerfestigkeit*
width across flats *Schlüsselweite*

8.2. drive *Antrieb*
to convey *übertragen*
axle *Achse*
shaft *Welle*
bearing *Lager*
coupling *feste Kupplung*
clutch *Schaltkupplung*
belt drive *Riementrieb*
chain drive *Kettentrieb*
gear drive *Zahnradtrieb*

8.2.1.1. funicular *Drahtseilbahn*
railway carriage *Eisenbahnwagen*
stepped *abgesetzt*

8.2.1.2. pulley *Riemenscheibe*
to refer to *zuschreiben, bezeichnen als*
roller *Walze*
to grind *zerkleinern*
cutter *Fräser*
to mill *fräsen*
collar *Bund*
shoulder *Ansatz*
locating bearing *Festlager*
loose bearing *Loslager*
to jam *klemmen*
crank shaft *Kurbelwelle*
piston *Kolben*
offset *Absatz*
crank *Kurbel*
crank pin *Kurbelzapfen*
throw *Kröpfung*
cam *Nocken*
drop forging *gesenkschmieden*

8.2.2. journal *Achszapfen, Wellenzapfen*
sliding bearing *Gleitlager*
lubricant *Schmiermittel*
rolling contact bearing *Wälzlager*
ball *Kugel*
roller *Rolle*
journal bearing *Querlager*
thrust *Axiallast*
thrust bearing *Längslager*

8.2.2.1. actual *wirklich, tatsächlich*
to build up *aufbauen*
clearance space *Zwischenraum*
to converge *sich verengen*
path *Bahn*
block *Kippsegment*
wedge shaped *keilförmig*
locus *geometrischer Ort*
semicircle *Halbkreis*
to coincide *zusammenfallen*
to starve *trocken laufen*
boundary lubrication *Grenzschmierung*
surface finish *Oberflächenbearbeitung*
property *Eigenschaft*
satisfactory *zufriedenstellend*
grease *Fett*
flash point *Flammpunkt*

point of self ignition *Selbstzündungspunkt*
setting point *Stockpunkt*
value of penetration *Penetrationswert*
point *Spitze*
dripping point *Tropfpunkt*
to drip *tropfen*
lubricator *Schmiergefäß*
force feed *Druckölschmierung*
splash feed *Spritzölschmierung*
oil cup *Öler*
wick oiler *Dochtöler*
wick *Docht*
sight feed lubricator *Tropföler*
to turn off *abschließen*
to adhere *anhaften*
oil scraper *Ölabstreifer*
spout *Abflußloch*
superfluous *überflüssig*
to drain *abziehen*
strainer *Sieb, Filter*
to circulate *in Umlauf setzen*
Stauffer screw-down greaser *Staufferbuchse*
grease gun *Fettpresse*
extrusion *Zusammenpressen*
bush *Buchse*
step *Lagerschale*
grit *feine Körner*
abrasive *Schmiergel*
remainder *Rest*
to anneal *glühen*
tough *zäh*
resilient *federnd*
web *Gewebe*
recess *Aussparung*
to suck *saugen*
groove *Nute*
to resort to *Zuflucht nehmen zu, lieber verwenden*
self-aligning bearing *selbst einstellbares Lager*
to tilt *kippen*
pedestal bearing *Stehlager*
solid *ungeteilt*
pedestal body *Lagerkörper*
housing *Gehäuse*
cap *Lagerdeckel*
wall bracket *Wandarm*
ceiling bracket *Hängebock*
foot step bearing or pivot bearing *Spurlager*
breast plate *Druckplatte*
pivot *Zapfen*
reliable *zuverlässig*
8.2.2.2. seal *Dichtung*
inner race *innerer Laufring*
outer race *äußerer Laufring*
cage *Käfig*
deep groove ball bearing *Rillenkugellager*
assembly *Zusammenbau*
offset *seitlich verschoben*
angular contact bearing *Schrägkugellager*
self-aligning ball bearing *Pendelkugellager*
appreciable *merklich*
angular misalignment *Winkelabweichung*
cylindrical roller bearing *Zylinderrollenlager*
end washer *Stützscheibe*
piston pin *Kolbenbolzen*
needle bearing *Nadellager*
taper roller bearing *Kegelrollenlager*
spherical roller bearing *Pendelrollenlager*
ball thrust bearing *Axialrillenlager*
inaccuracy *Ungenauigkeit*
alignment *Fluchten*
roller thrust bearing *Axialpendelrollenlager*
to grind *schleifen*

rating *Abschätzen, Bemessen, Dimensionieren*
service life *Lebensdauer*
to perform *ausführen*
failure *Unbrauchbarwerden*
speed *Geschwindigkeit, Drehzahl*
r.p.m. = revolutions per minute 1/min
stationary *feststehend*
circumferential load *Umfangslast*
point load *Punktlast*
arrangement *Anordnung*
fixture *Befestigung, Festlegen*
snap ring *Sicherungsring (Sprengring, Seegerring)*
adapter sleeve *Spannhülse*
withdrawal sleeve *Abziehhülse*
solid *einteilig*
split pedestal housing *geteiltes Lagergehäuse*
to dissipate *abgeben*
maintenance *Wartung*
cup grease *Maschinenfett*
lime *Kalk*
sodium *Natrium*
potassium *Kalzium*
leakage *Undichtwerden*
felt seal *Filzdichtung*
collar seal *Manschettendichtung*
sealing washer *Abdeckscheibe*
contaminant *Schmutz, Verunreinigung*

8.2.3. brake *Bremse*

8.2.3.1. device *Vorrichtung*
lineal alignment *geradliniges Fluchten*
split muff coupling *Muffenkupplung*
to tighten up *anziehen*
to dismantle *demontieren*
flanged coupling *Flanschkupplung*
split plate *geteilte Platte*
flexible *nachgiebig*
wear *Abnützung*
floating centre coupling *Kreuzscheibenkupplung*
lug *Ansatz*
to protrude *vorstehen*
coefficient of cyclic variation *Ungleichförmigkeitsgrad*
intermediate *dazwischenliegend*
to damp *dämpfen*
to space *mit Zwischenräumen anordnen*

8.2.3.2. positive *formschlüssig*
to effect *bewirken*
jaw *Klaue*
gear *Verzahnung*
sleeve *Hülse*
fabric *Gewebe*
jerk *Stoß*
wedge *Abschrägung*
multiple disc clutch *Lamellenkupplung*
to gripe *fressen*
toggle lever *Kniehebel*
to attach *befestigen*
to switch *schalten (elektrisch)*
current *Strom (elektrisch)*
impeller *Pumpenlaufrad*
runner *Turbinenlaufrad*
vane *Laufschaufel*
slip *Schlupf*
slip clutch *Rutschkupplung*
to retract *zurücknehmen*
overrunning clutch *Überholungskupplung*
free wheeling clutch *Freilauf*
coaster brake *Rücktrittbremse*
to jam *klemmen*

8.2.3.3. to slow down *verlangsamen*
dissipation *Vergeudung*
pivoted *glenkig gelagert*
wrap angle or angle of contact *Umschlingungswinkel*
ratchet mechanism *Sperrad*

pawl *Sperrklinke*
to notch *ausklinken*
8.2.4. competitive *wettbewerbsfähig*
maintenance *Wartung*
reliability *Betriebssicherheit*
availability *Verfügbarkeit*
timing *Regulierung*
8.2.4.1. adhesion *Anhaften*
gear ratio *Übersetzungsverhältnis*
layer *Schicht*
cambered *gewölbt*
textile fabric *Textilgewebe*
to glue *leimen*
tight (slack) portion *gespannter (loser) Teil*
belt tightener *Riemenspanner*
quarter turn drive *Halbkreuztrieb*
rim *Kranz*
guide pulley *Leitrolle*
hub *Nabe*
web *Scheibensteg*
to adapt *anpassen*
8.2.4.2. cord fabric *Cordfäden*
stiffness *Steifheit*
abrasion *Abnützung durch Reiben*
tangential force *Umfangskraft*
8.2.5. round iron chain *Rundeisenkette*
gearing chain *Gelenkkette*
lifting tackle *Flaschenzug*
8.2.5.1. link *Kettenglied*
rod *Stange*
to allow for *berücksichtigen*
8.2.5.2. spring clip *federndes Steckglied*
single strand roller chain *einfach Rollenkette*
multi-strand roller chain *mehrfach Rollenkette*
tightening device *Spannvorrichtung*
adjustable *nachstellbar*
link strand *Gliederreihe*
morse rocker joint chain *Zahnkette*
rocking bearing *Wiegelager*
seat pin *Lagerzapfen*
rocker pin *Wiegezapfen*
to rock *sich wiegen*
8.2.7. gear drive *Zahnradantrieb*
case *Gehäuse*
8.2.7.1. term *Bezeichnung*
symbol *Kurzzeichen*
gear train *Zahnradgetriebe*
stationary *feststehend*
planetary gear train *Planetengetriebe, Umlaufgetriebe*
to mesh together *zusammenarbeiten*
spur gear *Stirnrad*
pinion *Ritzel*
rack *Zahnstange*
bevel gear *Kegelrad*
skew gears *versetzte Kegelräder mit geraden Zähnen*
hypoid gears *versetzte Kegelräder mit Bogenzähnen*
screw gear *Schraubenrad*
worm *Schnecke*
worm wheel *Schneckenrad*
annular ring *innenverzahntes Zahnrad*
involute *Evolvente*
face *außerhalb des Teilkreises liegende Zahnflanke*
flank *innerhalb des Teilkreises liegende Zahnflanke*
straight spur wheel *geradverzahntes Stirnrad*
straight bevel wheel *geradverzahntes Kegelrad*
helical spur wheel *schrägverzahntes Stirnrad*
helical bevel wheel *schrägverzahntes Kegelrad*
herring bone spur wheel *Kegelrad mit Pfeilverzahnung*
spiral bevel wheel *Kegelrad mit Bogenverzahnung*
mesh *Eingriff*

pitch circle *Teilkreis*
addendum circle *Kopfkreis*
dedendum circle *Fußkreis*
addendum *Kopfhöhe*
dedendum *Fußhöhe*
fundamental law of gear tooth action *Verzahnungsgesetz*
to mate *zusammenarbeiten*
to trace *aufzeichnen*
generating line *Erzeugende*
base circle *Grundkreis*
angle of obliquity *Eingriffswinkel*
path of contact *Eingriffslinie*
arc of contact *Eingriffsbogen*
limiting number of teeth *Grenzzähnezahl*
undercut *Unterschnitt*
extended centre distance system *System mit Profilverschiebung*
rolling circle *Rollkreis*
mitre gear *Kegelrad mit 90° Kegelwinkel*
crown gear *Plankegelrad*
equivalent *gleichwertig*
stepped *versetzt*
helix angle *Steigungswinkel*
circular pitch *Stirnteilung*
normal pitch *Normalteilung*
overlap *Sprung*
contact ratio *Überdeckungsgrad*
speedometer *Tachometer*
distributor *Verteiler*
worm gearing *Schneckentrieb*

8.2.7.2. excessive *übermäßig*
cantilever beam *eingespannter Träger*
fatigue load *Dauerwechsellast*
crack *Bruch*
fillet *Hohlkehle*
to score *einkerben*
to scratch *zerkratzen*
to pit *Grübchen machen*
impact *Stoß*

8.2.7.3. to perform *ausführen*
to hob *wälzfräsen*
to shave *schaben*
to burnish *polieren*
form milling cutter *Formfräser*
generating milling cutter *Abwälzfräser*
pinion cutter *Stoßrad*
rack cutter *Zahnstangenwerkzeug*
hob cutter *Frässchnecke*
blank *Rohling*

8.2.7.4. single stage spur gear train *einstufiges Stirnradgetriebe*
spur bevel gear train *Kegel-Stirnradgetriebe*
geared motor *Getriebemotor*
planet carrier *Trägerscheibe*

8.3. slider crank mechanism *Schubkurbelgetriebe*
single (double) acting machine *einfach (doppelt) wirkende Maschine*
crank *Kurbel*
piston *Kolben*
connecting rod *Schubstange, Pleuelstange*
cross head *Kreuzkopf*
piston rod *Kolbenstange*

8.3.1. istantaneous centre *Momentanpol*

8.3.2. balance weight *Gegengewicht*
crown of piston *Kolbenboden*
crank pin *Kurbelzapfen*
flywheel *Schwungrad*
coefficient of cyclic variation *Ungleichförmigkeitsgrad*

8.3.3. journal *Wellenzapfen*
throw *Kröpfung*
web *Kurbelwange*
overhung crank *Stirnkurbel*
balanced crankshaft *Kurbelwelle mit Gegengewichten*
big end *unterer Schubstangenkopf*
big end bearing *unteres Schubstangenlager*
escape *Entkommen*

plunger piston *Plungerkolben*
packing *Dichtung*
packing spring *Kolbenring*
leakage *Leckwerden, Undichtsein*
gap *Spalte, Zwischenraum*
scraper ring *Ölring*
to clear *wegräumen*
surplus *überschüssig, Überschuß*
cylinder skirt *Zylinderwand*
crank case *Kurbelkasten*
gudgeon pin *Kolbenbolzen*
snap ring *Seegerring*
groove *Vertiefung*
stuffing box *Stopfbüchse*
case harden *einsatzhärten*
adjustable *einstellbar*
dead space *schädlicher Raum*
to drop forge *gesenkschmieden*
split bearing *geteiltes Lager*
eye bearing *Augenlager*

8.4. pipe *Rohr*
valve *Rohrschalter*
conveyance *Förderung*
to convey *fördern*
fluid *Medium*
bend *Bogenstück*
elbow *Kniestück*
expansion joint *Kompensator*
acid *Säure*
alkaline *Lauge*
nominal pressure *Nenndruck*
working pressure *Betriebsdruck*
proof pressure *Probedruck*
nominal diameter *Nennweite*
deviation *Abweichung*

8.4.1. centrifugal cast *Schleuderguß*
seamless *nahtlos*
conveying capacity *Fördermenge*
rust formation *Rostbildung*

8.4.2. socket joint *Muffenverbindung*
to embed *einbetten*
ground *Boden*
flexibility *Biegsamkeit*
to accomodate itself *sich anpassen*
slight *leicht, geringfügig*
settlement *Senkung*

8.4.3. fitting *Rohrformstück*
branch *Abzweig*
twin elbow *Zwillingskniestück*

8.4.4. to accomodate *aufnehmen*
alteration *Veränderung*
corrugated expansion joint *Wellrohrkompensator*
hinged expansion joint *Gelenkkompensator*
to stretch *dehnen*

8.4.5. to admit *freigeben*
to shut off *absperren*
operated manually *handbedient*
power operated *maschinell angetrieben*
globe valve *Ventil*
gate valve *Schieber*
plug cock *Hahn*
clack valve *Klappenventil*
to fulfil *erfüllen*
accessible *zugänglich*

8.4.5.1. shut-off valve *Absperrventil*
regulating valve *Regulierventil*
non-return valve *Rückschlagventi*
safety valve *Sicherheitsventil*
pressure reducing valve *Reduzierventil*
straight-way valve *Durchgangsventil*
angle valve *Eckventil*
slanted *geneigt*
slanted valve *Schrägsitzventil*
bonnet *Kopfstück*
guide frame *Aufsatz*
packing *Packung*
gland *Stopfbüchsenbrille*
valve disc *Ventilteller*
valve cone *Ventilkegel*
to lift off *abheben*
to lower on *herunterlassen*

orifice *Ventilöffnung*
extensively *in großem Umfang*
incline *Abhang*
to safeguard *schützen*
overloading *Überlastung*
sleeve *Hülse*
lock nut *Sicherheitsmutter*
to adjust *einstellen*
to release *entlassen*
to burst *zerplatzen*
jockey *Membrane*
8.4.5.2. gate *Schieberplatte*
sediment *Bodensatz*
sediment chamber *Schmutzkammer*
drain plug *Entleerungsschraube*
to deposit *ablagern*
collar *Bund*
to wedge *durch Keilwirkung pressen*
wedge *Keil*
wedge gate *keilförmige Schieberplatte*
gasket *Dichtungsring*
stuffing box *Stopfbüchse*
8.4.5.3. plug *Hahnküken*
spanner *Schraubenschlüssel*
to project *hervorragen*
washer *Scheibe*
8.4.5.4. suction pipe *Saugrohr*
pressure pipe *Druckrohr*
door *Klappe*
throttle valve *Drosselventil*
rate of flow *Durchflußmenge*
furnace *Feuerung*
carburettor *Vergaser*
8.5. spring *Feder*
helical spring *Schraubenfeder*
device *Vorrichtung*
cam follower *Nockenstößel*
buffer *Puffer*
scale *Meßgerät*
swing sieve *Schwingsieb*
sympathetic vibration *Eigenschwingung*
vibration damper *Schwingungsdämpfer*
superior *vorzüglich*
creep *Kriechen*
relaxation *Nachlassen*
spring stiffness *Federkonstante*
increment *Zunahme*
to plot *aufzeichnen*
graph *Schaubild*
work output *gewonnene Arbeit*
work input *aufgewendete Arbeit*
relief *Entlastung*
natural frequency *Eigenfrequenz*
8.5.1.1. helical cylindrical spring *zylindrische Schraubenfeder*
to buckle *knicken*
cyclic disturbance *periodische Störung*
to specify *im Einzelnen festlegen*
turn *Windung*
coil *gewundener Draht*
wire *Draht*
8.5.1.3. torsion bar *Torsionsstab*
hinge *Gelenk*
brush holder *Bürstenhalter*
door lock *Türschloß*
8.5.2.1. flat spring *Blattfeder*
farm machinery *landwirtschaftliche Maschinen*
8.5.2.2. laminated plate spring *geschichtete Blattfeder*
8.5.2.3. spiral spring *Spiralfeder*
8.5.2.4. counterbalance *Gegenzug*
8.5.3.2. mainly *hauptsächlich*
shock absorber *Stoßdämpfer*
to cushion *dämpfen*

Viewegs Fachbücher der Technik

Mathematik

Mathematik für technische Berufe von E. Gasse. **Band I: Arithmetik und Algebra,** DM 12,80 – **Band II: Geometrie,** DM 15,80
Mathematikaufgaben aus der Maschinentechnik von F. Riegel, DM 4,80
Einführung in die Nomographie von A. Bay, DM 5,80
Mathematik für Techniker von H. Simon, ca. DM 9,80

Physik

Physik – Grundlagen-Versuche-Aufgaben von A. Böge, DM 12,80
Physik für Ingenieure von H. Lindner, DM 24,80
Physikalische Aufgaben von H. Lindner, DM 8,80
Übungsaufgaben aus der Wärmelehre von W. Berties, DM 10,80
Technische Optik von H. Schade, DM 9,80

Elektrotechnik

Grundlagen der Elektrotechnik von J. Reth und H. Kruschwitz, DM 19,50
Aufgabensammlung Elektrotechnik von H. Kruschwitz, DM 10,80
Elektroaufgaben von H. Lindner. **Band I: Gleichstrom,** DM 9,20
Band II: Wechselstrom, DM 7,80
Elektronische Bauelemente der Nachrichtentechnik von A. Raschkowitsch, DM 24,8

Regelungstechnik

Einführung in die Regelungstechnik von H. Wittmers, DM 8,–

Mechanik und Festigkeitslehre

Mechanik und Festigkeitslehre von A. Böge, DM 17,80
Formeln und Tabellen zur Statistik, Dynamik, Hydraulik und Festigkeitslehre von A. Böge, DM 3,90
Aufgabensammlung zur Statik, Dynamik, Hydraulik und Festigkeitslehre von A. Bög und W. Schlemmer, DM 12,80

Maschinenelemente

Maschinenelemente von H. Roloff und W. Matek, DM 27,80
Aufgabensammlung Maschinenelemente von H. Roloff und W. Matek, DM 14,80

Fertigungstechnik

Schweißtechnik von A. Puhrer, DM 16,80
Stanztechnik von E. Semlinger, DM 16,80
Umformtechnik von K. Grüning, DM 16,80
Zerspantechnik von K.-Th. Preger, DM 14,80

Werkstoffkunde

Werkstoffkunde und Werkstoffprüfung von W. Weißbach, DM 14,80

Technisches Zeichnen

Technisches Zeichnen und Projektionszeichnen von G. Fey, DM 4,80

Betriebswirtschaftslehre

Arbeitsvorbereitung und Kalkulation von H. Sonnenberg, DM 19,80
Mitarbeiter Führen von G. Obst, DM 9,60
Wirtschafts- und Rechtskunde von R. Ott und M. Wendlandt, DM 12,80

uni–text

Einführung in die höhere Mathematik

Von Prof. Dr. Herbert Dallmann und Prof. Dr. Karl-Heinz Elster. uni-text / Lehrbuch. Braunschweig: Vieweg, 1969. DIN C 5. 620 Seiten mit 272 Abb. und 2 Tabellen. Paperback DM 36.–

(Best.-Nr. 3509)

Wechselströme und Netzwerke

Von Prof. Dr. Werner Leonhard. uni-text / Studienbuch. Braunschweig: Vieweg, 1968. DIN C 5. X, 215 Seiten mit 182 Abb. Paperback DM 9,80

(Best.-Nr. 3003)

Einführung in die Regelungstechnik

Von Prof. Dr.-Ing. Werner Leonhard. uni-text / Studienbuch. Braunschweig: Vieweg, 1969. DIN C 5. VIII, 233 Seiten mit 266 Abb. Paperback DM 9,80

(Best.-Nr. 3015)

Mechanik I

Grundbegriffe – Kinematik – Statik

Von Prof. Dr. Karl-August Reckling. uni-text / Studienbuch. Braunschweig: Vieweg, 1968. DIN C 5. VIII, 254 Seiten mit 195 Abb. Paperback DM 9,80

(Best.-Nr. 3002)

Mechanik II

Festigkeitslehre

Von Prof. Dr.-Ing. Karl-August Reckling. uni-text / Studienbuch. Braunschweig: Vieweg, 1969. DIN C 5. VIII, 215 Seiten mit 123 Abb. Paperback DM 9,80

(Best.-Nr. 3007)

» vieweg

»

Maschinenelemente

Normung – Berechnung – Gestaltung

Von Fachschuloberlehrer Ing. Hermann Roloff und Dipl.-Ing. Wilhelm Matek. VIEWEGS FACHBÜCHER DER TECHNIK. Braunschweig: Vieweg, 3., durchgesehene und verbesserte Auflage, 1968. DIN A5. XVI, 584 Seiten mit 416 Abb., 1 Anhang mit 94 Seiten und 44 Abb. Gebunden DM 27,80 (Best.-Nr. 4013).

Dieses Lehrbuch ist vorzüglich für die fachliche Ausbildun und die Berufspraxis des Technikers geeignet. Die beiden Verfasser haben ihre didaktisch gut gelungene Arbeit auf dem im gleichen Verlage erscheinenden bewährten Buch von A. Böge, „Mechanik und Festigkeitslehre", aufgebaut und damit zugleich erreicht, daß dem Erwerber der Neuerscheinung oft die Anschaffun oder Zuhilfenahme eines gesonderten Grundlagenbuches erspart wird.

»

Aufgabensammlung Maschinenelemente

Von Fachschuloberlehrer Ing. Hermann Roloff und Dipl.-Ing. Wilhelm Matek. VIEWEGS FACHBÜCHER DER TECHNIK. Braunschweig: Vieweg, 1967. VIII, 253 Seiten mit 342 Aufgaben und 22 Abb. Kartoniert DM 14,80 (Best.-Nr. 4015)

„Die Aufgabensammlung ist genau wie das Lehrbuch ‚Maschinenelemente' gegliedert, so daß vorteilhaft gleichzeitig mit beiden Büchern gearbeitet werden kann". (Schweißtechnik)

» **vieweg**

» **Physik für Ingenieure**

Von Studiendirektor Helmut Lindner. VIEWEGS FACHBÜCHER DER TECHNIK. Braunschweig: Vieweg, 1969. 8°. 591 Seiten mit 721 Abb., 64 Tabellen und 304 Beispielen und 1 Farbtafel. Halbleinen DM 24,80 (Best.-Nr. 4047).

Inhalt: Mechanik des Massenproduktes und des starren Körpers – Lehre von den Schwingungen – Mechanik der Flüssigkeiten und der Gase – Wärmelehre – Grundbegriffe der Wellenlehre – Geometrische Optik – Wellenoptik – Elektrizitätslehre – Atomphysik.

Dieses Lehrbuch ist für den werdenden Ingenieur und insbesondere für das Direktstudium an den Ingenieur- und Fachschulen bestimmt. Es kann aber auch in allen anderen Lehrgängen mit ähnlichem Bildungsziel sowie als Hilfsmittel beim Selbststudium verwendet werden.

» **Physikalische Aufgaben**

Von Studiendirektor Helmut Lindner. VIEWEGS FACHBÜCHER DER TECHNIK. Braunschweig: Vieweg, 11. Auflage, 1969. 8°. 320 Seiten mit 358 Abb. Halbleinen DM 8,80 (Best.-Nr. 4879).

Inhalt: Mechanik fester Körper – Mechanik der Flüssigkeiten und Gase – Wärmelehre – Optik – Elektrizitätslehre – Atom- und Kernphysik – Lösungen.

» **vieweg**

»

DAS MODERNE INDUSTRIEUNTERNEHMEN

Betriebswirtschaft für Ingenieure

Grundzüge des Wirtschaftsrechts

Von Dr. Rudolf Ott und Manfred Wendlandt. DAS MODERNE INDUSTRIEUNTERNEHMEN — Betriebswirtschaft für Ingenieure, herausgegeben von W. Zimmermann und S. Höwelmann. Braunschweig: Vieweg, 1969. DIN C 5. XI, 222 Seiten mit 7 Abb. Gebunden DM 16,8

(Best.-Nr. 4101)

Inhalt: Über das Recht — Schuldverhältnisse — Aus dem Sachenrecht — Aus dem Handelsrecht — Geld, Kredit, Zahlungsverkehr — Gerichtswesen — Arbeitsrecht — Versicherungen — Steuern — Gewerblicher Rechtsschutz.

Die Fachpresse urteilt:
Für den Ingenieur im Betrieb werden neben der Beherrschung seines Fachgebietes die Kenntnisse wirtschaftlicher Zusammenhänge immer wichtiger. Dieses Buch bietet ihm dabei eine Hilfe, durch kurzgefaßte Darstellung der wesentlichen Tatbestände schnell in die wirtschaftlich bedeutsamsten Gebiete des Rechts Einblick zu nehmen, da es keine besonderen Vorkenntnisse voraussetzt.

Die Berufsgenossenschaft, Bielefe

Planungsrechnung

Optimierungsrechnungen — Wirtschaftlichkeitsrechnungen — Netzplantechnik

Von Oberbaurat Dr.-Ing. Werner Zimmermann. DAS MODERNE INDUSTRIEUNTERNEHMEN — Betriebswirtschaft für Ingenieure, herausgegeben von W. Zimmermann und S. Höwelmann. Braunschweig Vieweg, 1968. X, 216 Seiten mit 42 Abb. und 69 Beispielen. Gebunder DM 16,80

(Best.-Nr. 4100)

Inhalt: Verfahrensvergleiche — Investitionsrechnungen — Operations Research — Optimierungsverfahren bei Ersatz-, Lagerhaltungs-, Wartezeit-, Konkurrenz-, Zuordnungs- und Zuteilungsproblemen in der industriellen Praxis — Grundlagen der Netzplantechnik — Besonderheiten spezieller Netzplantechniken.

Urteil der Fachpresse:
„Das moderne Industrieunternehmen", Betriebswirtschaft für Ingenieu ist eine neue Buchreihe, die die Eigenheiten der industriellen Unternehmung speziell für den Ingenieur und Betriebswirt erläutert. In dem erste nun vorliegenden Band werden Entwicklung und Verfahren der modernen Planungsrechnung dargelegt und anhand einer Vielzahl von Beispielen erklärt. Der Band soll den Studenten der Ingenieur- und Wirtschafts wissenschaften als Lehrbuch einen Überblick über die modernen Planur methoden und -techniken verschaffen.

Melliand Textilberichte, Heidelbe

» **vieweg**

GPSR Compliance

The European Union's (EU) General Product Safety Regulation (GPSR) is a set of rules that requires consumer products to be safe and our obligations to ensure this.

If you have any concerns about our products, you can contact us on ProductSafety@springernature.com

In case Publisher is established outside the EU, the EU authorized representative is:

Springer Nature Customer Service Center GmbH
Europaplatz 3
69115 Heidelberg, Germany

Batch number: 08566374

Printed by Printforce, the Netherlands